# INSECT PESTS

AND

# FUNGUS DISEASES OF FRUIT AND HOPS

# INSECT PESTS

# AND

# FUNGUS DISEASES OF FRUIT AND HOPS

A COMPLETE MANUAL FOR GROWERS

BY

PERCIVAL J. FRYER, F.I.C., F.C.S.

TECHNICAL DIRECTOR AND CHIEF OF RESEARCH DEPARTMENT,
YALDING MANUFACTURING CO., LTD.;
MEMBER OF SOCIETY OF ECONOMIC BIOLOGISTS

CAMBRIDGE
AT THE UNIVERSITY PRESS
1920

CAMBRIDGE
UNIVERSITY PRESS

University Printing House, Cambridge CB2 8BS, United Kingdom

Cambridge University Press is part of the University of Cambridge.

It furthers the University's mission by disseminating knowledge in the pursuit of education, learning and research at the highest international levels of excellence.

www.cambridge.org
Information on this title: www.cambridge.org/9781107544550



First published 1920
First paperback edition 2015

*A catalogue record for this publication is available from the British Library*

ISBN 978-1-107-54455-0 Paperback

# PREFACE

IN most scientific industries the practical worker has a choice of several standard books of reference by well-recognised experts to guide him in his operations. While this may also apply to certain sections of applied agricultural science, there is a most lamentable dearth of reliable and complete works of reference on the control of insect pests and fungus diseases for the fruit-growing industry. Such books of reference as exist either deal with a single aspect only of the subject, or have, on the other hand, too wide a scope. In any case they are not usually presented in a form which is available or adequate for the requirements of practical growers.

The present volume represents a careful and painstaking attempt to produce as complete a book of reference as possible, suited to the requirements of the fruit and hop grower, and presented in such a form that the information, while given with scientific precision, is also in a readily available form. Although the endeavour has been made to make the book as comprehensive and detailed as possible in every section of the subject, it has also been the author's aim throughout to exclude all matter which was not of direct scientific or practical interest. The subject matter, especially in the detailed descriptions of the pests, has been condensed as much as possible, and the significant facts presented in a tabular form for ease of reference, while all mere verbiage and unessential detail has been rigorously excluded.

The author's experience of fruit pests and diseases extends over some sixteen years, during which time he has been practically engaged in research and in contact with fruit growers all over the country.

Although his chief interest in the subject is from the chemical standpoint, it has been necessary in his researches to thoroughly familiarise himself with both its entomological and mycological aspects, and the reader may therefore have confidence in the accuracy and reliability of these sections of the book. As a short statement of the scope and arrangement of the subject matter and hints as to the use of the book follows it will not be necessary to refer further to this, and it only remains for the author to make acknowledgements to those persons to whom he is indebted in its preparation.

To his brother, Mr C. Henry Fryer, is due the entire credit for the many original photographs of insect and fungus pests with which practically each description is illustrated, and which contribute very greatly to the utility which it is hoped the book will achieve. A large number of diagrams and sketches, including the coloured plates of the fungus diseases, have been prepared by Mr Harry Ballard, who was also of assistance in the preparation of the index, and to whom the author expresses his sincere thanks. The Rev. F. M. Richards, M.A., who is an enthusiastic entomologist, has also rendered great assistance in this branch of the subject. The author especially desires to express his appreciation of the interest taken in the book by Mr A. C. Harradine, Fruit Farm Expert of Messrs Chivers and Sons, Histon, who made several useful suggestions on the practical treatment of pests.

Professor F. V. Theobald, M.A. of Wye College kindly supplied a corrected list of the scientific names for the aphis pests, and the author is also indebted to various published matter, especially that by Professors Salmon and Theobald. Mr Petherbridge has kindly sanctioned the copying of the excellent diagram of the development of the Capsid bug, shown on page 346. Other acknowledgements are also made in the appropriate places in the text of the book.

The author would like to take this opportunity of expressing his grateful thanks for many useful hints on the subject of the treatment of pests from fruit and hop growers, especially from Mr Selby Smith, Mr P. Manwaring, and Mr J. H. W. Best.

In a book of this character there are necessarily many differences of opinion, especially as regards recommendations concerning the application of remedies. The author will at all times welcome opinions or information from growers in this connection, or as regards any errors or mis-statements which, even when the greatest care is used, it is difficult entirely to avoid.

In conclusion, it has been the author's sincere desire throughout to assist the fruit grower in the continuous campaign waged against the multitude of insect and fungus pests which so persistently attack his plantations, and he will feel amply rewarded for his labours if the present work achieves this result to even a modest extent.

PERCIVAL J. FRYER.

RAVENSCAR,
TONBRIDGE.
*May* 1920.

# EXPLANATORY NOTE

THE **Index plates** are meant to be used by the practical grower in the following way:

When any species of pest is found attacking a fruit tree or hop in a grower's plantations, reference is made to the drawings on the index plates, which in the case of insects are placed just after the table of contents. The index plates of caterpillars and moths are coloured in their natural tints in order to assist in the identification. In other cases, it is hoped that some guide may at any rate be given by the sketches of the various beetles, etc. Underneath each pest is placed the *page number* where that particular insect is fully described and the grower may then turn to this page and learn the chief points as regards the life-history and the best means of dealing with the attack.

In the case of fungus diseases, coloured drawings of the appearance of the fungus on the tree are placed before each description of the diseases of that particular fruit. The page references are also given with insects.

The **arrangement** of the book is as follows:

At the commencement is a short INTRODUCTION on the general aspect of the subject, followed by the section on the habits of life or STRUCTURE OF THE PLANTS themselves, which is of a more or less simple botanical character. It was thought desirable to include this as there are several points which have a direct bearing upon spraying and other considerations.

**Part I** opens with Section III dealing with the CHIEF CHARACTERISTICS OF INSECTS and explains a good deal of the subject matter of Section IV, in which INSECT PESTS attacking fruit and hops are described in full detail. Each kind of pest has a special section to itself, and the author decided that this was a more convenient arrangement than classing them according to the kind of fruit they attack. The reason for this is that a good many insects attack more than one kind of fruit and any classification based upon this feature would therefore be unreliable.

At the end of each description is placed a *diagram* representing the different periods of the year in which the insect is in its different stages and pointing out the best time to apply the various remedies.

A full description of the INSECTICIDE MATERIALS which are available for combating insect pests is given in detail in Section V, and each material is placed in alphabetical order for ease of reference. It is also sub-divided into various headings for the same reason.

A short account of that very interesting class of insects which prey upon other insects and are thus to be regarded as FRIENDLY to the grower is given in Section VI, while some brief remarks on their employment on the practical scale will be found in Appendix IX on page 708. This completes the first part of the book.

**Part II** of the book deals with the FUNGUS DISEASES OF FRUIT and their control. Section VII opens with a short INTRODUCTION on the subject of fungi and the diseases they cause and the remarks on preventive measures to be employed. This is succeeded by various sub-sections, dealing with the DISEASES applying to each kind of fruit. At the end of each description is also placed, as in the case of those of the insects, a diagram showing what time of the year the fungus is best treated. Section VIII deals with SPRAYING MATERIALS to be used in combating fungus diseases which may be termed FUNGICIDES.

**Part III** deals with the ART OF SPRAYING, and contains descriptions of spraying appliances and methods and a calendar for general spraying work. Section IX opens with an INTRODUCTION discussing the various kinds of SPRAYING NOZZLES available, followed by a detailed description of the various types of SPRAYING APPLIANCES, ranging from the small syringe to the powerful mechanically operated power sprayer, suitable for dry spraying. This is followed by a short chapter on SPRAYING METHODS and another on the use of combined or MIXED SPRAYS. Section X contains a SPRAYING CALENDAR in tabular form, which is really a spraying introduction to the rest of the book, and which the author trusts will serve as a reminder to growers of the various pests, for the first appearance of which they should keep a sharp look-out. The next section contains a few REFERENCE TABLES and APPENDICES, including four DIAGRAMS showing in a concise manner the particular stage in which each insect exists in a given week in the year. This is really a composite diagram made up of a collection of each of those given at the end of the descriptions of individual pests.

A very full and detailed INDEX will be found at the end of the book.

# CONTENTS

# PLATES



The author wishes to make acknowledgment to Messrs Frederick Warne & Co. for permission to reproduce the figures in Plates V—VIII from their book *The Moths of the British Isles* and to use other illustrations from the same book in the preparation of Plates I—IV.

*PLATE I*

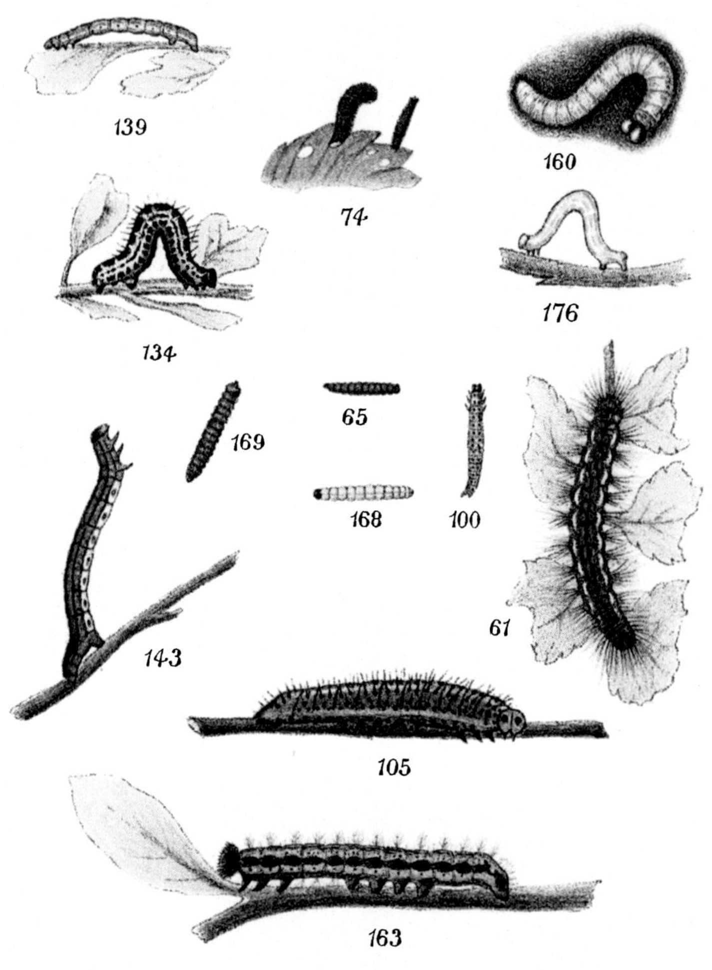


| | | | | | |
|---|---|---|---|---|---|
| 61 | Brown-tail | 105 | Figure of Eight | 160 | Swift |
| 65 | Bud (*Heyda ocellana*) | 134 | Magpie | 163 | Tortoiseshell (large) |
| 74 | Case-bearers | 139 | March | 168 | Green bug caterpillar |
| 100 | Ermine (small) | 143 | Mottled umber | 169 | Tortrix |
| | | 176 | Winter | | |

chromo-lith Cambridge University Press

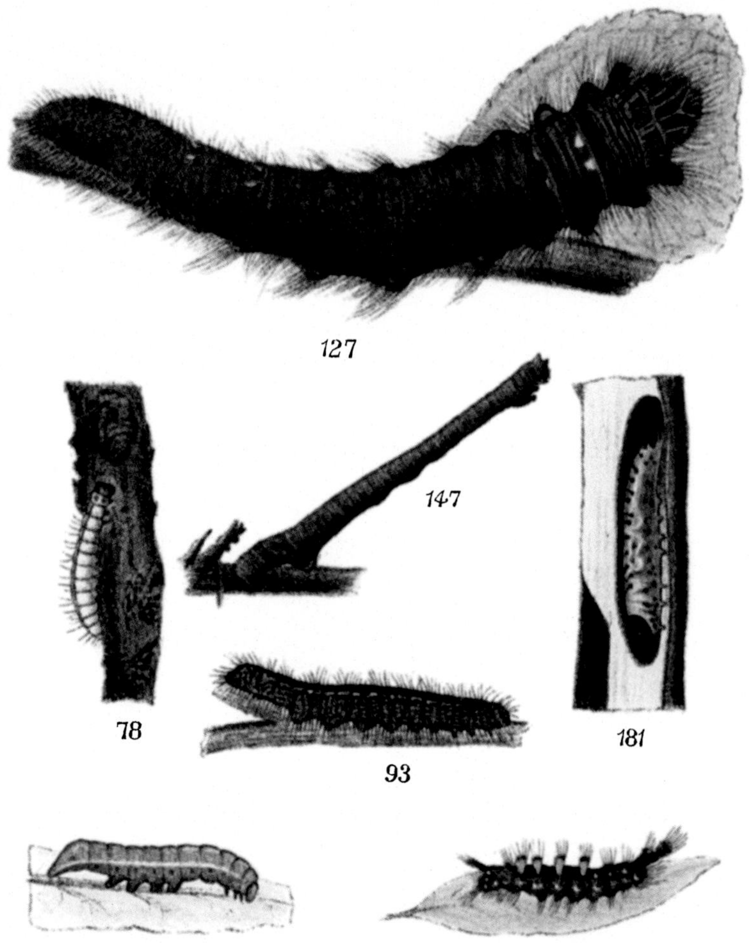


| | | | | | |
|---|---|---|---|---|---|
| 78 | Clear Wing | 84 | Clouded Drab | 93 | December |
| 127 | Lappet | 147 | Peppered | 172 | Vapourer |
| | | 181 | Wood Leopard | | |

chromo-lith. Cambridge University Press

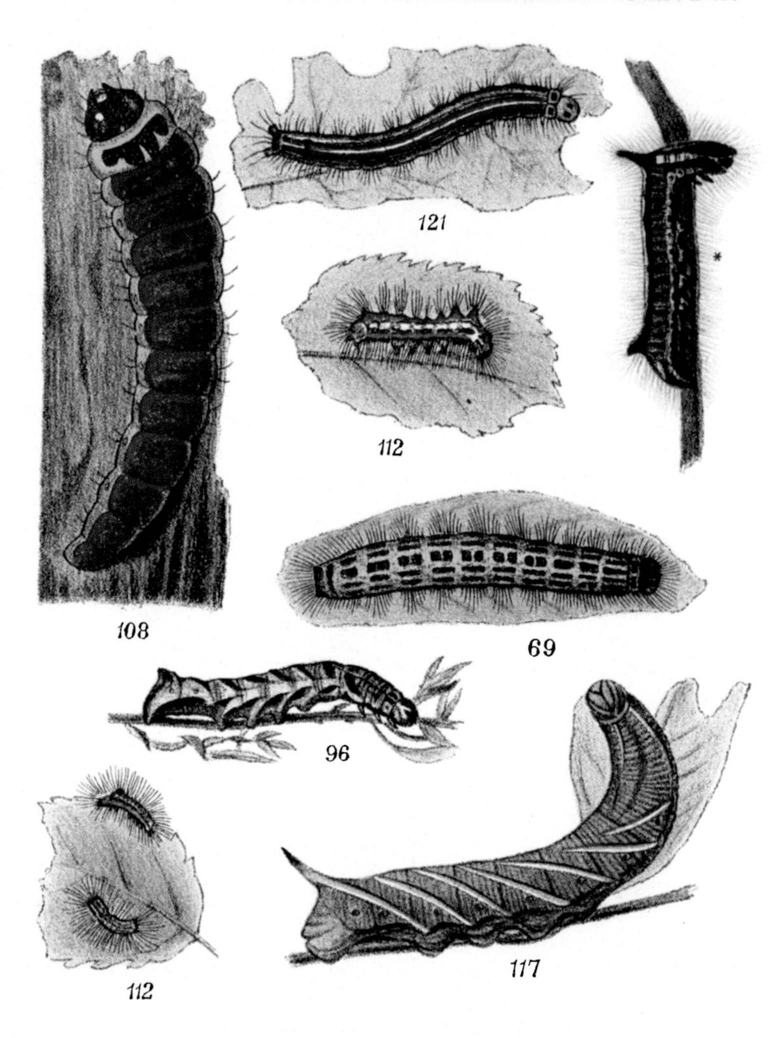


69 Buff-tip 96 Dot 108 Goat
112 Gold-tail 117 Hawk (eyed) 121 Lackey
* Grey dagger (occasionally found in fruit)

chromo-lith. Cambridge University Press

*PLATE IV*

| | | | |
|---|---|---|---|
| 66 | Bud and allied bud moths | 100 | Ermine (small) moth |
| 74 | Case-bearer moth | 155 | Plum fruit caterpillar |
| 88 | Codling caterpillar | 168 | Green bug moth |
| 88*a* | Codling moth | 169 | Tortrix moth |

chromo-lith. Cambridge University Press

*PLATE V*

69

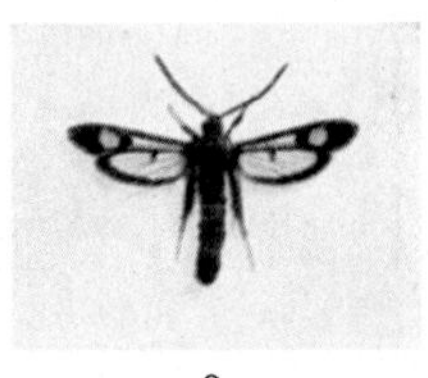

78

81

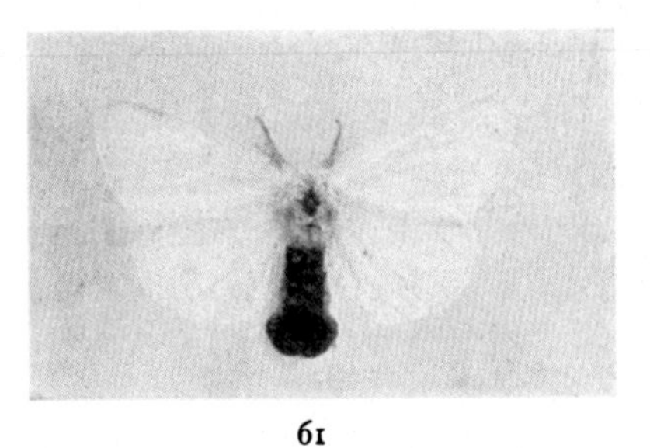

61

84

93

96

61 Brown-Tail Moth
69 Buff-Tip Moth
78 Apple Clearwing Moth
81 Currant Clearwing Moth
84 Clouded Drab Moth
93 December Moth
96 Dot Moth

108

105

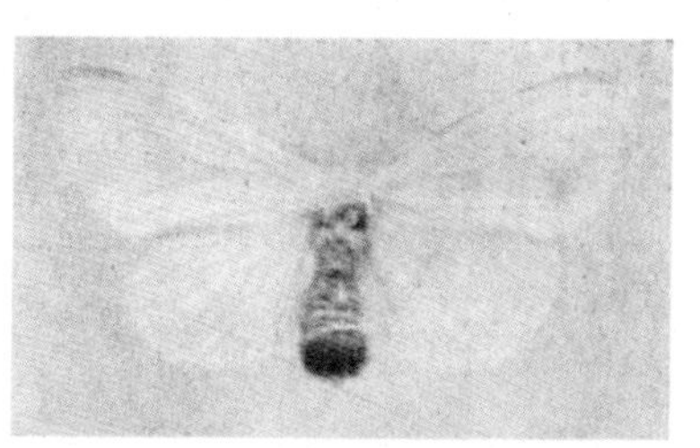

113

117

105 Figure of Eight Moth
108 Goat Moth

113 Gold-Tail Moth
117 Eyed Hawk Moth

*PLATE VII*

121

139

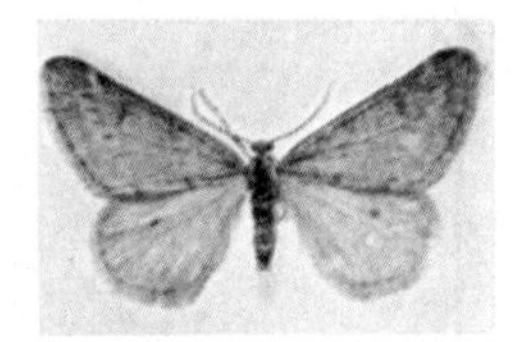
139

134

143

127

143

121 Lackey Moth
127 Lappet Moth
134 Magpie Moth
139 March Moth
143 Mottled Umber Moth

176

147

176

182

172

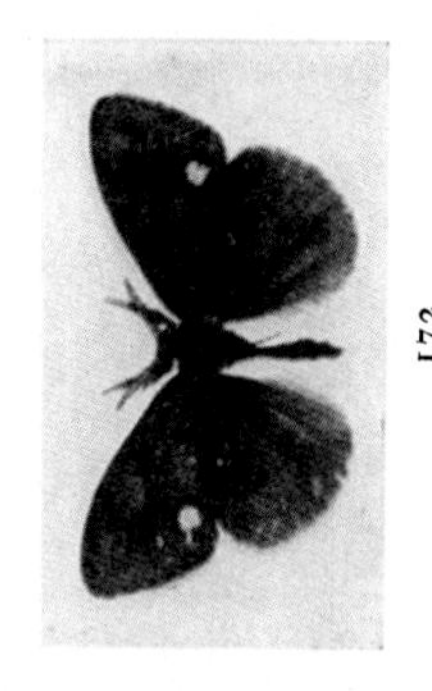

172

163

147 Peppered Moth
163 Tortoiseshell Butterfly
182 Wood Leopard Moth
172 Vapourer Moth
176 Winter Moth

*PLATE IX*

## BEETLE PESTS

### (including Weevils and Chafers)

Showing also the injurious grubs

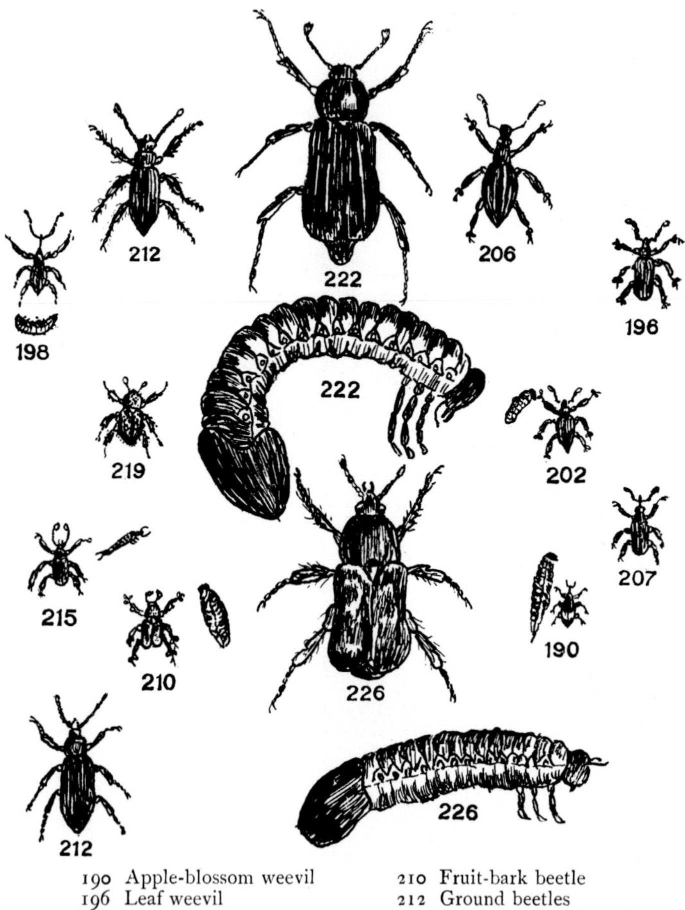


| | | | |
|---|---|---|---|
| 190 | Apple-blossom weevil | 210 | Fruit-bark beetle |
| 196 | Leaf weevil | 212 | Ground beetles |
| 198 | Nut weevil | 215 | Raspberry beetle |
| 202 | Raspberry weevil | 219 | Shot-borer-beetle |
| 206 | Red-legged weevil | 222 | Cockchafer |
| 207 | Twig-cutting weevil | 226 | Rose chafer |

*The numbers refer to pages where the pest is described*

## MAGGOT PESTS

### (Larvae of Sawfly and Midges)

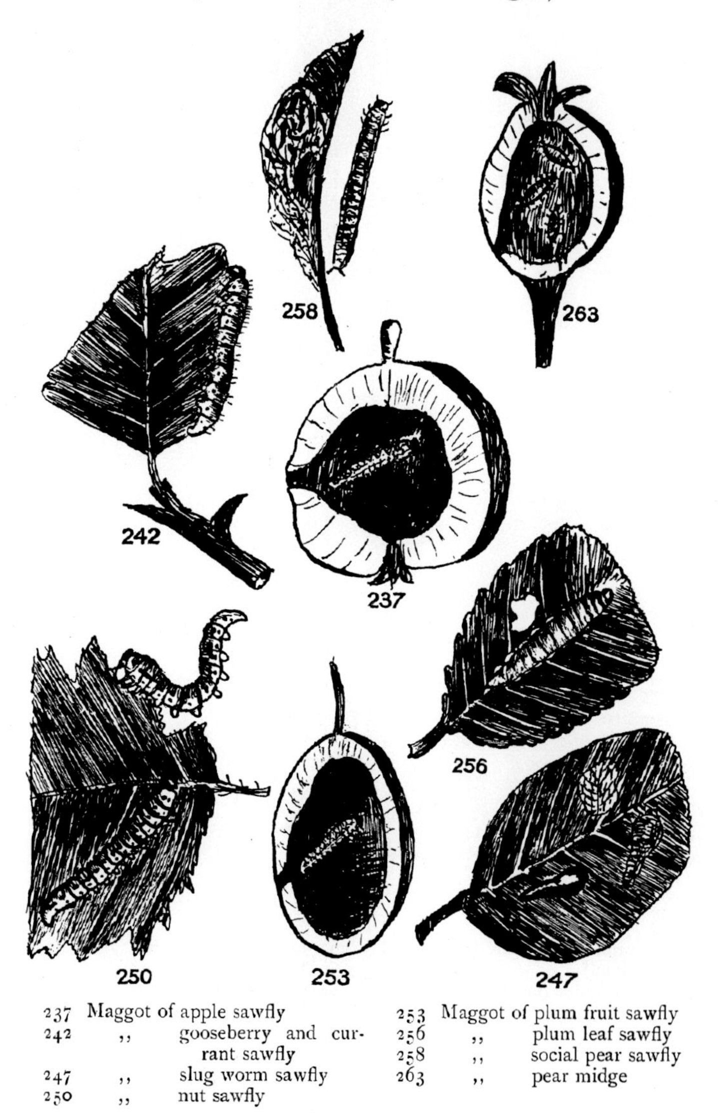


237 Maggot of apple sawfly
242 ,, gooseberry and currant sawfly
247 ,, slug worm sawfly
250 ,, nut sawfly
253 Maggot of plum fruit sawfly
256 ,, plum leaf sawfly
258 ,, social pear sawfly
263 ,, pear midge

*The numbers refer to pages where the pest is described*

*PLATE XI*

## APHIS PESTS

276 Blue apple aphis (on leaves and magnified)
281 Apple-oat aphis (magnified)
283 Green apple aphis (natural size and magnified)
290 Black cherry aphis (natural size)
293 Leaf-bunching currant aphis (magnified)
297 Leaf blister currant aphis (natural size and magnified)
302 Currant-lettuce aphis (magnified)
305 Hop aphis (nat. size and magnified)
313 Plum leaf-curling aphis (on leaves)
317 Mealy plum aphis (on leaves)
321 Raspberry aphis (magnified)
327 Woolly aphis (natural size)

*The numbers refer to pages where the pest is described*

## BUGS, HOPPERS, MITES, ETC.

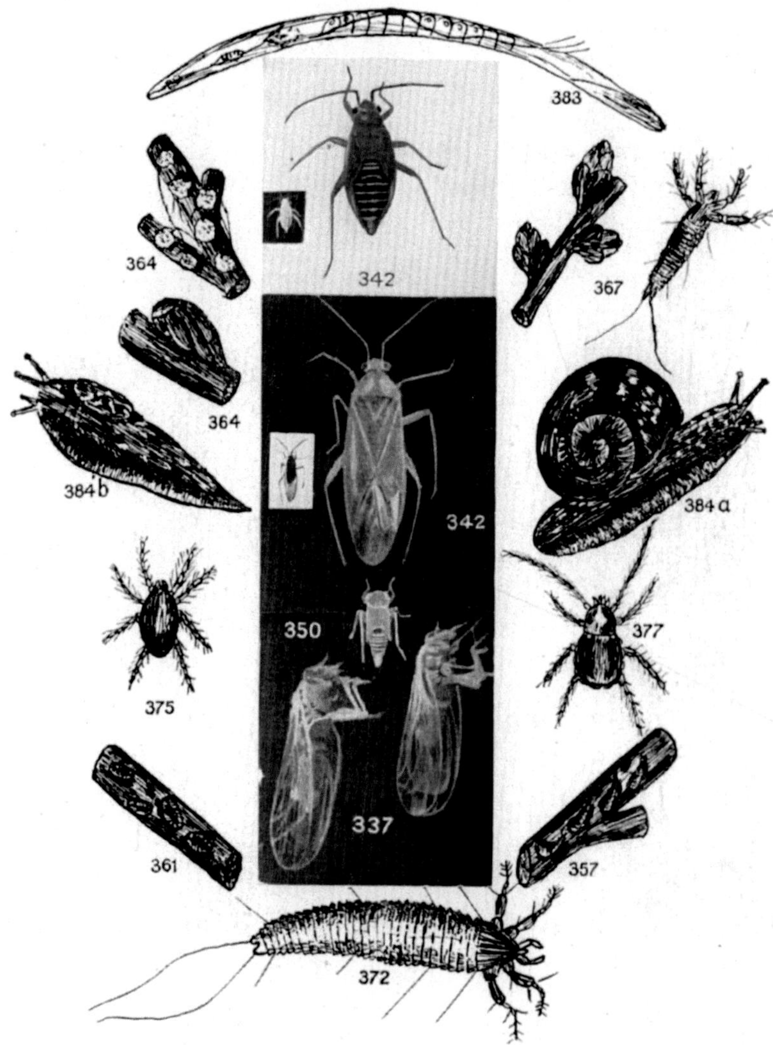


337 Apple suckers (Psylla) magnified
342 Capsid bugs: adult and nymph (magnified and natural size)
350 Leaf hopper (magnified)
357 Mussel scale
361 Brown scale
364 Woolly currant scale
367 Big bud (buds and mite magnified)
372 Leaf blister mite (highly magnified)
375 Hop red spider
377 Gooseberry red spider
383 Eelworm (highly magnified)
384*a* Snail
384*b* Slug

*The numbers refer to pages where the pest is described*

# SECTION I

## INTRODUCTION

# SECTION I

## INTRODUCTION

In common with almost all forms of life, fruit trees are subject to **attacks of disease** of various kinds, the effect of which is to weaken the vitality of the tree or plant, and to prevent to a greater or less degree its various activities, especially the bearing of healthy and abundant fruit.

The diseases which fruit growers have to contend with are practically all **parasitic** in character, that is, they are produced by the attacks of other animals or plants which feed and live upon the juices or tissues of the plants.

A **parasite** may be defined as an animal or plant which continuously derives its nutriment from another individual without immediately destroying it, since such destruction would also involve the death of the parasite.

The animal or plant upon which a parasite feeds is called its **host**. The extra demand made upon a host plant by its parasite almost always results in a weakening of the former, producing finally in many cases death or decay.

The parasites which attack fruit trees belong mainly to two classes, the well-known forms of animal life called collectively **insects**, and a low order of plant life termed **fungi**.

The insects form a very large class in the great division of animals called **invertebrata** or **boneless animals**, distinguished by having **no internal skeleton**. The higher types of insects possess considerable powers of intelligence, or, more strictly "instinct," but most of the parasitic forms are very lacking in this respect.

The **red spiders** (mites) and certain minute "worms" as well as slugs and snails all attack plants, and belong to lower forms of animal life than insects.

The **fungi** are a low class of non-flowering plants which contain no green colouring matter (*chlorophyll*) and are consequently unable, like the green plants, to live upon mineral matters alone, but are obliged to have their food ready prepared for them.

Many fungi are able to live upon decaying animal or vegetable matter, and these (termed *saprophytic*) are usually harmless to plants. Some fungi which are plant parasites are able also to live upon dead organic matter at certain stages of their existence.

As in the case of human beings, a **lowered vitality** produced by unhealthy surroundings is a common predisposing cause of attack. Thus many fungus diseases are only able to attack plants when these are in an unhealthy condition, and "blight" attacks of *aphis* are notably greater when the general health of the plants is poor.

In addition to the effect of the health of the plant on the prevalence of disease, there is the important influence of **weather conditions**. Thus many fungi are only able to maintain a vigorous growth on plants during favourable weather (usually damp heavy atmospheric conditions). This is probably usually due to the effect of such conditions on the plant cells, these becoming charged with excess of sap or nutritious liquid, but such weather conditions may be actually necessary for the activities of the fungus itself, as in the case of the fungus producing "potato blight."

As regards the effect of the weather on insect attack, it is a fact very often observed by growers that **aphis** ("fly") attack frequently follows upon a spell of easterly winds. In common parlance, the east wind "brings the fly." This is specially noticeable in the case of hops.

The effect of **frost**, especially in early spring when the sap is rising in the plant, is often very similar in its effects to actual disease, and the damage noticeable as the result of fungus attack is often attributed to frost, and occasionally vice versa (see page 485).

Although plants are often attacked independently by fungi and by insects, these may work **in combination**. Thus the wounds in the leaves, etc. of the plants produced by the incisions of insects open the way to the entrance of the fungus. A notable instance of this occurs in the case of apple canker following upon attacks of woolly aphis (see page 501). The wounds made by the insect are often so minute as to be invisible to the naked eye. They are quite large enough however to afford a means of entrance to the fungus.

As every grower knows, the **loss** occasioned by insect and fungus attacks is frequently very heavy, and may involve an entire absence of fruit. More frequently the fruit produced is stunted in size, deformed in shape, and rendered unsightly by spots, cracks, or surface markings. Another very common injury is revealed on cutting open the fruit, when the interior is found eaten away by a grub, leaving more or less dark brown "frass" (the excrement of the grub) in its place. In such cases the grub often remains in the apple or other fruit after the latter has been marketed, though a "wormy" apple can always be told by the hole made by the insect in entering or leaving the fruit. Another common source of loss due to disease is the premature falling of the fruit before it is ripe. In the wide-spread "brown rot" disease the fruit continues to hang on the trees in a hard "mummified" condition.

Owing mainly to the orchards in this country being scattered over considerable areas, and to the diverse varieties of fruit planted, English growers are not subjected to attacks of such virulence as in the case of some other countries, notably America, where attacks on fruit assume often the nature of **epidemics**, rapidly spreading over large areas. Huge sums of money have been, and are being spent annually by the State in combating various pests which have obtained a firm hold, as, for example, the caterpillar of the gipsy moth.

Although at a first glance it may appear that many insects and fungi are very similar in their methods of attack, closer study has shown that many important differences exist of which an accurate knowledge is essential if a successful method of treatment is to be employed. In particular it is necessary to know at what period in the **life-history** of the pest it is most open to attack and to take advantage of the information thus gained.

Thus it is found that during a considerable part of the life of most insects and fungi these are either very resistant to any destructive agents we can employ, or that they are completely inaccessible, being securely hidden away, or in some manner protected from attack.

An important point to bear in mind in regard to the question of devising remedies for insect and fungus attacks, is that the tissues of the plant itself consist of living cells like those of the parasite which it is desired to destroy. It follows that the choice of remedies which, while being harmless to the plant, are effective in killing the parasite is **very restricted**, and also that in any of the remedies employed there will always be a certain risk of some degree of injury to the plant. A further point to note is the importance of using exact and uniform strengths in the application of remedies, since if certain concentrations are exceeded, damage to the plant is bound to result, the appropriate strength being obviously a carefully ascertained mean between what will prove harmless both to the plant and the parasite, and such a strength as will injure them both.

The **remedies** employed in practice are various, and depend upon the character and situation of the pest in question. A very effective means of control is by "**spraying**" or "**dressing**" with solutions or suspensions of various substances. The material is thus brought in contact with the whole exposed surface of the pest which would be difficult of accomplishment in any other manner. An alternative method is the blowing on to the plant of the material in the form of a *fine dry dust* or powder. Pests which live in the soil are dealt with by *digging-in* various poisons, or by *injecting* poisonous volatile liquids near the parts attacked. Where circumstances permit of its employment—as in the case of small indoor plants—*fumigation* with poisonous gases is very effective.

In the case of insects it is necessary to distinguish between those which *eat* the tissues of the plant, and those which *suck* the juices only, having no means of "eating" as ordinarily understood. The first class of insects can be killed by means of poisons placed on the surface of the plant, but this method is of no use at all to the insects which suck the sap of the plant. The only means of reaching this class of offender is by absorption of poisonous solutions through the soft body of the insect, or by acting on the breathing pores which may be made to absorb the poison, or may become choked up, thus causing suffocation.

Spraying has long been employed as a **standard practice** in fruit growing abroad, especially in America, where it is regarded as essential to the production of clean healthy fruit, and good crops. It is only in comparatively recent times however that the importance of spraying has been recognised in this country, and in many ways we are still very backward in this direction.

A fact which greatly helps the American fruit grower is the *interest in the industry taken by the State*. A government department is responsible not only for giving free advice on all subjects connected with fruit culture, but undertakes free analyses of materials, soils, manures, etc., and has State Research stations established in every district for carrying on experimental work in the interests of the growers. Analyses of advertised remedies are given, together with hints as to their relative efficiency, so that profiteering at the growers' expense is effectively prevented, while legitimate manufacturers whose products merit recommendation are encouraged.

It cannot be too often emphasised that unless spraying is intelligently carried out at the right times with suitable materials, accurately compounded and with efficient appliances, it is often **worse than useless.**

It will therefore amply repay every grower of fruit, be he small or large, to make a **careful study** of the pests by which his fruit is attacked and to take full advantage of the information which has been gained by careful and patient research through many years. Much has yet to be learnt, and new and interesting facts are constantly coming to light, but it is now possible by careful treatment to "control' or limit the activities (and therefore the damage) of most of the pests to which fruit trees are subject, and to ensure a clean and healthy crop of fruit. The actual yield of fruit is obviously dependent on other factors besides attacks of parasites, such as variety of tree, fertilisation of the blooms, frost, unsuitable weather conditions and other somewhat obscure influences.

A matter which requires serious attention is the question of making the spraying of fruit and other crops **compulsory** by the introduction of the necessary legislation. This has been already done in the United States with most beneficial results for the industry. Until such procedure is adopted in this country it will be possible for growers to take all possible pains in the culture of their fruit in keeping it free from disease and yet fail entirely to eradicate serious pests because a neighbour allows his plantations to become breeding grounds for the pests and thus form a centre of infection in the whole surrounding district. At the present time this is the chief deterrent to the proper control of disease and it is extremely discouraging to intelligent, up-to-date growers who, after having done their utmost in this direction, find their labours nullified by the carelessness or indifference of unprogressive neighbours. What is needed is united action to compel the government to give the necessary consideration to the question and to give facilities for introducing into parliament the necessary legislation[1].

[1] Signs are not wanting that this important matter will soon receive full consideration. Compulsion has been adopted in the case of Silver Leaf disease (see page 577), certain measures to prevent its spread being now enforced.

# SECTION II

## HOW FRUIT TREES LIVE

# SECTION II

## HOW FRUIT TREES LIVE

In order to have a correct understanding of the diseases which attack fruit, it is necessary to start with some knowledge of the botanical aspect of the subject, that is, of the **structure** and **use** of the various parts of the plant. As an illustration of this, if in the past a close study of the human body had not been made, the physicians would have had little knowledge of the diseases to which it is subject and of successful methods of cure. It is not possible however to do more, within the limits of a book of this character, than to indicate in a very brief and simple manner some of the more important considerations. For further details and for a more scientific treatment of the subject, reference should be made to the many excellent works available on the botany of plants.

In the first place it is well to emphasise that a fruit tree is a **living being**, exactly in the same sense as we ourselves are. It was (or might have been) produced from a seed, the product of two parents of different sexes. It lives by eating and drinking, having "mouths" to devour, digest and assimilate the food supplied. The organs which correspond with the digestive organs of animals are the leaves, whose business it is to catch the minute floating particles of carbon dioxide from the air around, and to "suck out" from them the carbon, which is the principal food and mainstay of plant life. Further, fruit trees "drink," by means of their roots, the liquid constituent from the soil, along with that other necessity of all living organisms, namely, nitrogen.

Fruit trees also "marry" and rear families, having **two distinct sexes**, male and female, but each existing upon the same plant. The reproductive parts of the tree are the flowers, whose function it is to produce the seeds by means of which the plant is propagated. The male portions of the tree are known as *stamens*, and these produce and disperse a yellow powder, the "pollen," which is able to fertilise the young seed or *ovule*; the female portion is called the *pistil* and it contains the undeveloped ovules in an *ovary*, and these can only develop and grow after fusion with a pollen grain from the male portion of the plant. The ovules are thus like the eggs of animals, and after fertilisation the pistil develops into the fruit.

Like all the higher forms of life, the tree is made up of millions of minute living *cells*, these forming the "bricks" of which the structure is built. Every plant cell is constructed in an exactly similar manner. First there is the *cell-wall* which separates one cell from another, and within the cell is a semi-fluid substance called **protoplasm**. This material is the essence of all life; all living beings are built up of it and it is the only vital material with which we are acquainted. Chemically, it consists mainly of four elements combined together, namely, Oxygen, Hydrogen, Carbon and Nitrogen, but the chemical composition explains very little of its living activities, the important point being that this remarkable substance is endued with life and it is owing to its life activities that the plant is able to feed and grow.

In all the higher plants the protoplasm of many of the cells contains a green colouring matter called **Chlorophyll**, and it is this substance which is able, in conjunction with the protoplasm, to feed upon the carbon dioxide of the air (see p. 15). Under the action of light, it splits up the latter into *carbon* which it absorbs and builds up into the nutrient material and *oxygen* which is set free. Carbon is used, together with oxygen and hydrogen, for the manufacture of sugar and starch, the latter being a reserve food material for the plant. Cells which contain no green chlorophyll are unable to split up the carbon dioxide in the air in this manner—thus, the cells of fungi, which have none of this substance, are obliged to obtain their supply of carbon from other sources; those which attack other plants, from the cells of their host.

Trees have a **skeleton** which supports them just as have animals. This skeleton is made up of ribs or bundles of woody material which, commencing in the smallest rootlets, run up the stem or trunk through the branches, and ramify into every leaf. The cells contained in these bundles, which form the supporting part of the tree, have walls specially thickened with woody material, while the tissue of the remaining cells is comparatively soft and yielding. This hardened or *vascular tissue*, as it is called, consists partly of very long cells called *vessels* which form a rapid means of communication with all parts of the plant.

We are now in a position to consider in greater detail the different portions or organs of the tree and their special structure and use.

# CHAPTER 1

## The Leaves

The leaves of a plant correspond with its **digestive organs**. It is not generally recognised that the chief food of every plant is derived, not from the *soil*, but from the *air*. The amount of carbon in a tree is easily seen to be a large proportion of its weight (as, e.g., from the amount of charcoal, which is mainly carbon, formed on burning), and this is all derived from the air by means of the leaves of the plant. Thus every leaf is, during the daytime and under the influence of light, feeding the tree by separating the carbon from the air and by digesting this in cells of which it is composed.

In order that each leaf may intercept a maximum amount of **light**, the leaves are extended horizontally, and in order that there may be a very large surface exposed to the air, each leaf is very thin and flat. They are also not rigid, but are placed upon flexible stalks which are able to move with the slightest breeze and thus come into contact with fresh "particles" of atmosphere.

Each leaf consists of a supporting skeleton of a **vascular tissue** previously mentioned and this ramifies throughout the leaf forming numbers of small enclosed spaces. These are filled in with softer tissue composed of several layers of cells, most of them containing the green *chlorophyll* referred to above.

To the naked eye the under surface of the leaf is seen to vary from its upper surface; thus, if an apple leaf is examined it will be seen that the upper surface is more or less clear and of a dark green tint, while the under surface is lighter and usually quite dull, and the skeleton of veins is much more clearly seen than from above. If the leaf is examined under a powerful microscope the under surface will be seen to have a great number of small openings (see fig. 2). These pores or **stomata** allow free entrance of air and of water vapour to the cells of the leaf. The upper surface of the leaf is usually devoid of any such openings.

If we take a thin slice through a leaf, and magnify it many times under a microscope, we shall find that it presents the appearance shown in fig. 1.

Each surface of the leaf consists of a thin layer of colourless cells, known as the **cuticle**, which protects the softer cells below from injury. The inner skin or **epidermis** consists of a row of regular cells all coloured dark green with chlorophyll. Then come several layers of densely packed cells followed by a spongy tissue consisting of cells very loosely packed, with many interspaces. The under surface of the leaf has a similar "skin" layer of cells but in this case the layer is interrupted at close intervals with the pores or

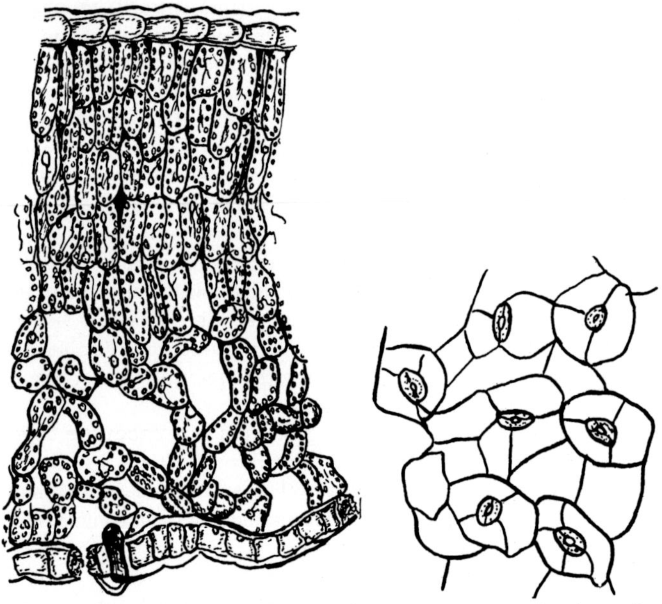

Fig. 1. Appearance of a thin slice through a leaf (highly magnified).

Fig. 2. Diagram of under surface of leaf showing "pores" (stomata) highly magnified.

"stomata," each of these admitting air to the spaces in the spongy tissue before mentioned. These stomata have dark-coloured "guard cells" which open or close the apertures under certain conditions.

Leaves are often described as the *breathing organs* of the plant, and to some extent this is true, but it would be more correct to call them the **digestive organs** of plants. Thus quite one-half of the dry weight of any plant is made up of a substance *carbon* which is entirely derived through this digestive action of the leaves. This fact is even more remarkable than it appears at first sight, since the carbon (readily seen in the form of charcoal on burning) is obtained from a constituent of the air in such a dilute form that it only exists in the proportion of 2—3 parts per 1000 of air. Further, of this substance, which is termed carbonic acid, or more correctly *carbon dioxide*, less than one-third consists of carbon, the rest being oxygen, the well-known gas which forms the active constituent of the atmosphere. When the leaves of the plant, by the action of the light on the chlorophyll, assimilate or "digest" the carbon, they set free the oxygen; consequently all green plants are *oxygen producers*, although the actual amount set free is of course relatively small. The form in which the carbon appears on its separation from the air is either as *starch* or *sugar*. Starch is the form in which the food is stored up for future consumption, while in the form of sugar it can be conveyed to all parts of the plant.

All fruit trees are *deciduous*, that is, the leaves only last for a single season, falling off the trees and dying every autumn, to be followed by a new growth the succeeding spring. Thus each leaf has a **definite length of life**. Towards the end of the summer the leaf cells become less active and the starch granules in the leaves are transformed into soluble sugar which is conveyed to the roots of the plant where it is again stored as starch. With the ripening of the fruit the immediate necessity for the activity of the leaves is over, but in many cases the leaves do not at once cease their work, but continue to "extract" the carbon from the air, and to store this in the roots. This action of the leaves is undoubtedly very important, as it results in a rich sap ascending the trunk and branches to supply nourishment to the young shoots in the following spring.

### Practical Considerations

It follows from what has been stated above how important it is to **avoid injury** of all kinds to the leaves of plants, since with every leaf or part of a leaf which suffers injury, there results a diminished food supply resulting ultimately in loss or damage to the fruit crop. Also, that even after the fruit is mature, it is important that there shall be no loss or injury of leaves until their natural fall in the autumn, because the vitality of the tree is thereby lessened, and there is a decreased food supply for the following spring.

#### CATERPILLARS, AND OTHER LEAF-EATING INSECTS

1. Caterpillars usually attack leaves from the under surface, chiefly because of the greater protection afforded against their natural enemies.

2. Young caterpillars generally "skeletonise" the leaves, eating out the soft tissue from between the harder "veins" of the leaf.
3. When caterpillars are very numerous on a tree, they can soon reduce the entire leaf surface by one-half, and this implies that only one-half the nutriment in the form of sugar or starch is being manufactured. This must inevitably result in a great drain on the vitality of the tree, and emphasises the need of dealing very promptly with a caterpillar attack.
4. Even if the measures taken to deal with a caterpillar attack do not suffice to save the fruit for the current season, a very great amount of good is effected if the leaves are saved from injury, since the storage of food is not interfered with, and the tree will greatly benefit in the following year.
5. Conversely, when a caterpillar attack is allowed to continue unchecked, resulting in a great loss of effective leaf surface, the general health of the tree is bound to suffer, and the effect may continue for several years before the consequent loss of vitality is recovered.

PLANT LICE AND OTHER SUCKING INSECTS

1. Aphides and other plant lice invariably attack the under surface of the leaves and by removing the sap from the cells often cause the leaves to curl and so form a protected enclosure. The under surface is also probably easier of penetration by the "suckers" (proboscides) of the insects.
2. The immense amount of harm done by a bad attack of plant lice is seldom fully realised; it may be seen however from the above considerations, that:
   (*a*) Free access of air is interfered with by the lice blocking the stomata.
   (*b*) In the case of the leaf curling lice, access of light is also prevented.
   (*c*) The cells are continuously exhausted of the sugary sap which it is their function to produce.
   (*d*) To make up for the loss of liquid in the leaves, sap flows from other parts, producing a general exhaustion of the plant.

Plant lice are often so small that growers are apt to underestimate the harm which is produced by their attack. It is not however the *actual* size, but the *relative* size, as compared with the thickness of the leaf which they attack, that is the point to consider.

Thus, if a thin section of the leaf is made and the thickness of the leaf compared with the size of the lice which attack it, it will be seen that the latter often greatly exceed the total leaf thickness, that is, their bodies are larger than the total number of cells which they attack.

# CHAPTER 2

## The Trunk and Branches

The trunk or stem of a tree or plant serves primarily as its support and for this purpose it contains bundles of thickened or **woody material**. Surrounding this is a softer tissue made up of more delicate cells. These are protected by the outer bark of the tree, and if this outer bark is injured or destroyed there is grave risk to the life and health of the plant. The trunk also contains special long cells, called *vessels*, along which the sap passes to and from the root and the branches and leaves.

The outer bark of some trees is very frequently rough and scaly, especially at certain times of the year, and other trees shed some of their bark each year. Moss and lichenous growths are apt to occur on fruit trees in neglected orchards.

The branches may be regarded as sub-divisions of the trunk for the purpose of supporting and feeding the shoots, and consequently the leaves, blossom and fruit.

### Practical Considerations

1. It is necessary to protect the bark of the tree from injury, in order that the food supply from the roots may not be hindered or cut off. Thus the use on the trunk or branches of any corrosive or poisonous material which may penetrate into the inner tissues must be avoided.
2. Beetles, which bore into the bark, and certain sucking insects, such as woolly aphis, scale insects, red spiders, etc., do a great deal of unsuspected injury and must be regarded as dangerous pests.
3. The bark of the trees affords shelter and winter quarters for many different kinds of pests, and the trunk and branches must be kept clean and not allowed to become covered with lichen and moss, or still more protection is provided for pests which would otherwise find no harbourage.
4. In winter, during the dormant stage of the tree, there is a good opportunity to apply cleansing sprays, as the trunk and branches are then far less liable to injury from a more or less corrosive substance than during the spring and summer months.

# CHAPTER 3

## The Root

The root of a tree has broadly two functions. It maintains a firm hold on the ground, and it is the organ whereby **nourishment** is obtained **from the soil**. The root furnishes all the nutritious substances which are required for the life and growth of the tree or plant with the single exception of the *carbon*.

The root of a tree is a wonderfully complicated system, the primary root forming very numerous branches ramifying in all directions, and the final endings or rootlets being often hair-like in fineness. Such a system ensures that very thorough contact is maintained between the root and the soil particles.

Roots differ a good deal in their modes of growth, some spreading fan-like near the surface of the ground, and others growing deep into the subsoil.

As previously mentioned, the root not only supplies the plant with moisture and nutriment, but it also acts as a storehouse for the food manufactured by the leaves in the form of starch.

The roots of trees are not protected by bark as are the trunk and branches and the tissues of which they are formed are much softer and more delicate. The cells are also not thickened with woody material as, being buried in the soil, the root is practically immune from injury from ordinary physical causes.

### Practical Considerations

1. The roots of trees and plants are liable to attack from several species of insects and fungi. Since the attack is unseen it often becomes very serious before the nature of the trouble is suspected.
2. Fungus attacks are very difficult to eradicate. In the case of attacks by insects, recourse is had to the injection of poisonous liquids, such as *carbon bisulphide*, which readily vaporises and so diffuses through the soil. Care must be taken that such liquids do not come into actual contact with the root itself.
3. During the dormant season it is possible, though often not practicable, to lay bare the root and apply spray fluids containing insecticides in solution.

# CHAPTER 4

## The Flower (*Blossom*)

The flowers of plants correspond with the **reproductive organs** of animals, and are the means, in the wild state, of carrying on the plant from one generation to another. From the point of view of the botanist, therefore, they are perhaps the most important, and certainly the most highly specialised parts of the tree.

To the grower they are also of extreme importance, since they are the necessary means by which the fruit is produced—"no blossom, no fruit."

All flowers consist of certain well-defined, distinct parts, or **organs**, and those which are commonly best known, the conspicuous, often highly coloured *petals*, are perhaps the least important parts of the flower—their chief function in point of fact being undoubtedly to attract bees and other insects (see below).

The figure shows a section or cut through an **apple blossom**.

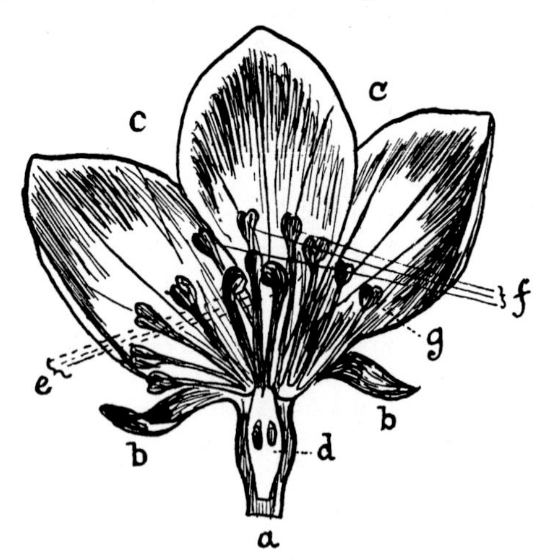


Fig. 3. Cut through an apple blossom, showing its various parts.

There is first the flower stem (*a*), then the green *sepals* (*b*) which form a protective sheath around the unopened flower, together termed the *calyx* (or "cup").

Then comes the circle of *petals* (*c*) forming the *corolla* which besides attracting the bee and other insects by its conspicuous appearance (and perhaps in some cases scent also) acts as a protection to the delicate portions within.

Now come the parts of real significance, but often overlooked by the casual observer.

Of primary importance is the chamber, revealed by cutting through the flower, which contains the future seeds, now termed *ovules*. This is the **ovary** (*d*) and is the part of the flower which subsequently develops into the fruit.

The ovary extends upwards into one or more stalks, or *styles* (*e*), ending usually in a more or less sticky cap—the *stigma*—which is designed to retain the *pollen grains* (see below).

These parts of the flower are collectively termed the *pistil* and form the FEMALE ELEMENT.

The MALE ORGANS of the flower consist of the **stamens** (*f*). These are formed of stalks of various sizes in different plants, which carry hollow cases or capsules, the *anthers* (*g*). Inside the anthers are produced hundreds of fine floury grains, the *pollen grains*, which are liberated when the anthers become ripe.

The importance of the pollen grains is well known. Without their assistance no development of the fruit is possible.

The object of the flower therefore is to **produce the seed** so as to carry on the plant from generation to generation. Before this can happen, each *ovule*, or immature seed, must be "fertilised" by a pollen grain. The manner in which this is carried out is very interesting and remarkable. The pollen, on being liberated from the anthers, is carried by the wind, or by bees or other insects, until a grain happens to alight on the sticky surface of the stigma of another flower. The pollen grain now produces a long tube, which passes down the stalk or "style" of the pistil, enters the ovary, and penetrates into one of the ovules therein contained. Complicated and imperfectly understood changes now occur in the latter, resulting finally in the production of the seed and the **formation of the fruit**.

Usually flowers require the pollen from another plant for fertilisation, i.e. they are "cross-fertilised." Occasionally flowers are capable of self-fertilisation.

After successful fertilisation, the ovary enlarges, the petals fall off, the stamens and stigma wither up, and a young fruitlet is produced. A portion of the calyx remains in the apple and pear forming the "eye" of the fruit.

### Practical Considerations

A number of practical points suggest themselves from the foregoing description. A few of these may be mentioned.

1. It is of the first importance to preserve the flowers from damage either by insect or fungus attacks, or from the use of sprays which may injure the blossom.
2. Injury to the *petals* only is a very minor consideration. The important thing to guard against is burning or other DAMAGE TO THE STAMENS AND THE PISTIL (or stigmata). In the case of spraying being necessary to control insect or fungus attack while the tree is in blossom, these parts of the flower should be carefully examined for browning or signs of scorching with a powerful hand glass or microscope.
3. Bees in an orchard are often of great advantage in ensuring cross-fertilisation, since the time during which the pollen grains are liberated from the anthers may be very short, and it is unwise to rely on the agency of wind alone to distribute them. Since the blossom, when fully expanded, is very exposed and very susceptible to injury, it is important to confine spraying with substances of a corrosive nature to the time before the *bursting of the buds.* At a short interval of time just before this happens it is possible to kill many of the forms of insect attack before the bud is entered and before they are protected.
4. Considering the profusion of blossom which often occurs, it is quite impossible for more than a small proportion to develop, as otherwise the tree would be greatly overloaded with fruit.
5. The appearance of a profusion of blossom on a tree cannot be taken as ensuring a proportionate yield of fruit.

# CHAPTER 5

## The Fruit

The fruit is the final product of the tree, and it is produced by the development of the ovary. The important part of the fruit from a botanical point of view is the **core** which contains the seed.

The flesh (pericarp) of the fruit, which is the only part of practical importance to the grower, was probably designed by nature to attract birds and other animals to devour the seed, and to pass the latter through their system and so provide for its propagation in other localities.

The fruit is protected by a fairly tough outer skin, any injury to which while the fruit is in an immature stage resulting in *dwarfing* or *distortion*.

### Practical Considerations

1. It is necessary to see that the amount of fruit on a tree is not in excess of what it can bring to maturity without too great a drain on its vitality.
2. A small degree of insect or fungus attack resulting in the loss of a proportion of the fruit may thus not be a disadvantage.
3. Poisonous sprays (such as the arsenic compounds) should not be applied when the fruit is near maturity.
4. A careful watch should be kept for those pests which cause slight injuries to the skin of the fruit when immature, so that distortion may not afterwards result.

# PART I

## INSECT PESTS AND THEIR CONTROL

## SECTION III

### ABOUT INSECTS

# SECTION III

## ABOUT INSECTS

## CHAPTER 6

### The Structure of Insects

Insects belong to a class known as *Arthropoda* or **jointed limbed animals**. Besides the insects, this group contains also the spiders and mites (Arachnoidea) a few of which are of importance to fruit growers. The only other orders of any interest are the **worms** (Vermes), of which the strawberry eelworm is a destructive member, and the snails and slug family (Mollusca).

True insects are distinguished by having

1. A distinct division of the body into three parts
   (*a*) Head,
   (*b*) Thorax or chest,
   (*c*) Abdomen.

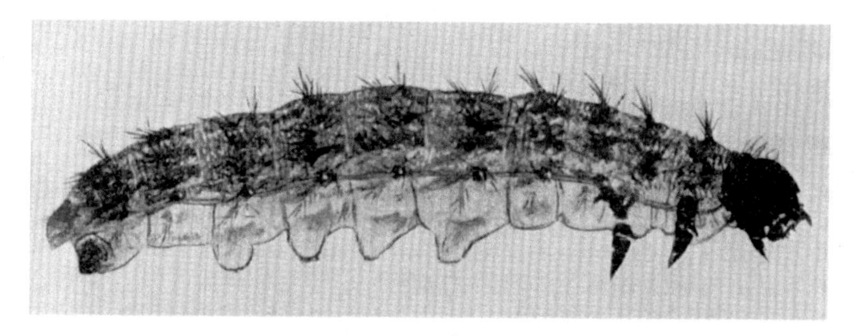

Fig. 4. Caterpillar of the Large Tortoiseshell Butterfly, showing true and false feet (prolegs) also main breathing tube and spiracles (see p. 30). Magnified ( × 3).

2. Neither more nor less than **six legs**.

Instead of having a backbone and internal skeleton as in the higher animals, insects have an external protective covering, formed of a hard or elastic substance, called *chitin*, which is originally liquid but hardens on exposure to air. To this the muscles of the body are attached.

## The Head

The head of an insect is very variable in size and shape, according to its habits. It possesses a pair of **eyes** which differ from those of mammals in being *compound*, that is, made up of hundreds of eyes grouped together. For this reason it is probably able to distinguish objects in motion very readily. Besides these compound eyes, some insects are also furnished with one or more simple eyes. Near the eyes are the **feelers** or *antennæ*, and these vary very much in size and shape.

The **mouth** is however, from the growers' point of view, the most important of the structures of the head. According to the type of mouth which insects possess they may be divided into two classes:

1. Biting insects.
2. Sucking insects.

The former are provided with powerful jaws for biting and chewing, with an upper and lower lip, and a "tongue." The latter have none of these structures, but in place of them there is a stout beak which is hollow and formed like a tube. This is for insertion into the tissues of the plant to draw up the juices.

Many insects of this type have also two or three sharp spines or lancets which are used to rasp or prick the tissues so as to cause a greater flow of juicy sap.

As will be seen in a later chapter it is most important to distinguish between these two classes, since insects of the first class (those with jaws as e.g. beetles) may be killed by placing poisonous materials on the leaves or other portions of the plant, and this will be eaten by the insect along with its food. In the case of insects of the other group (without jaws but having beaks, such as *aphides* or plant lice) no effect is obtained by placing poisons on the leaf, and the insect must be killed by the action of a corrosive poison on its body, by suffocation, or some other means.

## The Thorax

The thorax is the middle part of the insect's body. It has usually three distinct rings or *segments*. On each of these is a pair of jointed legs, and on each of the last two, a pair of wings, except in the case of the two-winged flies, which have only one pair on the middle segment.

The **legs** of insects consist usually of four joints :

1. A short stout joint next the body (*the trochanter*),
2. A large and heavy joint (*the femur*),
3. A slender joint (*the tibia*),
4. A foot or *tarsus* of five joints, and often having a claw and pad or *pulvillus*.

The **wings** of insects vary enormously in shape, size, appearance and texture. They form the basis of classifying many insects, as described in a later chapter. Thus in the beetles, the front pair of wings is hard and horny and serves as wing cases for the hind pair. In moths and butterflies the wings are covered with coloured scales, which form many diverse and beautiful patterns.

### The Abdomen

The third and hindmost part of the body is termed the abdomen. In many cases, it consists of ten rings or segments, though some of these may be joined together. It has no legs or other appendages. It contains the greater part of the digestive organs and the sexual or reproductive apparatus. The abdomen is usually very muscular, and can be twisted almost double, which is often of great use to the insect, e.g. in egg-laying. See also Appendix III, pp. 698—700 (figs. 296, 297).

## CHAPTER 7

### How Insects Live

**Breathing** is necessary to insects as to higher animals. Their method of obtaining the air they require is however different. No air is taken in through the mouth parts or head. There are no lungs. In place of these there are a number of small openings, called *spiracles*. These form two rows, one on each side of a segment of the abdomen.

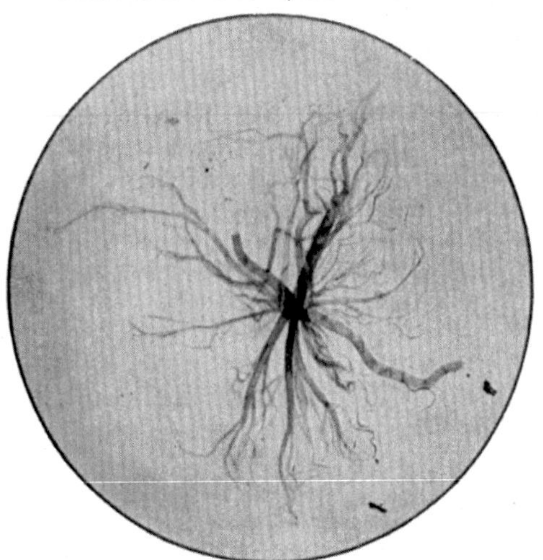

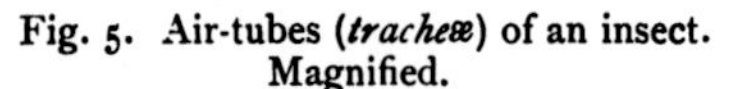

Fig. 5. Air-tubes (*tracheæ*) of an insect. Magnified.

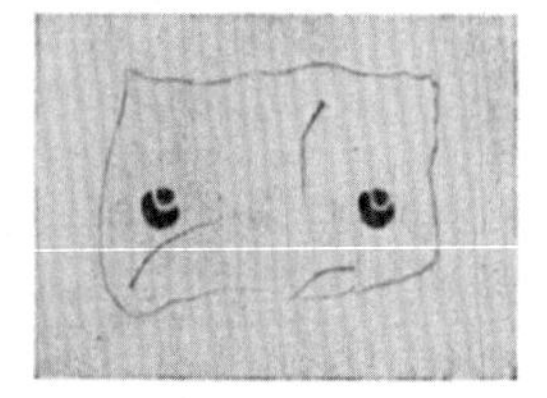

Fig. 6. Spiracles of an insect. Magnified.

Each of these openings communicates with a main air-tube running lengthwise on each side of the body. From these tubes, many smaller branches are given off which branch again many times, till every part of the body is fed with them even to the tips of the feelers. Air is thus distributed to all parts of the tissues. The tubes are termed *tracheæ*, and the openings are guarded by hairs or other structures. If these are blocked up, the insect is sooner or later suffocated. It is also susceptible to the influence of poisonous gases mixed with the air, and upon this the success of *fumigation* for the destruction of insect pests depends. The tracheæ are strengthened by a spiral thread of tough material, by which means they are kept expanded.

**Blood** as we know it is absent, but there is a greenish yellow fluid surrounding the internal organs, which answers to it. There is no closed system of arteries and veins, as in the higher animals. The fluid is however propelled by a species of "heart" consisting of a pulsating structure just beneath the upper surface of the body. It contains chambers and valves.

**Digestion** of food is carried out in a series of organs answering

Fig. 7. Diagram of digestive tract of an insect. Magnified.

to those of the higher animals. These vary greatly in many species according to their food habits. There is often a gullet, crop, gizzard, and intestine, though these may be more or less continuous and similar in structure in the simpler types. The undigested food is expelled at the end of the body through the *anus*. Organs that answer to kidneys are also often present.

**Nerves** are supplied in abundance. These are probably necessary on account of the great activity of many insects. There is a central nervous system, passing lengthwise just over the lower surface of the body. In the head, thorax and each segment of the abdomen are swellings of the main nerve, termed *ganglia*, and from these, numerous branches diverge to all parts.

**Sensory organs** are in many cases highly developed. The *sight* of insects has already been referred to. In many instances, *hearing* also exists, the position of the "ear" having been located. In addition there is a sense of *taste* and a very finely developed sense which we may describe as *smell*. This is probably situated in the feelers or antennæ, and guides the insect to its food or to the plant suitable for egg-laying. Its chief purpose, however, is to enable the male insect to find the female at the mating season, and so acute is its sense in this respect that the male insect has been proved to be able to discover a female from long distances, or within closed doors.

The sense of *touch* is also specially developed. It is associated not only with the "feelers" but with various hairs, connected with nerves, which often occur freely over the body.

### Reproduction

The male and female are distinct in practically all insects. Highly developed organs of generation exist. Pairing takes place immediately preceding egg-laying, the season of the year varying a good deal in different species. Eggs are laid by the female insect in all cases of sexual reproduction, but in some of the lower forms, e.g. the "green-fly" or *aphis*, "unsexed" forms exist which are capable of producing living young for many generations during the summer months. In some of the higher forms (e.g. bees and wasps) a "worker" insect is produced, along with the sexual insects, which is incapable of reproduction. Definite communities, "hives," are formed in which the "workers" attend upon the female insect and rear the young.

In order that the egg shall be laid in the most suitable surroundings for its development, the female is provided in many cases with an egg-depositor or *ovipositor*. This may be very long and sharp, and capable of puncturing the tissues of the plant in order to place the egg therein. In the case of insects which are parasitic upon other insects, the eggs are laid inside the body of the caterpillar or chrysalis (see Section VI, p. 471). The curious manner of egg deposition adopted by the apple blossom weevil is described on page 192.

Male insects are invariably fully winged, but many females are without wings or have only very imperfectly developed ones. In the case of the moth family, such females ascend the trunks of trees to lay their eggs and may readily be caught by means of bands of sticky composition placed around the trunk (see pages 180, 404).

## CHAPTER 8

### The Changes which Insects Undergo

The great majority of insects, during their growth, go through **distinct changes of form.** This process is termed *metamorphosis.*

Take an example at random. The Winter Moth female, after pairing with the male, lays EGGS, which in course of time hatch out to minute CATERPILLARS; these grow until mature, during which time the skin is cast or "moulted" several times. They then grow listless and inactive, enter the ground, and commence to shrink up, the skin finally becoming quite horny; each caterpillar thus changing into a PUPA or "chrysalis." In due time, the pupa skin breaks open, and the fully developed MOTH or adult insect appears.

The foregoing is an example of *complete metamorphosis.* It involves four distinct changes of form, viz.:

**egg** form (or **ova**),
**larva** ,, in this case a "caterpillar,"
**pupa** ,, ,, ,, "chrysalis,"
**adult** ,, ,, ,, "moth."

The higher classes of insects, e.g. moths and butterflies, beetles, bees, wasps, and sawflies, two-winged flies, and most of the parasitic insects, all pass through these four stages.

In some of the lower classes, the larva (grub, maggot, or caterpillar) form is indistinguishable from the pupa condition. For example the adult female capsid bug lays an egg in the shoot of an apple tree, which in the spring hatches into a minute "bug" or wingless insect. As this grows, several moultings of the skin occur, and at each moult (or *instar*) slight differences in the appearance result. Wing-buds appear, and at the last moult the adult winged insect is formed, but the bug remains active in all stages. In such cases the term "pupa" is not applicable, so the word NYMPH is employed to distinguish the stage when the wing-buds first appear till the adult insect is formed. The word "larva" may be used (though not strictly correct) to cover the period from the hatching of the egg till the appearance of the wing-buds.

In order to distinguish the young stages of the various classes of insects, the following words are used throughout this book:

| | Moths and Butterflies | Beetles, also weevils, chafers, etc. | Flies and Sawflies | Aphides ("green flies") and plant bugs |
|---|---|---|---|---|
| egg stage | ova (eggs) | ova | ova | ova |
| larva ,, | caterpillar | grub | maggot | louse |
| pupa ,, | chrysalis | pupa | pupa | nymph |
| adult ,, | moth | beetle | fly | adult |

Fig. 8. The four insect forms or *metamorphoses* of a moth (Scarlet tiger). Natural size.

Thus, when the word "maggot" is employed, it is always intended to refer to the larva stage of a fly or sawfly and never to that of a moth or butterfly. Similarly the word "grub" is reserved for use when a beetle (including weevil or chafer) is spoken of. "Louse" and its plural "lice" mean, in this book, the young stages of the aphis, psylla (sucker), capsid bug or similar insect.

## The Egg Stage

Great diversity is shown in the shape, size, colour, number, arrangement, position, time of laying and duration of insect eggs.

SHAPE. This is usually round (spherical) or oval, but in some cases it assumes many curious forms. Thus the eggs of the capsid bug are tubular in form, while those of the psylla (sucker) have a curious curved "process" or "appendage" (see pages 338, 346). In the moth class, the figure of 8 moth lays a curious umbrella-shaped egg (see page 106).

SIZE. Eggs laid by the larger insects are usually bigger than those of the smaller kinds, though there is no direct proportion.

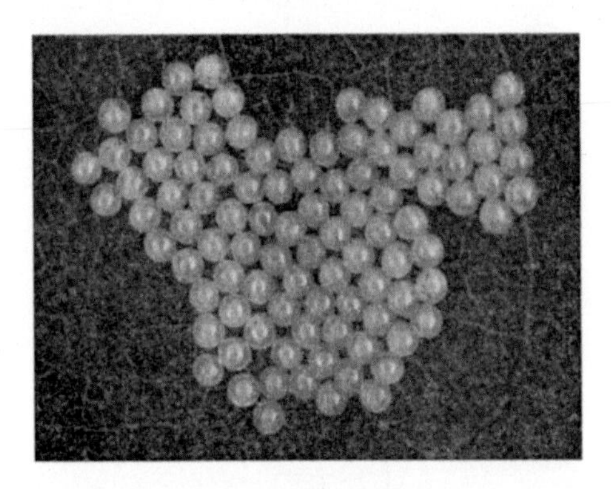

Fig. 9. Typical eggs of a moth which will hatch into caterpillars, magnified.

COLOUR. This is very variable. Many eggs change colour on exposure to the air. Thus the egg of the winter moth when first laid is a pale green, but this changes in course of time to a rusty red tint. Aphis eggs are usually shiny black; Psylla (apple sucker) yellowish, becoming reddish. Beetle eggs are usually white or cream coloured, those of flies transparent. The eggs of moths and butterflies are very varied in tint. Many have characteristic rings of colour around them, as e.g. Vapourer and Lackey. The Lappet Moth lays eggs with very curious wavy markings and rings.

NUMBER. The number of eggs laid by the female insect in a given position may vary in different species from one to several hundreds. Thus the Apple Blossom Weevil lays a single egg in each blossom: the

Eyed Hawk Moth a single egg on each leaf it visits, while the Lackey and many others lay hundreds in one position. Generally speaking, if the eggs are protected from injury, only a few are laid, but where exposed, or placed in a conspicuous position, many more are deposited.

ARRANGEMENT. Eggs may be laid in crevices without special arrangement, as in the case of the Winter Moth, the Aphides, Psylla, etc. In many cases they are placed with wonderful order and precision. Instances of regular egg-bands placed on the branches are those of the Lackey and March Moths. The Vapourer Moth lays a single layer of eggs placed close together. The eggs of the Lady-bird beetle are in little square clumps, placed on end in a characteristic manner (page 469).

In some cases the eggs are covered with hairs from the tail tuft of the insect (e.g. March Moth, Gold Tail Moth).

POSITION. The position of the eggs is subject to great variation. They are usually placed near to the suitable food supply for the young larva on hatching out. Thus leaf-eating caterpillars, aphides, sawflies, etc., hatch from eggs placed on the leaves or branches of the trees. Maggots and grubs which infest fruit, from eggs placed by the insect inside, or on the surface of the young fruitlet. The Swift Moths, the larvæ of which feed upon roots, drop their eggs during flight on to the ground. Some weevils, beetles and chafers deposit their eggs under the surface of the ground, where the maggots will live till mature. Parasitic insects lay their eggs inside the body of the attacked insect.

In some cases the female insect flies long distances before depositing her eggs. Others are wingless and crawl up the trunks of trees to the smaller branches. The Vapourer Moth never leaves her cocoon and pupal case, but lays the eggs on the cocoon itself (page 174). Incidentally, it has frequently been noticed, that where the adult female is sluggish in her habits, the caterpillars are often extremely active, and can themselves travel long distances, and so spread from plantation to plantation.

Most eggs are placed in sheltered and inconspicuous places, such as crevices of the bark, axils of branches, etc., obviously for protection, but some are conspicuously deposited in exposed situations, and in this case, there is usually a large number laid in one position to ensure survival.

TIME OF LAYING. This varies very greatly in different species. It also varies very considerably in the same species according to the climatic conditions. Thus the Winter Moth may lay her eggs from

October to January. Some insects have more than one brood in a year, and the eggs are then laid at different seasons for the different broods. Eggs which are laid by insects which have passed the winter in the mature state, will appear sooner or later, according to the seasonal conditions, a cold spring delaying the laying operations. The approximate time of the appearance of the eggs is given under the descriptions of the various pests (Chapters 12—24) and also in the table, Appendix I, page 692 (fig. 289).

DURATION. Many insects pass the winter in the egg stage and hatch in the spring or early summer. In many other cases, hatching takes place in a few days. In others again, it may be delayed for some weeks. In the case of aphides, reproduction takes place throughout the summer without any egg laying at all. In this case it is probable that the hatching occurs in the body of the insect. Pairing however is apparently unnecessary under such conditions. See also the table on page 692.

## The Larva Stage

The larvæ of insects show almost as great variety in character as the eggs. As previously mentioned, a larva is, for the sake of distinction, referred to as a caterpillar, grub, maggot or louse, according to whether it belongs to the moth, beetle, fly or "bug" orders respectively. The following is a brief summary of the chief variations:

SIZE. This varies roughly in relation to the size of the adult insect, and great variations are shown in each order of insects. Thus there are minute caterpillars like the leaf miners, and some over four inches in length, e.g. the Lappet. Of beetle grubs we have the small Apple Blossom Weevil and the Cockchafer larva. Only in the "bug" family (*Hemiptera*) is there little variation shown.

APPEARANCE. Caterpillars may be naked or covered with hairs, almost colourless or of brilliant hues. They all, however (with the exception of the "loopers") possess *four "sucker" feet*, in a middle position, and a tail pair of "claspers" in addition to three jointed legs near the head.

Beetle grubs are normally pale cream or flesh coloured, often with brownish or blackish heads. The skin is often much wrinkled, and feet are usually *absent altogether*, except in the Chafers, which possess three pairs of jointed legs.

The maggots of Flies and Sawflies show a fair amount of variation. They may be almost colourless or distinctly marked, as in the Social

Pear Sawfly, without feet or with *six middle pairs* of sucker feet (e.g. Apple Sawfly).

The "lice" of the bugs and aphides differ entirely from the larvæ of the other orders. They have bodies similar in shape to the adult insects, and not of the elongated "caterpillar" form. They are commonly green or yellowish green, often with red eyes, but bluish, purple and black forms are also frequent. They have normally three pairs of jointed legs and prominent feelers (antennæ). The Scale insects show very great variation from the rest, and are quite an exceptional class.

Fig. 10. A typical caterpillar on apple leaf. Natural size.

LOCATION. The great majority of the larvæ are found upon the surfaces of the leaves. There are however many exceptions in all the orders of insects. Thus there are caterpillars which infest the interior of the leaves, the shoots, the flowers and the fruit, and some which feed under the soil. Many pass different periods of their growth in different situations.

The beetle grubs are usually found in the ground and many feed upon the roots of plants. There are however several notable exceptions. Thus the Apple Blossom Weevil larva lives in the blossom, the Shot-borer Beetle grub in the interior of the trunk or branch, etc.

The Fly and Sawfly maggot resembles the caterpillar in being often on the surface of leaves, but may infest the fruitlets or pass part of its life in the ground.

Those which are parasitic (the Ichneumons, etc.) live in the interior of the insects they infest until mature.

The lice of aphides and bugs are normally found on the under surfaces of leaves, or in the shoots. The Woolly Aphis however is also found in the ground. The Scale insects appear upon the young branches.

TIME OF APPEARANCE AND DURATION. Various forms of insect larvæ appear at all times of the year, according to the nature of their food. The great majority of the leaf-eating larvæ make their appearance in the spring when the leaves are young and succulent. The

Fig. 11. Caterpillars (Mottled Umber) about to enter the ground to pupate. Natural size.

caterpillar of the Peppered Moth however does not hatch out till much later and feeds upon the old leaves, and there are many which appear in the early summer. Almost all of the aphis family occur in early spring, though a large number leave their original hosts and fly to other plants during the summer.

Beetle grubs which live in the soil often appear in late summer.

As regards duration this is also very variable. Very many caterpillars hibernate during the winter, under shelter or in the ground, ready to feed upon the young leaves in the following spring. In other cases, their duration in this stage is comparatively short. An exception is the Goat Moth caterpillar which lives for three years in the interior of the tree.

Many caterpillars spin cocoons in which they pass the winter, and leave these in the spring: others change into the chrysalis form in the same cocoons.

Beetle grubs may have a very short period, as in the Apple Blossom Weevil, or a very long one like the Chafers. They mostly pass the winter in the soil in this stage.

The Fly and Sawfly maggots show almost as great variations in duration as the caterpillars, some hibernating throughout the winter.

The length of life of the larvæ of the parasitic flies depends upon that of the host they infest.

In the case of aphides and bugs, the duration of the larval stage is always short and may last only a few hours under suitable weather conditions. See also fig. 290 (p. 693).

## The Pupa Stage.

This stage shows much less variation in the different orders than the others. Except in the case of the *nymph* stage of the aphides and bug family (*Hemiptera*) which is not a true pupal stage, no insect is injurious to plants as a pupa, but is in a resting and almost motionless condition. An insect is therefore liable to escape notice, in this stage, but it is often important from a practical standpoint, since many pests can be successfully dealt with as pupæ.

Fig. 12. A typical pupa (chrysalis). Natural size.

APPEARANCE. Most of the moths give rise to a pupa or *chrysalis* varying in size in the different species, but very similar in colour and appearance.

Pupæ of moths are usually light or dark brown, either naked, or in cocoons of grey silk, and have a shiny surface. Some have spines along the abdomen, and many a sharp spine at the tail end. In a large number of cases, the abdomen is capable of a twisting movement when disturbed, and some are able to take high leaps into the air in this manner. The body is always devoid of a distinct head, eyes or limbs.

Beetle pupæ are usually paler in colour and differ in the limbs being distinct, though incapable of use. In the case of many weevils, the long snout is free and folded under the head.

Fly and Sawfly pupæ in general resemble the chrysalides of the moths. They are usually enclosed in silken cocoons. The parasitic flies form a number of cocoons of yellow or white silk near to the dead body of their host.

In the case of aphides and plant bugs, the pupal stage is termed the *nymph*. In this condition, the insect is freely active, though not to the same extent as in the larval condition. No resting stage occurs, and the only distinguishing mark of this stage is the appearance of the *wing-buds*.

LOCATION. Most commonly, in all the orders, the pupæ are found in the ground. Chrysalides of moths are also commonly to be seen on fences, among the leaves, or in curled-up leaves, under rubbish, etc. The pupa of the Apple Blossom Weevil is found in the interior of the "capped" blossom.

Aphides and other members of this family pass the nymph stage on the leaves or in the buds.

TIME OF APPEARANCE AND DURATION. Pupæ normally appear as soon as the food of the larvæ is exhausted or become unsuitable. This may be at all times of the year except during the winter months (excluding larvæ living in the soil).

Many pupæ last during the winter: others hatch out into adult insects in the course of a few days (see Appendix I, fig. 291, p. 694).

The *nymph* stage of the aphides and plant bugs is comparatively short, often only a few days, or even hours.

### Adult Insect

This is the final or "perfect" stage in the development of insects. Sexual differences, occasionally slightly shown in the immature stages, now become fully marked.

The adult stage is not, in general, a feeding stage, and little or no growth occurs. In many cases, therefore, the adult insects do no damage to the crops. They are almost entirely concerned with the reproduction of the species.

Thus, the *moths* do not attack fruit trees, and the same is true of the saw-flies and flies. Exceptions however occur in the case of some leaf-eating weevils, and the adult females ("unsexed") of the aphis class.

In the case of some beetles, injury is done, not by eating, but by boring into the trunk. Some sawflies and weevils make incisions in the fruit for depositing their eggs.

Fig. 13. A typical adult insect (Wasp Fly) showing the various parts: magnified ( × 6). Inset natural size.

SIZE. Insects vary greatly in size, from almost microscopic dimensions to large moths, like the Hawk and Lappet, and large beetles such as the Cockchafer. Between these extremes there are an almost infinite series of gradations.

APPEARANCE. The *male* insect, with scarcely a single exception, is winged in the adult stage. This is to ensure mating with the female. The feelers (antennæ) are commonly more highly developed, and are the seat of the sense which enables the male to discover the location of the female of the species.

The female is often larger than the male. It possesses in many cases a specially formed organ, the *ovipositor*, for placing the eggs in the most suitable position.

In a number of cases the female is almost or wholly devoid of wings. Conversely, in the aphis group, certain of the females (the so-called "migrants") are winged. These are however not "egg-laying," and are to that extent "non-sexual" (*unsexed*[1]) although capable of giving birth to living young.

The COLOUR of the adult insects is, of course, extremely variable. Often it is protective in character, that is, it so closely resembles the normal surroundings of the insect as to render the latter difficult to distinguish. This is frequently the case when the insect is at rest, although with the wings spread in flight, it makes quite a conspicuous object.

LOCATION. The winged insects fly from tree to tree. Those females without wings are found upon the trunk or branches laying their eggs. The Woolly Aphis and some beetles live in the ground, at any rate during a portion of the time.

TIME OF APPEARANCE AND DURATION. Adult insects appear mainly in summer and late summer, others in autumn, while a few hatch out from the pupæ in winter (e.g. Winter Moths) or spring (e.g. March Moths).

The duration of the moths is normally very short, and merely covers the period of egg-laying. The Tortoiseshell butterfly and some moths however frequently live over the winter in the adult stage. Beetles have a much longer life. Some weevils pass the winter in shelter or under the ground (see Appendix I, fig. 292, p. 695).

Most of the aphides and bugs pass the winter in the egg stage, but a few each season hibernate in sheltered crevices of the trees.

[1] See note on page 271.

# CHAPTER 9

## How Insects are Classed

There are more than three hundred thousand distinct species of insects which have been recognised and studied, and at least as many more remain to be classified. Amongst such a large number there are obviously many which have features in common, and can be grouped together.

The popular names of the insects fairly accurately distinguish the main groups or **orders**. Below are given these names, the scientific names and the chief distinguishing feature of the orders which contain insect pests.

1. Orders in which *complete transformation* occurs, i.e. which have an egg, larva, pupa and adult stage.

| | | |
|---|---|---|
| MOTHS AND BUTTERFLIES | **Lepidoptera** | insects with coloured scales on the wings. |
| BEETLES | **Coleoptera** | insects with hard and horny wing cases. |
| TWO-WINGED FLIES | **Diptera** | insects with one pair of membranous wings. |
| FOUR-WINGED FLIES | **Hymenoptera** | insects with two pairs of membranous wings. |

2. Orders in which *incomplete transformation* occurs, i.e. in which there is no true pupal or resting stage.

| | | |
|---|---|---|
| APHIDES AND PLANT BUGS | **Hemiptera** | small insects with jointed sucking beak. |
| THRIPS | **Thysanoptera** | small insects with long fringed wings and mouths adapted for piercing and sucking. |

There are other orders of insects, including two in which the members undergo no transformations at all, and never acquire wings. The above however include all which contain pests of fruit of any importance, in this country.

In addition to the above, there are pests belonging to the following orders which are *not true insects.*

| | | |
|---|---|---|
| SPIDERS AND MITES | **Acarina** | usually with four pairs of legs, no distinct division into head, thorax and abdomen, mouth often suctorial. |
| EELWORMS | **Vermes** | with elongated, ringed bodies, no limbs, and no distinct head. |

Remarks on each order of insects will be found preceding the descriptions of pests in the next section.

# SECTION IV

## INSECT PESTS

# SECTION IV

## INSECT PESTS

## CHAPTER 10

### Introduction to Insect Pests

Insect pests take an annual large toll of the fruit grown in this country. Despite the great extension of the practice of systematic spraying, these pests show no signs of decreasing in numbers and virulence. This matter has been touched upon previously (page 7) but it may be well to emphasise that this is due to two main causes:

1. The laxity as regards spraying of many fruit growers, whose plantations form centres of infection each year.

2. The accumulation in compact areas and districts of the food plants on which the pests live.

Although some pests, such as leaf eating caterpillars and leaf sucking insects like the aphides and plant bugs, always do harm by injuring the vitality of the tree, others which directly reduce the fruit crop are not always actually harmful. This is because no tree can normally bear the fruit corresponding to the whole of the blossom in a normal year without being seriously weakened in vitality. Pests therefore like the apple blossom weevil, bud moth, pith moth, and to some extent, the apple sucker, are only really serious when the percentage of blossoms destroyed exceeds a certain figure. Unfortunately it is impossible to forecast beforehand what extent of damage is likely to occur, and in recent years all these pests have in places well nigh ruined the crop of fruit.

Experience has shown that systematic spraying is the best and safest policy and the one which is cheapest in the long run. In order however to spray with maximum effect, a THOROUGH KNOWLEDGE OF THE PEST is necessary. The following descriptions are intended to give this information in a concise and readily available form.

The same order has been observed throughout in all the descriptions. Information as to the APPEARANCE OF THE VARIOUS STAGES of the pest is first given to ensure its identification. In most cases, also, an **index sheet of pests in natural colours** has been carefully prepared (see frontispiece), so that, after having identified the pest by reference to the figure, the grower may at once turn to the detailed description given on the page indicated under each pest.

Next follow brief statements of the TIME TO EXPECT THE PEST and the PERIOD OF ITS DURATION, and also its exact LOCATION ON THE ATTACKED PLANT. Also any special details which are of interest or importance.

A brief statement of its DISTRIBUTION follows and then a short and simple account of the LIFE-HISTORY.

Under the heading "**Economic**" is given information of the TREES ATTACKED, the FREQUENCY OF THE PEST, the NATURE AND DEGREE OF THE DAMAGE and the natural enemies of the pest, if any.

METHODS OF PREVENTION meeting with success are stated. These may not always be applicable, but are given so that they may be employed if the occasion permits.

For instance, it is advised to collect and destroy the capped blossoms in the case of apple blossom weevil attack; but this may not be feasible if the plantation is large, or if labour conditions are difficult. In small orchards however it is often quite possible. The same remarks apply in the case of remedies (below).

**Remedies of proved benefit** are then stated. In this connection it must be remembered that

1. There are no absolutely successful remedies yet found for a few of the worst pests.

2. As a result of active research and investigation, fresh means of treatment are constantly being devised.

The author strongly advises EVERY GROWER who can, to HIMSELF EXPERIMENT on portions of his plantations, and record the results obtained. *Untreated trees* and plants should *in all cases* be left as *controls* to show what happens when the treatment is not applied. This is *most important.* The reason why individual trials are so useful is because every case has its own peculiar circumstances, e.g. the variety of tree, soil, situation, surroundings, available water supply, available appliances and labour, etc.

Finally, at the end of each description a **Calendar of Treatment** is given. EACH SPACE in the vertical column represents ONE MONTH, and the thickened lines opposite show in *what particular stage* the pest exists during those periods. Opposite is indicated the correct treatment as near as is known, and the RIGHT TIME to apply it.

It is confidently hoped that this device will be of great value to growers, as summarising in a pictorial and readily accessible form the information required. At the end of the book (Appendix I) will also be found a condensed calendar showing in the same manner the time and duration of insect pests in each of the four stages (i.e. egg, larva, pupa and adult).

## Authorities

Our present knowledge of the life-history and treatment of insect pests has been gradually built up by the careful observations of many workers. In this country, and especially amongst fruit growers, Miss Ormerod will always be remembered as one of the most indefatigable of these. In more recent years the work of Professor Theobald of Wye College and his collaborators has been of great value, especially in connection with the habits and life-history of the various species of aphides. From the point of view of treatment, Mr Spencer Pickering of Woburn has covered a very large field, and in many respects laid the foundation of our present-day methods of spraying, especially in this country. The various agricultural colleges and experiment stations, both here and especially in the States, as well as several independent workers are now engaged in pursuing many lines of research in regard to insect pests, and our knowledge of the subject is constantly being revised or added to.

The following is a list of most of the authorities concerned:

| | | | |
|---|---|---|---|
| Awati | Fitch | Miall | Rep |
| Bourcart | Fletcher | Morris | Sanderson |
| Buckler | Fryer, J. C. F. | Murray | Slingerland |
| Buckton | Howard | Newman | Smith |
| Cameron | Kaltenbach | Newstead | Stainton |
| Carpenter | Kollar | Ormerod | Taschenberg |
| Collinge | Lefroy | Petherbridge | Theobald |
| Curtis | Lees | Pickering | Warburton |
| Edwards | Marchal | Porrit | Wood |

It is almost needless to add that the author will be very grateful to receive any comments on this part of the subject, and any suggestions for more efficient remedies or methods of application.

The various insect pests are placed in **alphabetical order** in their particular classes. This ensures facility in reference. The more important pests are given in *larger type*, and all the descriptions are illustrated with original photographs or diagrams.

# CHAPTER 11

## Moths and Butterflies

(LEPIDOPTERA)

## and their larvæ

## CATERPILLARS

### 1. General and Economic

A good many of the characteristics of this class have already been referred to (page 38). The great majority of the caterpillars are **leaf eaters**, and are pests on this account. The seriousness of the attack is therefore in proportion to the number of caterpillars present, and their size, since a large caterpillar will eat more in a given time than a small one. Many of the pests here described are not normally very destructive because they do not occur on fruit in sufficient numbers. Many however are liable to become serious plagues locally, when for some reason they have increased beyond the usual limits.

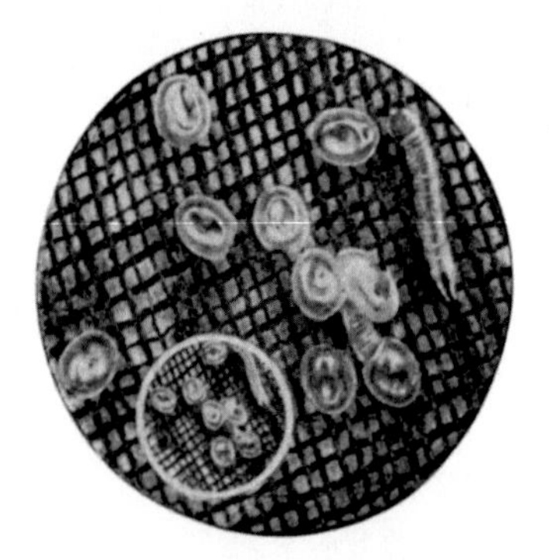

Fig. 14. Eggs (of Eyed Hawk Moth) just hatching out. Magnified (× 3). Inset natural size.

There are however two caterpillars, the larvæ of the BUD MOTH, and of the CODLING MOTH, which are annually responsible for great damage to the fruit crop. This is not due to their eating the leaves, but to their injury to the *opening buds* and *the fruit.* From the growers' point of view, therefore, they come into a separate class.

The same remarks apply, though in a less degree, to the larvæ of the Clearwing, Goat, Wood Leopard, Pith, Plum fruit, and Swift moths. The first three *bore into the branches* or trunk of the tree; the Pith moth *destroys the blossom trusses* by tunnelling in the pith of the shoots; the Plum fruit moth caterpillar *infests the plums*, and the Swift moth larva *attacks the roots* of plants.

Of the caterpillars which attack the leaves of fruit trees, that of the **winter moth** is without question the most serious pest. It occurs almost every year all over Britain in large numbers and damages the foliage and young shoots at an early and tender stage. The female of this moth is fortunately wingless, and is obliged to ascend the trunks of trees to lay her eggs in the branches. This fact places a ready means of prevention in the growers' hands, that of banding the trees with a sticky composition in the autumn.

Other pests, the adult females of which ascend the tree trunks and can be caught in a similar manner are:

March moth,
Mottled Umber moth.

Each of these is fairly common in the midlands and south of England and does a considerable amount of damage.

Another class of caterpillars from the growers' point of view is comprised of those which form **nests** or **webs** ("tents") amongst the branches. The most serious of these is the LACKEY MOTH CATERPILLAR. The LITTLE ERMINE also causes a large amount of trouble, and the BROWNTAIL caterpillar is serious at times, though not of very frequent occurrence in this country.

The BLISTER MOTH and LEAF MINER caterpillars work in the interior of the leaves, but are not often serious pests.

The MAGPIE MOTH caterpillar is very destructive to currants and gooseberries but does not attack the larger fruit trees. The RASPBERRY MOTH larva is equally destructive to raspberries and loganberries, attacking the buds and shoots.

Of the remainder, which are all leaf eaters, and defoliate the trees, several occur as destructive pests locally when in large numbers, while a few are not commonly very harmful, *but should be able to be recognised at sight*, so that prompt remedies may be applied if called for.

PREVENTIVE MEASURES.

The chief of these may be summarised as follows:

1. Grease bands for Winter Moth, March Moth and Mottled Umber Moth.

Fig. 15. A moth (Eyed Hawk) just hatched out from pupa case. Natural size.

2. Run fowls into the orchard in winter or remove and bury or bake the top spit of soil to destroy pupæ of Buff Tip, Clouded Drab, March, Mottled Umber, Peppered, and Winter moths.
3. Remove egg-bands when seen on the branches, of the Lackey, March, and Vapourer moths, and of the Tortoiseshell butterfly.
4. Provide shelter for the wintering caterpillars of the Codling Moth.

Remedies.

A stomach poison is required to kill caterpillars. One of the most efficient of these is *Arsenate of Lead*[1] which does no damage to the leaves if properly manufactured. A fine spray is used, and the leaves are covered with the poison[2].

*Paris green*[3] and *London purple*[4] both have a tendency to severe scorching of the foliage.

*Arsenate of calcium*[5] is being used largely in the States, and in many cases with successful results.

An efficient caterpillar poison, harmless to animals, is also available.

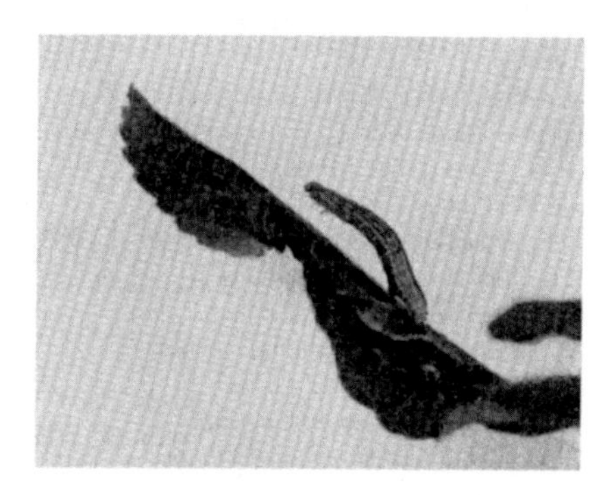

Fig. 16. Caterpillar of the Winter Moth. Most destructive of caterpillar pests on fruit.

In addition to the above, *Nicotine*[6], especially with soft soap, will kill caterpillars probably by absorption through the spiracles (breathing apertures). For this purpose the spray must wet each caterpillar, and this is not always an easy result to accomplish, as many protect themselves in curled leaves, blossoms, etc. or within webs.

In the case of *tent caterpillars*, the tent should be removed, burnt or blown away with a shot gun and a little powder.

Other special remedies apply to particular pests and will be found in detail under the descriptions.

It is very essential to spray as *early as possible* when the caterpillars

[1] See page 397. [2] See page 402. [3] See page 442. [4] See page 431.
[5] See page 395. [6] See page 436.

are young. Less damage is then done to the tree, and the caterpillars are killed with a much smaller dose of poison.

The dangers of neglecting to spray at all are discussed elsewhere (pages 7, 16).

## 2. Scientific

There are several systems in vogue of classifying the moths, but the following has fair acceptance. Those caterpillars only are mentioned which occur as pests on fruit.

Order **LEPIDOPTERA.**

*Rhopalocera*, Butterflies.

Large tortoiseshell, *Vanessa polychloros.*

*Heterocera*, Moths.

CLASS I. SPHINGES (Hawk Moths).

Large moths with thick bodies, and rather narrow and long fore-wings. Caterpillar cylindrical with a dorsal horn at back extremity of body.

Eyed Hawk, *Smerinthus ocellatus.*

CLASS II. BOMBYCES (silk spinners).

Body thick and short. Antennæ comb-like, and larger in male. Caterpillars hairy, spinning cocoons.

Family, Lasiocampidæ.

Lackey, *Clissiocampa neustria.*

Lappet, *Lasiocampa quercifolia.*

Clouded Drab, *Tæniocampa incerta.*

December, *Pœcilocampa populi.*

Family, Notodontidæ.

Buff-tip, *Phalera bucephala.*

Family, Liparidæ.

Vapourer, *Orygia antiqua.*

Brown-tail, *Euproctis chrysorrhœa.*

Gold-tail, *Porthesia similis.*

CLASS III. NOCTUÆ (owlets).

Night-flying moths. Colours usually grey or brown. The fore-wing often marked in a characteristic manner.

Figure of 8, *Diloba cæruleocephala.*

Dot, *Mamestra persicariæ.*

CLASS IV. GEOMETRIDÆ (loopers).

Body slender. Caterpillars have less prolegs than in other classes, and progress by *looping the body.* When resting they project out like twigs.

Winter, *Cheimatobia brumata.*
March, *Anisopteryx æscularia.*
Mottled Umber, *Hybernia defoliaria.*
Magpie, *Abraxas grossulariata.*
Peppered, *Amphidasys betularia.*

CLASS V. TORTRICES.

Small moths. The caterpillars generally roll up the leaves tightly; others live in buds and fruits. They wriggle backwards when disturbed and often eject a brown fluid.

Bud, *Hedya ocellana.*
Codling, *Carpocapsa pomonella.*
Little Ermine, *Hyponomeuta malinella.*
Tortrix moths, *Ribeana*, *rosana*, etc.

CLASS VI. XYLOTROPHA (wood-borers).

Larvæ, which burrow in the wood of trees, are smooth skinned.

Family, Sesiidæ (clearwings).
Apple clearwing, *Ægeria myopiformis.*
Currant ,, *Ægeria tipuliformis.*

Family, Hepalidæ (Swifts).
Garden Swift, *Hepialus lupulinus.*

Family, Cossidæ (goat moths).
Goat, *Cossus ligniperda.*
Wood leopard, *Zeuzera pyrina.*

CLASS VII. TINEIDÆ.

Small moths, with long and narrow fringed wings.

Blister moths, *Ornix*, *Lithocolletis*, etc.
Case-bearers, *Coleophora*, etc.
Leaf miner, *Lyonetia clerckella.*
Pith, *Blastodacna hellerella.*
Plum fruit, *Opadia funebrana.*
Raspberry, *Tinea rubiella.*

# CHAPTER 12

## Caterpillar Pests

### BLISTER MOTHS (on leaves)

Names *Ornix petiolella* *Lithocolletis coryli* *Cemiostoma scitella*, etc. Class *Tineidæ* Order *Lepidoptera*

### 1. General

**Description**

**Larva (Caterpillar)**

| | |
|---|---|
| APPEARANCE. | About $\frac{1}{4}$ inch when mature: has four pairs of stiff hairs on side. |
| COLOUR. | White at first, then greenish. |
| LOCATION. | Inside leaves. |
| APPEARS IN | May to July. |
| REMARKS. | The caterpillar enters the tissues of the leaf from the under surface and when mature crawls out. |

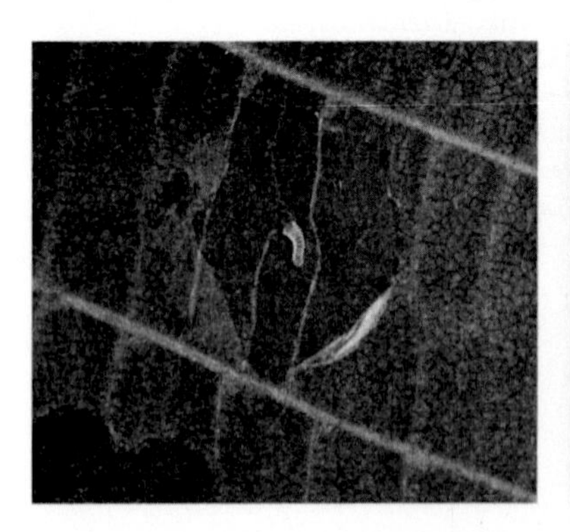

Fig. 17. On right, leaf blister on nut leaf. Natural size.
On left, same blister opened to show caterpillar. Magnified ( × 3).

**Pupa (Chrysalis)**

| | |
|---|---|
| APPEARANCE. | In silky cocoons. |
| COLOUR. | Cocoons white—chrysalis pale brown. |
| LOCATION. | On the trees in crevices. |
| DURATION. | Throughout the winter. |

**Adult Insect (Moth)**

| | |
|---|---|
| SIZE ACROSS WINGS. | About ½ inch. |
| COLOUR (WINGS). | Grey, hind-wings with grey fringe. |
| APPEARS IN | End of April to May. |

**Ova (Eggs)**

| | |
|---|---|
| LOCATION. | On under surfaces of leaves. |
| APPEAR IN | May. |

**Distribution**

Widespread in Britain.

**Remarks**

The caterpillar passes its life in the interior of the leaf, producing the blisters by eating away the soft interior tissue.

## 2. Economic

**Trees Attacked**

Apples, pears, peaches and nuts [also sloe, hawthorn and ash] by different kinds of blister moths.

**Frequency of Pest**

Not common.

**Nature of Attack**

The leaves are blistered and often blackened by the action of the caterpillars.

**Degree of Damage**

Occasionally serious.

**Remedies**

Against this pest the usual remedies are of little use, since the caterpillar is effectively protected in the interior of the leaf.

Spraying, to be of any service, must be done when the moths are first seen with ARSENATE OF LEAD[1].

HAND PICKING the attacked leaves is effective where it is practicable. These should be at once burnt.

Winter spraying with LIME-SULPHUR[2] at full strength or strong

[1] See page 397. [2] See page 612.

caustic emulsion[1] may corrode the skin of the pupæ and destroy them, and in any case exposes them to frost and to birds by its cleansing action on the bark.

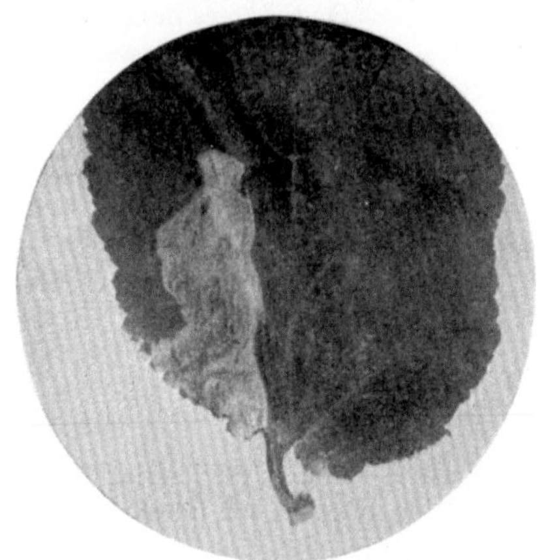

Fig. 18. Leaf blister, showing further damage caused by extension of blistered area. Natural size.

## Calendar of Treatment

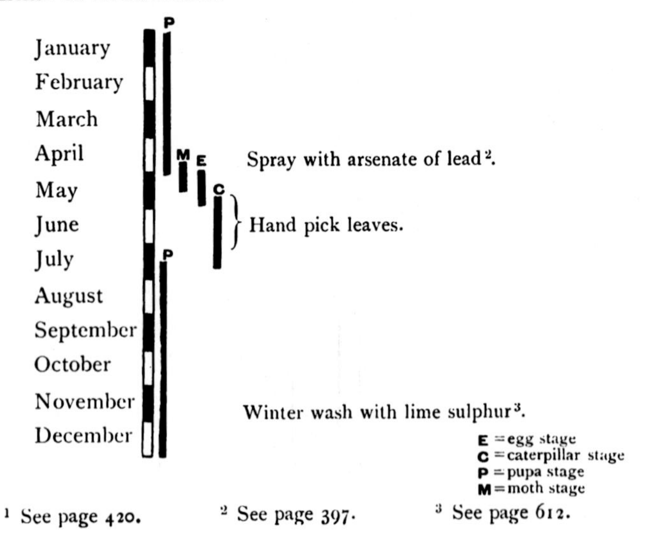


[1] See page 420. [2] See page 397. [3] See page 612.

## BROWN-TAIL MOTH

Name *Euproctis chrysorrhœa*[1] Family *Liparidæ*
Order *Lepidoptera*

### 1. General

**Description**

**Larva (Caterpillar)**

| | |
|---|---|
| SIZE. | About $1\frac{1}{2}$ inches when mature. |
| COLOUR. | At first, yellowish: later, deep brown with a row of white spots, and two red "humps." |
| LOCATION. | Amongst the leaves in "tents" or "nests," which later on are abandoned. |
| APPEARS IN | Beginning of August. |
| DURATION. | Throughout the winter. |
| REMARKS. | The young caterpillars form shelters of leaves, spun together with silk, in which they winter. |

Fig. 19. Diagram of caterpillar of Brown-tail Moth. Natural size.

[1] Recently re-named *Nygmia phæorrhœa.*

**Pupa (Chrysalis)**

| | |
|---|---|
| APPEARANCE. | In cocoons. |
| COLOUR. | Grey cocoons, containing a dark-brown chrysalis. |
| LOCATION. | Among the leaves of the trees. |
| DURATION. | About a month (July). |

**Adult Insect (Moth)**

| | |
|---|---|
| SIZE ACROSS WINGS. | $1\frac{1}{2}$ to $1\frac{3}{4}$ inches. |
| COLOUR (WINGS). | Glistening white; female moth has dark spots on fore-wings. |
| „ (BODY). | Downy white, with a *brown tail-tuft.* |
| APPEARS IN | Late July to August. |

Fig. 20. Brown-tail Moths: female (above), male (below). Natural size.

**Ova (Eggs)**

| | |
|---|---|
| COLOUR. | Dull yellowish. |
| ARRANGEMENT. | In compact masses of a hundred or more, covered over with hairs from the tail-tuft. |
| LOCATION. | On under surfaces of leaves. |
| APPEAR IN | August. |

**Distribution**

Mainly in the south of Britain, on the continent, and in America.

**Life-history**

The moths, male and female, appear in late July and August, flying at night and resting in the open during the day, when they are very inert. The female moth uses the hairs of the tail-tuft to conceal the eggs from observation. The eggs hatch early in August and the young caterpillars spin themselves up in leaves. Later on they make nests, in which they pass the winter, being unaffected by the sharpest frosts. The nests are left in the spring and the caterpillars feed on the young leaves. In early summer they spin large nests or "tents" and in these they change to dark brown pupæ.

**Remarks**

It occurs as a very destructive pest in AMERICA.

When the caterpillar is handled, the hairs stick to the skin, and produce great irritation (sometimes an actual rash).

Carbolised zinc ointment has been found a good cure for this.

## 2. Economic

**Trees Attacked**

Chiefly apples and damsons.

**Frequency of Pest**

Liable to occur locally in great numbers for a single season.

**Nature of Attack**

The foliage is devoured by the caterpillars, the attack being similar in most respects to that of the Lackey (page 121).

**Preventive Measures**

Where practicable, the winter nests should be hunted out and burnt to prevent attack in the following spring.

**Natural Enemies**

Ichneumon[1] flies attack the caterpillar on the continent and a Tachina fly[2] has been reported as a parasite in this country.

**Remedies** (see also "Lackey Moth" page 126)

1. CUT OUT THE TENTS, which are easily seen, during the summer, taking care that no caterpillars escape by dropping to the ground. These are then burnt, or placed in buckets of lime.

[1] Kollar, *Treatise on Insects*. (See also page 471.)

[2] Theobald, *Report on Economic Zoology* 1907, 25. (See also page 472.)

2. Destroy the tents by blowing them to pieces with a gun and a little powder and paper wad.
3. With a coarse nozzle sprayer force into the nests SOFT SOAP[1] solution with 7—8 ozs. NICOTINE[2] per 100 gallons.
4. Spraying during the dormant period with strong LIME-SULPHUR[3] wash or caustic emulsion[4] probably destroys a percentage of the young caterpillars and exposes others to attack by its cleansing action on the bark.

**Calendar of Treatment**

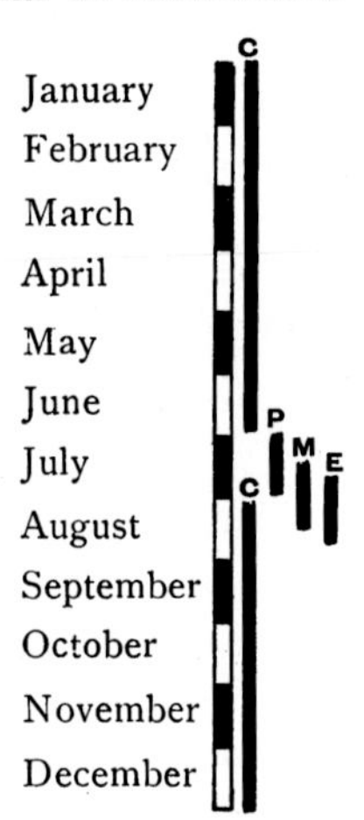


Remove or destroy tents, as above.

In winter tents—spray with strong lime-sulphur[3] or caustic emulsion[4].

**E** = egg stage
**C** = caterpillar stage
**P** = pupa stage
**M** = moth stage

[1] See page 452.
[2] See page 436.
[3] See page 612.
[4] See page 420.

## BUD MOTH

Name *Hedya ocellana* Class *Tortricidæ* Order *Lepidoptera*

### 1. General

### Description

#### Larva (Caterpillar)

| | |
|---|---|
| APPEARANCE. | When mature, nearly $\frac{1}{2}$ inch. |
| COLOUR. | Green until spring, when it becomes brown, with black head and dark first segment and legs. |
| LOCATION. | Summer. Under leaves in silken tubes. |
| | Winter. At base of buds in silken cases. |
| | Spring. (*a*) In the leaf and blossom buds. |
| | (*b*) Between the leaves (before pupation). |
| APPEARS IN | July. |
| DURATION. | Throughout winter, till following June. |

Fig. 21. Caterpillar of Bud Moth. Magnified (×3). Inset natural size.

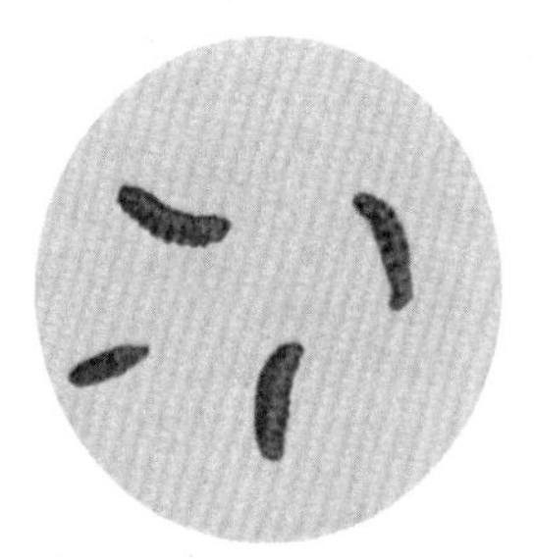

Fig. 22. One pupa and three caterpillars of Bud Moth. Natural size.

### Pupa (Chrysalis)

| | |
|---|---|
| APPEARANCE. | Small, with spines on each segment. |
| COLOUR. | Bright brown. |
| LOCATION. | In pockets of dead leaves. |
| APPEARS IN | June. |
| DURATION. | 1—2 months. |

### Adult Insect (Moth)

| | |
|---|---|
| SIZE ACROSS WINGS. | $\frac{1}{2}$ to $\frac{2}{3}$ inch. |
| COLOUR (WINGS). | Front pair: dark grey with grey band: back pair: grey. |
| APPEARS IN | June and July. |

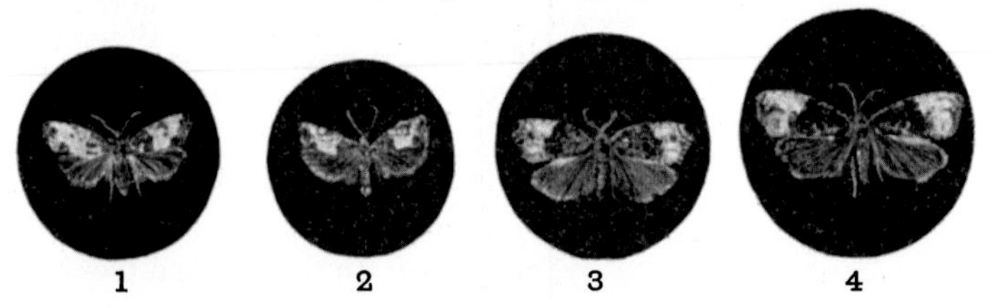

Fig. 23. 1, 2, Bud Moths. 3, 4, Allied Bud Moths. Natural size.

### Ova (Eggs)

| | |
|---|---|
| APPEARANCE. | Round and flattish. |
| COLOUR. | Transparent (gelatinous). |
| ARRANGEMENT. | Overlapping like the tiles of a house. |
| LOCATION. | Upper surface of leaf—singly, or in clusters. |
| APPEAR IN | June and July. |
| HATCHING PERIOD. | After 7 to 10 days. |

## Distribution

Widespread in Europe and America.
In Britain, occurs chiefly in South.

## Life-history

The small moths are on the wing during June and July after dusk, egg laying taking place at night on the upper surface of the leaves. The young caterpillars commence to feed in a little tube of silk on the under surface of the leaves. On the coming of autumn, the caterpillars find hiding places at the base of the buds and cover themselves with silk and small particles of bark, etc., being, at this stage, less than $\frac{1}{8}$ inch in length.

In the following spring, the caterpillars leave their winter houses, and enter the buds, which they destroy. Later on they feed on the leaves, spinning them together, and pupate in pockets of dead leaves.

## 2. Economic

### Trees Attacked

All fruit trees, but especially apples and cherries.

### Frequency of Pest

Common in south of England.

### Nature of Attack

The opening buds are entered and destroyed, turning brown, the leaves and blossom being *spun together*.

### Duration of Attack

During spring and autumn, but chief attack on buds in spring.

### Degree of Damage

Often great.

### Natural Enemies

On continent, numbers of Ichneumon flies; the author has also found specimens in Kent (see fig. 24).

The Blue and Great Tits, also the Sparrow, all feed on them in the winter.

Fig. 24. Ichneumon parasite of Bud Moth found in Kent. Magnified (× 3). Inset natural size.

## Remedies

Spraying in the late summer with ARSENATE OF LEAD[1] will kill the young larvæ.

In spring, just when the buds are bursting, the caterpillars are also open to control with ARSENATE OF LEAD[1].

CAUSTIC WINTER WASHES[2], if thoroughly used, should kill some of the hibernating caterpillars. Late washing in the early spring with LIME-SULPHUR[3] is of proved benefit.

## Calendar of Treatment

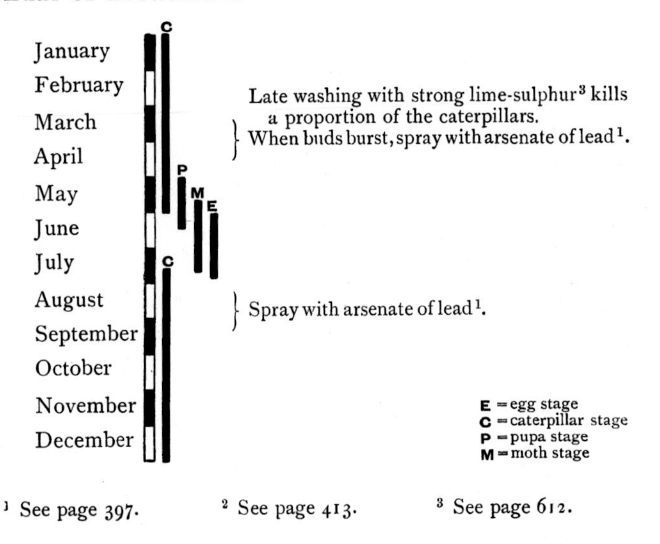


[1] See page 397. [2] See page 413. [3] See page 612.

## BUFF-TIP MOTH

Name *Phalera bucephala* Class *Notodontidæ*
Order *Lepidoptera*

### 1. General

#### Description

**Larva (Caterpillar)**

| | |
|---|---|
| APPEARANCE. | About 2 inches when mature—conspicuous. |
| COLOUR. | Dark yellow and black. |
| LOCATION. | On the leaves. |
| APPEARS IN | Middle of June and July. |
| DURATION. | Till middle September or October. |
| REMARKS. | The young larvæ feed together in parallel groups, see fig. 25. |

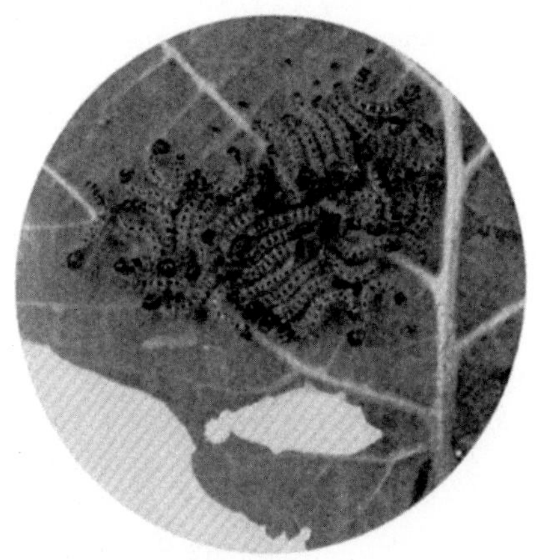

Fig. 25. Young Buff-tip caterpillars on nut leaf. Natural size.

**Pupa (Chrysalis)**

| | |
|---|---|
| APPEARANCE. | About 1 inch long. It has two spines at the tail end. |

| | |
|---|---|
| COLOUR. | Dark brown. |
| LOCATION. | Under surface of ground, or beneath fallen leaves. |
| DURATION. | From September, during winter and till following June—July. |

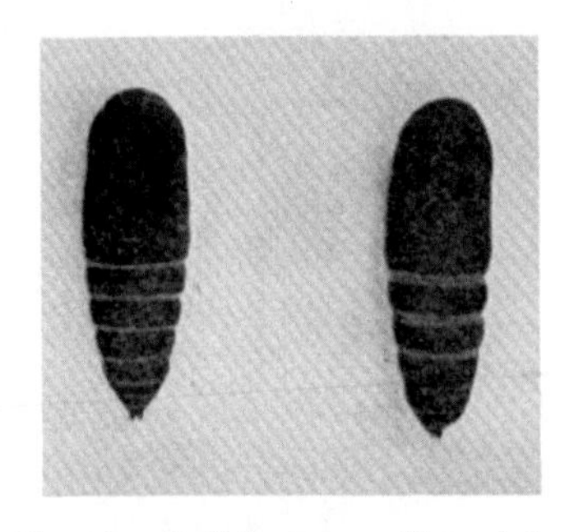

Fig. 26. Buff-tip pupæ. Natural size.

**Adult Insect (Moth)**

| | |
|---|---|
| SIZE ACROSS WINGS. | $2\frac{1}{2}$ to $2\frac{3}{4}$ inches. |
| COLOUR (WINGS). | Grey with cross streaks of brown, and a large buff spot at the tips of the wings. |
| ,, (BODY). | Buff. |
| REMARKS. | The male is distinguished by its feathery antennæ (feelers). |
| APPEARS IN | June and July. |

**Ova (Eggs)**

| | |
|---|---|
| APPEARANCE. | Rounded above, flat beneath: fairly large and conspicuous. |
| COLOUR. | White with dark spot in centre. |
| ARRANGEMENT. | In groups of one to four score. |
| LOCATION. | Firmly attached to the shoots or the under surfaces of leaves. |
| APPEAR IN | June. |
| HATCHING PERIOD. | 10 to 14 days. |

Fig. 27. Buff-tip Moth. Natural size.

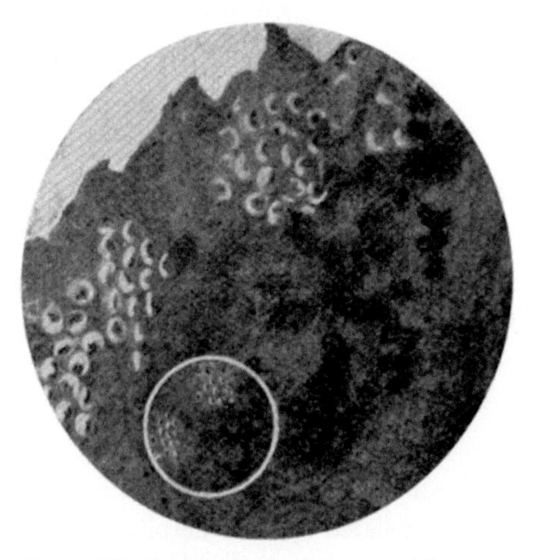

Fig. 28. Eggs of Buff-tip Moth showing caterpillars hatching out. Magnified ($\times 3$). Inset natural size.

### Distribution

General over Great Britain.

### Life-history

The moths, which appear in June and July, are of striking appearance, but while at rest with wings partially folded on the bark of trees, etc., they are difficult to distinguish (see fig. 29). The eggs are also readily seen and the young caterpillars group themselves on the leaves in parallel rows in a characteristic manner (fig. 25). They grow to a large size, and are conspicuous objects, maturing in September, when they fall to the ground and change into their chrysalis state under the soil or under leaves, remaining thus till the following summer.

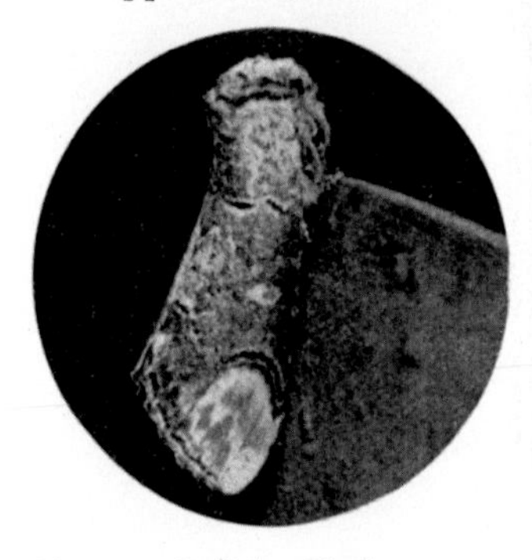

Fig. 29. Buff-tip Moth at rest with wings folded. Natural size.

## 2. Economic

### Trees Attacked

Nuts and occasionally apples and plums; also the elm, lime, sallow, beech, birch, oak, etc.

### Susceptible varieties

Cobs and filberts.

### Frequency of Pest

Common.

### Nature of Attack

The trees are defoliated by the caterpillars.

### Duration of Attack

Till the maturity of the caterpillar in September.

**Degree of Damage**

Very variable.

**Preventive Measures**

Poultry feeding under the trees during the winter will keep down the attack by destroying the pupæ. Removing the top spit of soil under trees is also effective.

**Remedies**

1. Spray the trees with ARSENATE OF LEAD[1].
2. Jar the trees and destroy the caterpillars which readily drop to the ground. A good method is to band the trees with sticky-composition[2] before doing this and so catch the caterpillars as they re-ascend the trunk.

**Calendar of Treatment**

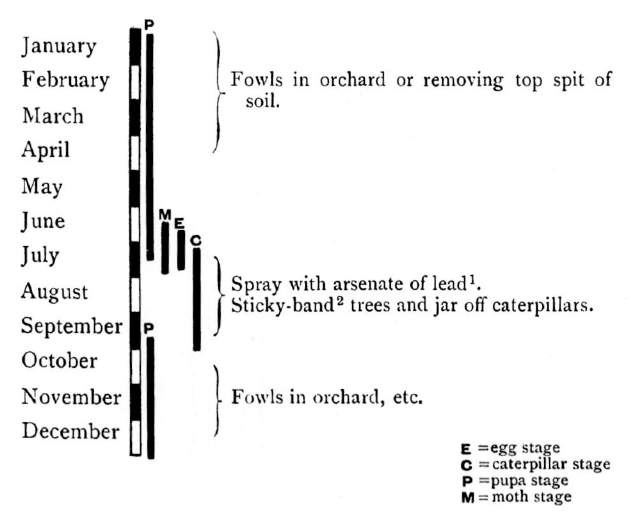


[1] See page 397. [2] See page 404.

## CASE-BEARER MOTHS

Name *Coleophora anatipernella* (Pistol-case-bearer) and other varieties Class *Tineidæ* Order *Lepidoptera*

### 1. General

**Description**

**Larva (Caterpillar)**

| | |
|---|---|
| APPEARANCE. | About ½ inch: after a few days entirely enclosed in brown or black "case" with white border. |
| COLOUR. | Reddish brown with shiny head. |
| LOCATION. | On the leaves projecting from the surface of the leaf, head downwards. |
| PROGRESSION. | Very slow. |
| APPEARS IN | August. |
| DURATION. | August till the following June (throughout winter). |

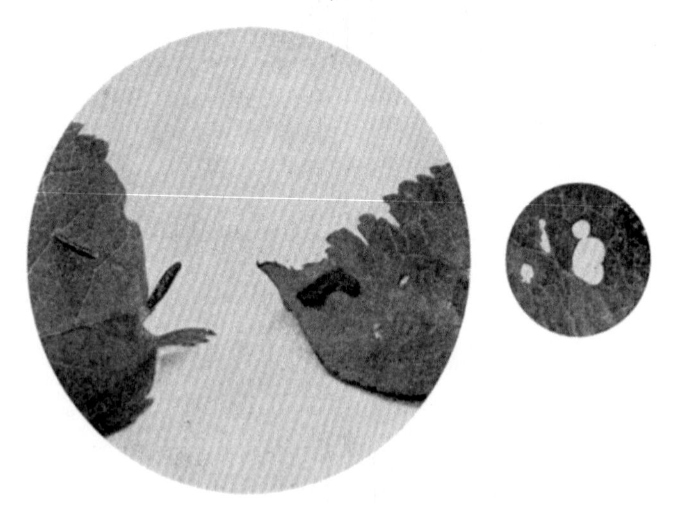

Fig. 30. Case-bearer caterpillars on apple leaf. Damaged leaf on right. Natural size.

**Pupa (Chrysalis)**

| | |
|---|---|
| APPEARANCE. | Enclosed in case. |
| COLOUR. | Pale brown. |
| LOCATION. | Lying flat upon the leaves. |
| DURATION. | 3 to 4 weeks. |
| REMARKS. | The chrysalis leaves the case before the moth emerges. |

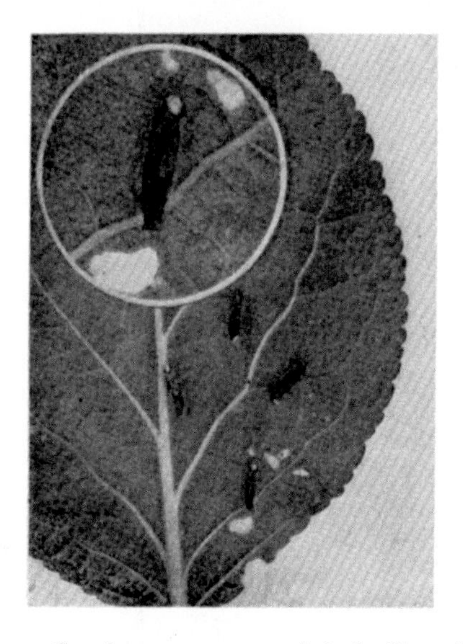

Fig. 31. Case-bearer pupæ on apple leaf. Natural size. Inset magnified (× 3).

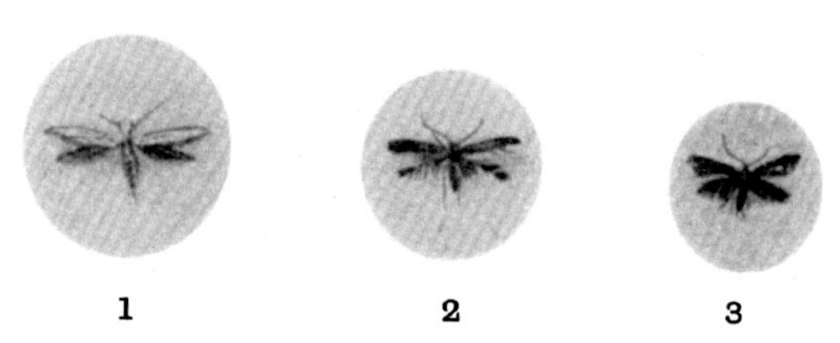


Fig. 32. 1, Pistol-case-bearer. 2, 3, Case-bearer Moths. Natural size.

**Adult Insect (Moth)**

| | |
|---|---|
| SIZE ACROSS WINGS. | About ½ inch. |
| COLOUR (WINGS). | Creamy white. |
| APPEARS IN | July till August. |

**Ova (Eggs)**

| | |
|---|---|
| LOCATION. | On the under surface of leaves. |
| HATCHING PERIOD. | 2 to 3 weeks. |

### Life-history

This small inconspicuous moth appears during July, flying at dusk. The eggs hatch in about a week and the young caterpillars soon become covered with the curious-shaped "cases." They feed upon the leaves, head downwards, and move gradually over the surface, often removing only the upper surface of the leaf, leaving a patch of the lower skin (cuticle) like the slugworms. They continue feeding as long as there is foliage, and winter by attaching themselves with threads in their cases to the shoots of the trees. On the arrival of spring they commence to feed on the buds and young leaves, pupating during June in their cases.

Fig. 33. Pupa of a variety termed lichen-case-bearer. Shelters on rough bark. Magnified (× 3). Inset natural size.

## 2. Economic

### Trees Attacked

Cherries, apples, sloes.

### Frequency of Pest

Common in the southern counties.

### Nature of Attack

The young buds and leaves are attacked in early spring and this is the most serious damage. When the leaves are larger the harm caused is not usually great.

**Duration of Attack**

Throughout spring, early summer and autumn.

**Degree of Damage**

Occasionally serious in spring.

Fig. 34. Pistol-case-bearer Moth. Magnified (× 3).
Inset natural size.

**Natural Enemies**

Case-bearers are attacked by several small flies (hymenoptera).

**Remedies**

1. Spray in the autumn with ARSENATE OF LEAD[1].
2. Or in the spring (if other caterpillars are present).
3. Spray in dormant season with strong LIME-SULPHUR[2] or other winter wash. (This exposes them to attacks of frost or birds.)

[1] See page 397. [2] See page 612.

**Calendar of Treatment**

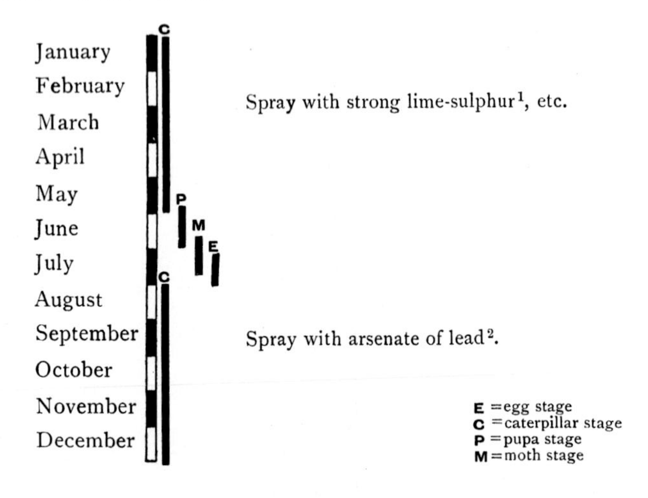


## **CLEARWING MOTH** (Apple)

Name *Ægeria myopiformis* Family *Sesiidæ*
Order *Lepidoptera*

### 1. General

**Description**

**Larva (Caterpillar)**

| | |
|---|---|
| APPEARANCE. | About $\frac{3}{5}$ inch when mature, with bristly spines. |
| COLOUR. | Dull yellowish white with shining brown head. |
| LOCATION. | In tunnels under the bark. |
| APPEARS IN | End of June and July. |
| DURATION. | Throughout the winter till following May. |
| REMARKS. | When mature, the caterpillar makes a hollow lined with grey silk and covered with wood chips. |

[1] See page 612. [2] See page 397.

**Pupa (Chrysalis)**

| | |
|---|---|
| APPEARANCE. | $\frac{1}{3}$ to $\frac{1}{2}$ inch, curved body with spines. |
| COLOUR. | Light brown. |
| LOCATION. | In hollow chamber formed by the caterpillar. |
| DURATION. | May and June. |
| REMARKS. | The empty pupal cases are left projecting out of the bark. |

**Adult Insect (Moth)**

| | |
|---|---|
| SIZE ACROSS WINGS. | About $\frac{3}{4}$ inch. |
| COLOUR (WINGS). | Transparent with dark borders. |
| „ (BODY). | Black with bright red band above and white below in male. |
| REMARKS. | The body ends in a black "fan-tail." |
| APPEARS IN | May—July. |

**Ova (Eggs)**

| | |
|---|---|
| LOCATION. | On trunks of trees. |
| APPEAR IN | May—July. |

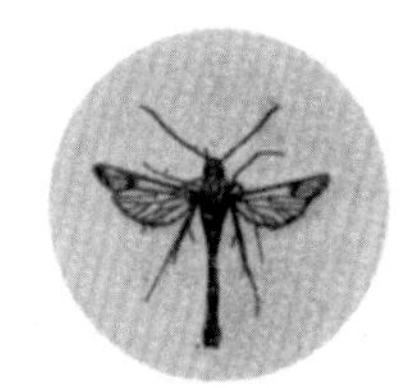

Fig. 35. Apple Clearwing Moth. Natural size.

### Life-history

The adult moths appear, flying in the daytime, in May, June and July, and lay their eggs on the trunks of apple trees. The larvæ burrow under the bark and remain in their tunnels throughout the winter until mature, when they hollow out a small chamber and line it with silk, covering the outside with chips of wood. Pupation takes place during May and June, the pale brown pupæ working their way (assisted by the spines) out of the cocoons, and leaving the empty pupal cases projecting from the trees.

## 2. Economic

### Trees Attacked

Apples.

### Susceptible Varieties

Apparently most susceptible to attack when about 6 years old.

### Frequency of Pest

Fairly common.

## Nature of Attack

The caterpillars eat out hollows in the trunk and larger branches of the trees.

## Degree of Damage

Variable.

## Remedies

The caterpillars are difficult to reach, being effectively protected in the interior of the tree.

Sealing up the holes with a sticky composition[1] in May stops the moth from emerging, and so prevents further attacks.

## Calendar of Treatment

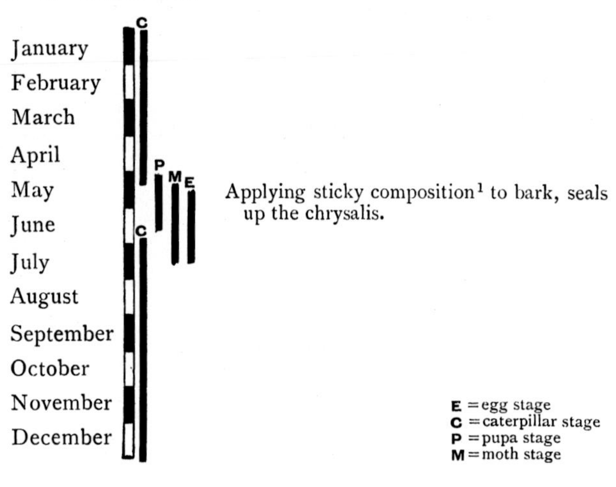


[1] See page 404.

## **CLEARWING MOTH** (Currant)

Name *Ægeria tipuliformis* Family *Sphingidæ*
Order *Lepidoptera*

### 1. General

**Description**

**Larva (Caterpillar)**

| | |
|---|---|
| Colour. | Creamy white with brown shiny head and last segment, and with black spots on each segment. |
| Location. | Inside pith of shoot. |
| Appears in | June or July. |
| Duration. | Until following April (throughout winter). |

**Pupa (Chrysalis)**

| | |
|---|---|
| Appearance. | Has a series of spines on back. |
| Colour. | Brown. |
| Location. | Inside the branch. |
| Appears in | April or May. |
| Remarks. | At the end of the period, the chrysalis forces itself partly out of the hole formed by the caterpillar, ready for the moth to emerge. |

**Adult Insect (Moth)**

| | |
|---|---|
| Size across wings. | About $\frac{2}{3}$ inch. |
| Colour (wings). | Transparent, edged with black and orange. |
| „ (body). | Shining bluish black with 3 yellow stripes (female) or 4 (male). |
| Remarks. | The body ends in a "fan-tail." |
| Appears in | June. |

**Ova (Eggs)**

| | |
|---|---|
| APPEARANCE. | Small, oval. |
| COLOUR. | Yellowish white. |
| ARRANGEMENT. | Singly. |
| LOCATION. | On the stem, near buds. |
| APPEAR IN | June. |
| HATCHING PERIOD. | About 7—10 days. |

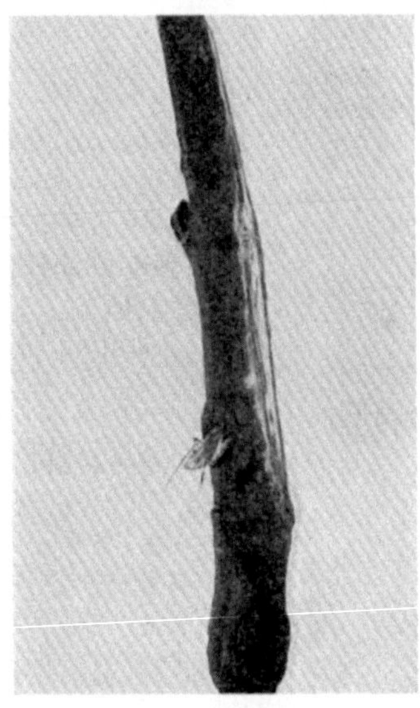

Fig. 35*a*. Currant Clearwing Moth—empty pupa-case projecting from currant stem. Natural size.

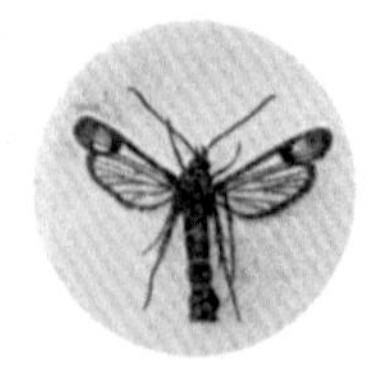

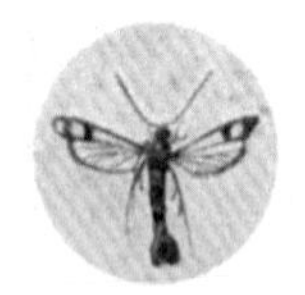

Fig. 36. Currant Clearwing Moths. Natural size.

**Distribution**

Widely distributed in England, the Continent and America.

**Life-history**

The moths emerge from the pupal cases in June, being seen on the wing in the early morning on fine, sunny days.

The eggs hatch out in a few days and the small white grub first eats its egg-shell, and then bores its way into the pith of the branch, where it remains till the following spring, when it pupates. At the end of the pupal period, the insect forces an exit half through the hole made by the caterpillar, by means of its spines, and the empty pupal cases may be seen in the summer protruding from the branches (see fig. 35*a*).

**Remarks**

The presence of the grub in the currant branch may be suspected by the withering of the shoots in the spring, or if the leaves and fruit are dwarfed.

In winter the hole made by the grub may be noticed by the appearance of "frass" at the opening.

## 2. Economic

**Trees Attacked**

Red, black and white currants.

**Susceptible Varieties**

Chiefly black currants.

**Frequency of Pest**

Fairly common, mainly in garden plots.

**Nature of Attack**

In early summer the shoots die off, owing to the activities of the grub inside. If they do not die, the leaves and fruit are dwarfed.

**Degree of Damage**

Often serious.

**Remedies**

Prune off all the shoots showing holes, cutting back till all sign of tunnelling in the pith disappears.

**Calendar of Treatment**

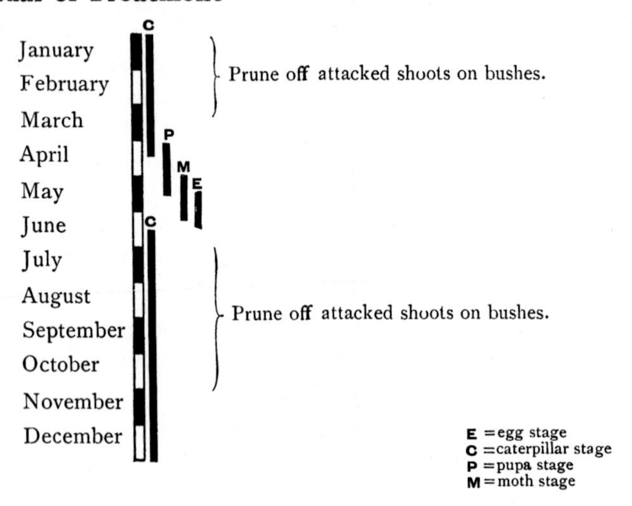


## CLOUDED DRAB MOTH

Name *Tæniocampa incerta* Family *Lasiocampidæ*
Order *Lepidoptera*

### 1. General

**Description**

**Larva (Caterpillar)**

| | |
|---|---|
| APPEARANCE. | About 1¼ inches when mature. |
| COLOUR. | Green dotted with black with one middle and two side ye lowish stripes. |
| LOCATION. | On leaves. |
| APPEARS IN | May. |
| REMARKS. | Said to also attack the fruit. The caterpillars sometimes devour each other. |

**Pupa (Chrysalis)**

| | |
|---|---|
| APPEARANCE. | Naked, or an earth cocoon. |
| COLOUR. | Deep brown. |
| LOCATION. | In the soil. |
| APPEARS IN | July. |
| DURATION. | Through the winter to March. |

Fig. 37. Caterpillar of Clouded Drab Moth. Natural size.

**Adult Insect (Moth)**

| | |
|---|---|
| SIZE ACROSS WINGS. | $1\frac{1}{4}$ to $1\frac{1}{2}$ inch. |
| COLOUR (WINGS). | Pale dull grey—3 wavy lines on fore-wings. Hind-wings all one colour, but veins prominent. |
| „ (BODY). | Front downy and like fore-wings in colour: rest of body paler. |
| APPEARS IN | March and April. |
| REMARKS. | For illustration, see coloured plate. |

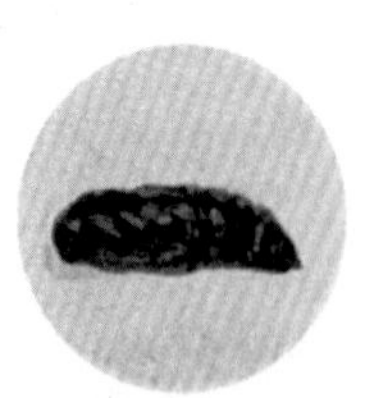

Fig. 38. Clouded Drab pupa. Natural size.

**Ova (Eggs)**

APPEAR IN — March and April.

**Distribution**

Common in Britain.

**Life-history**

The moth appears in March and April and lays her eggs, which hatch out in a few days, on the leaves. The caterpillars mature about July and fall to the ground, pupating in the soil, and passing the winter in this state.

## 2. Economic

**Trees Attacked**

Apples, but more commonly the sloe, willow and oak.

**Frequency of Pest**

Not uncommon.

**Nature of Attack**

The leaves eaten and occasionally the fruit.

**Degree of Damage**

Not usually great.

**Preventive Measures**

Keeping fowls in the orchard during the winter will account for some of the pupæ.

**Natural Enemies**

Its own species, the caterpillars often devouring each other in preference to vegetable food.

**Remedies**

1. Spray the trees with ARSENATE OF LEAD[1].
2. Shake caterpillars off young trees, and prevent re-ascending by bands of sticky composition[2].

[1] See page 397. [2] See page 404.

**Calendar of Treatment**

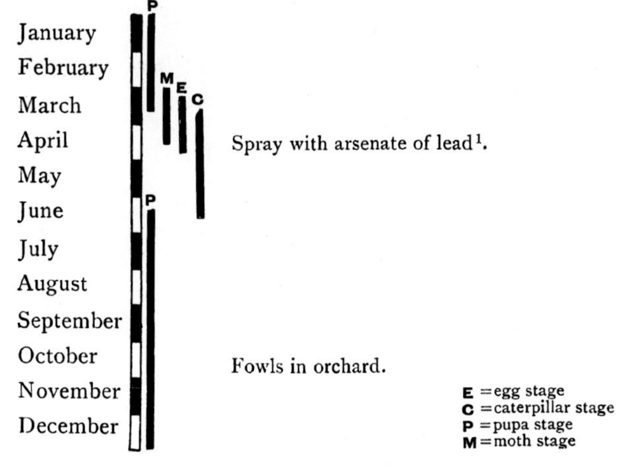


[1] See page 397.

## **CODLING MOTH** ("Apple maggot")

Name *Carpocapsa pomonella* Class *Tortricidæ*
Order *Lepidoptera*

### 1. General

**Description**

**Larva (Caterpillar)**

| | |
|---|---|
| APPEARANCE. | About ½ inch when mature: almost naked. |
| COLOUR. | Creamy white or pink, with brown head. |
| LOCATION. | Inside the fruit, which it afterwards leaves and finds shelter under the bark, etc. |
| APPEARS IN | May. |

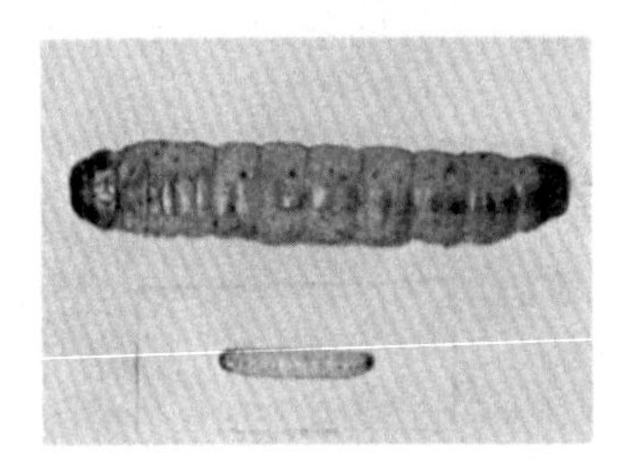

Fig. 39. Codling Caterpillar magnified (× 3) and natural size.

| | |
|---|---|
| DURATION. | Throughout the winter till the following spring. |
| REMARKS. | The caterpillar has three pairs of *true legs*, four pairs of *claspers* and one "tail pair." It is thus a true caterpillar (see page 39) and is distinguished from the larva of the sawfly (see page 237). |

**Pupa (Chrysalis)**

| | |
|---|---|
| APPEARANCE. | Small, smooth, shiny. |
| COLOUR. | Brown. |
| LOCATION. | In sheltered places in the caterpillar cocoons. |
| APPEARS IN | March—April. |
| DURATION. | 2 to 3 weeks. |
| REMARKS. | Pupation may occur in October as well as in spring and thus two broods be produced in a single year. |

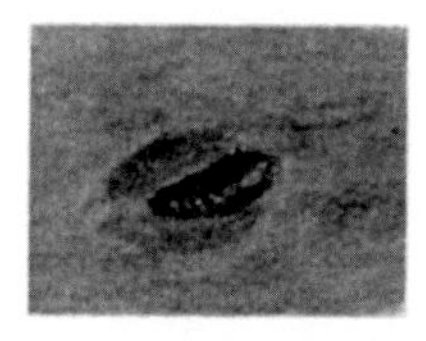

Fig. 40. Pupa of Codling Moth. Natural size.

Fig. 41. Codling Moth. Natural size.

**Adult Insect (Moth)**

| | |
|---|---|
| SIZE ACROSS WINGS. | About $\frac{1}{2}$ inch. |
| COLOUR (WINGS). | Greyish, with silky sheen. The male has a dark line on the under surface of the front wings. |
| APPEARS IN | April—May. |

**Ova (Eggs)**

| | |
|---|---|
| APPEARANCE. | Small, and glistening—somewhat flat. |
| COLOUR. | Transparent. |
| ARRANGEMENT. | Usually singly. |
| LOCATION. | On the sides of the fruit: occasionally on the leaves. |
| APPEAR IN | April—May. |
| REMARKS. | Only one egg is generally laid on each fruit—one moth may deposit over a hundred eggs on separate apples. |

**Distribution**

Widespread.

**Life-history**

The moths emerge from the caterpillar cocoons at the fall of the blossom and fly from fruit to fruit, laying one egg on each. The minute grub crawls over the apple till it arrives at the "eye" when it feeds a little time here and then enters the fruit, reaching the core and ejecting its excrement ("frass") through the opening, the appearance of which shows the nature of the attack.

(*In the case of Sawfly attack the entrance hole is usually on the side of the fruit.*)

Later on the grub burrows to the side of the fruit, and forms another opening at which more "frass" appears. Before the grub is mature, the apple often falls off the tree. If this has not occurred, it leaves the fruit and, crawling down the trunk, seeks any shelter it may find. It there spins a rough cocoon and passes the winter in the larval state, changing to a brown pupa, after casting its skin.

**Remarks**

The grub is liable to be mistaken for the SAWFLY larva (see page 237), but the latter has 6 pairs of "claspers" (instead of 4).

## 2. Economic

**Trees Attacked**

All varieties of apples.

Pears and plums are also attacked.

**Frequency of Pest**

Extremely common; but the moth is not often observed, and a very few can do an immense amount of damage. See above

**Nature of Attack**

The fruit is disfigured in the well-known manner ("maggoty fruit"). Often a great part of the pulp is devoured, sometimes only the core, and the tunnels. The grub often remains

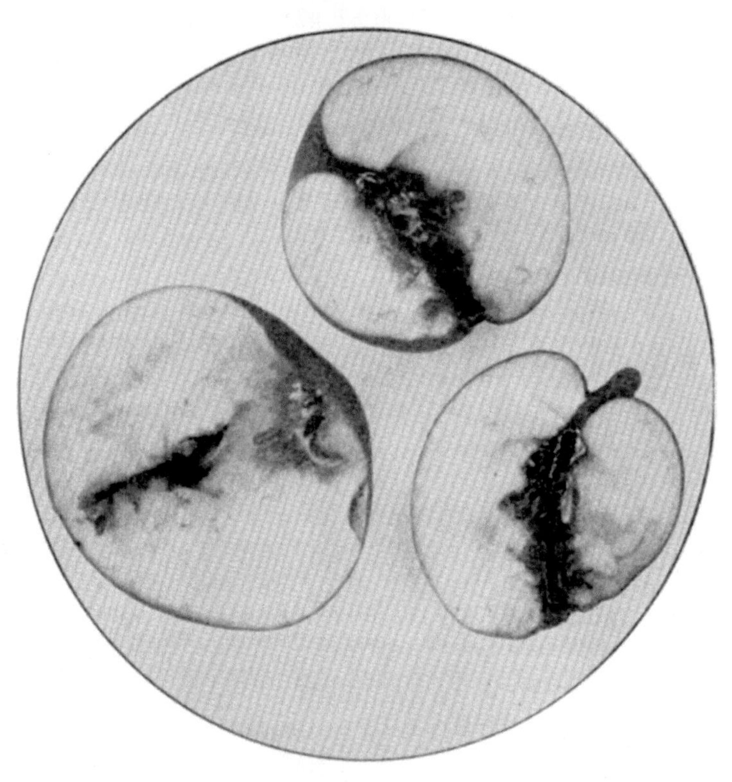

Fig. 42. Apples attacked by the Codling Caterpillar (reduced).

in the apple when marketed and is found inside the fruit when it is eaten. The attacked apples usually fall off the tree before they are ripe.

**Degree of Damage**

Often very serious, almost every apple in an orchard being attacked.

**Preventive Measures**

1. If maggoty fruit is kept it should be fed to pigs or destroyed, and should not be left on the ground.
2. Keep ground under trees free from grass or weeds.

**Natural Enemies**

In America, parasites are reported attacking the caterpillars. The tits greedily devour them in this country, but these birds are rather scarce.

**Remedies**

1. Where practicable (e.g. in small plantations) provide shelter for the hibernating caterpillars by tying bands of sacking round the trunk and destroy those which shelter there.
2. Keep tree-bands sticky during summer and autumn.
3. Spray with ARSENATE OF LEAD[1] immediately the blossom has fallen. A spot of arsenate is left in the eye of the fruit, and this poisons the caterpillar at his first meal. A fine misty spray is best, and *this time must be adhered to.*
4. Winter spraying with strong LIME-SULPHUR[2] or other winter wash may kill caterpillars sheltering under bark, and in any case exposes them to frost and to birds by the peeling of the lichen and rough outer bark.

**Remarks on Remedies**

If spraying with ARSENATE OF LEAD is decided on, it must be done *within a week of the falling of the blossoms.* Otherwise the labour may be in vain.

[1] See page 397. [2] See page 612.

## Calendar of Treatment

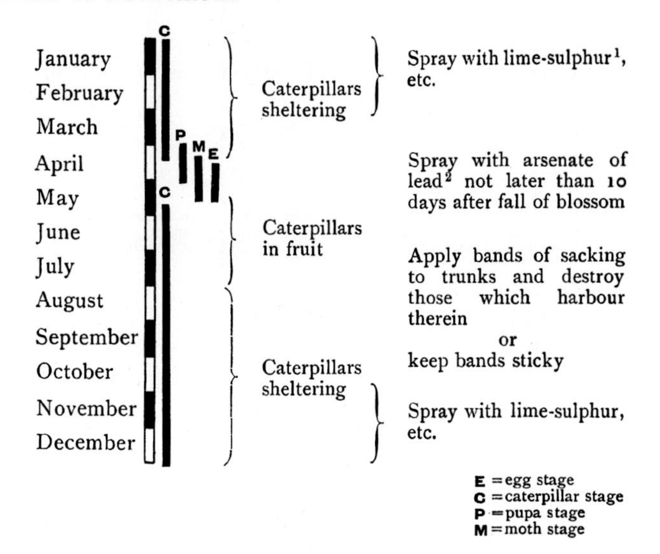


## DECEMBER MOTH

Name *Pœcilocampa populi* Family *Lasiocampidæ*
Order *Lepidoptera*

### 1. General

#### Description

**Larva (Caterpillar)**

| | |
|---|---|
| APPEARANCE. | About 1¾ inches when mature. |
| COLOUR. | Variable. |
| LOCATION. | On leaves. |
| APPEARS IN | March and April. |
| DURATION. | About 3 months. |

[1] See page 612. [2] See page 397.

**Pupa (Chrysalis)**

| | |
|---|---|
| APPEARANCE. | In grey cocoons. |
| COLOUR. | Reddish brown. |
| LOCATION. | At the foot of trees, or in the soil. |
| APPEARS IN | June and July. |
| DURATION. | Till December. |

Fig. 43. Caterpillar of December Moth. Natural size.

**Adult Insect (Moth)**

| | |
|---|---|
| SIZE ACROSS WINGS. | About $1\frac{1}{2}$ inches. |
| COLOUR (WINGS). | Purple-red: hind-wings paler. |
| „ (BODY). | Thick and hairy. |
| REMARKS. | Flies at night, being attracted by lights. |
| APPEARS IN | December. |

Fig. 44. December Moths. Natural size.

**Ova (Eggs)**

| | |
|---|---|
| LOCATION. | On branches. |
| APPEAR IN | December. |
| HATCHING PERIOD. | 4 to 5 months. |

**Distribution**

Widespread.

**Life-history**

The moth emerges from the ground in *December*, hence its name. It lays eggs at night on the bare branches, which hatch in the following spring. The caterpillars, which grow to a large size, mature about June, and then pupate in cocoons at the foot of the trees, or on the ground.

## 2. Economic

**Trees Attacked**

Occasionally apples and other fruit; normally the forest trees.

**Frequency of Pest**

Not common on fruit.

**Nature of Attack**

Defoliation of the trees by caterpillars.

**Remedies**

When the caterpillars occur in sufficiently large numbers, spray with ARSENATE OF LEAD[1].

[1] See page 397.

**Calendar of Treatment**

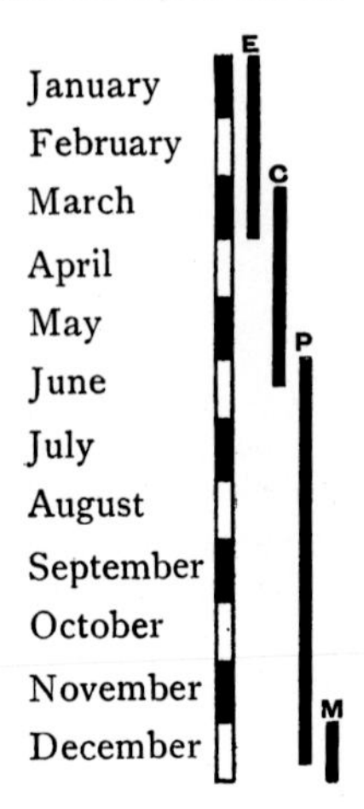


Spray with arsenate of lead[1].

E = egg stage
C = caterpillar stage
P = pupa stage
M = moth stage

## DOT MOTH

Name *Mamestra persicariæ* Class *Noctuæ*
Order *Lepidoptera*

### 1. General

**Description**

**Larva (Caterpillar)**

| | |
|---|---|
| APPEARANCE. | About $1\frac{1}{2}$ inches when mature. |
| COLOUR | Greenish or grey with characteristic markings. |
| LOCATION. | On leaves. |
| APPEARS IN | July and August. |
| DURATION. | Till end of September. |
| REMARKS. | The caterpillar is variable in colour according to the tint of the foliage on which it feeds. It is difficult to detect. |

[1] See page 397.

**Pupa (Chrysalis)**

| | |
|---|---|
| COLOUR. | Brown. |
| LOCATION. | In ground. |
| APPEARS IN | End September and October. |
| DURATION. | Till following summer. |

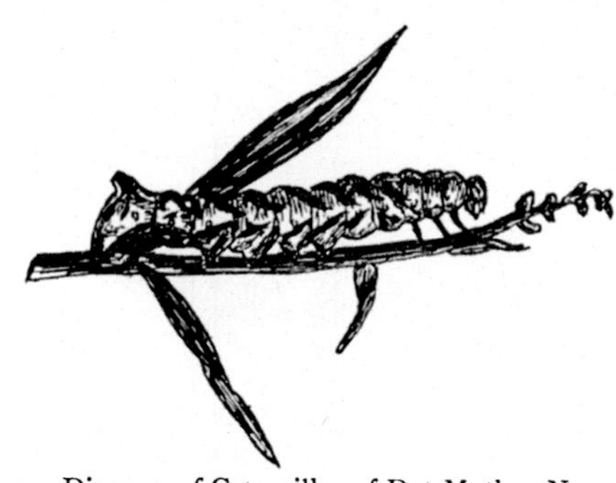

Fig. 45. Diagram of Caterpillar of Dot Moth. Natural size.

**Adult Insect (Moth)**

| | |
|---|---|
| SIZE ACROSS WINGS. | $1\frac{1}{2}$ to $1\frac{3}{4}$ inches. |
| COLOUR (WINGS). | Fore-wings dark with kidney-shaped dot. Hind-wings dark grey. |
| ,, (BODY). | Dark and hairy. |
| APPEARS IN | June and July at night. |

**Ova (Eggs)**

| | |
|---|---|
| LOCATION. | On leaves. |
| APPEAR IN | June and July. |

**Life-history**

The moths appear in June and July, and lay their eggs on the leaves of the plant. The caterpillars live on the leaves and mature about the end of September, pupating in the ground till the following summer

## 2. Economic

**Trees Attacked**

Gooseberries, and occasionally apples, plums, raspberries and currants.

Occurs most frequently on garden plots, and on cabbages etc.

**Frequency of Pest**

Common on gooseberries locally.

Fig. 46. Dot Moths and pupæ. Natural size.

**Nature of Attack**

The caterpillars devour the foliage and rapidly strip the bushes.

**Remedies**

1. In gardens, hand picking the caterpillars is advisable.
2. Spray the bushes with ARSENATE OF LEAD[1].
3. Run fowls or ducks over ground in winter.

[1] See page 397.

**Calendar of Treatment**

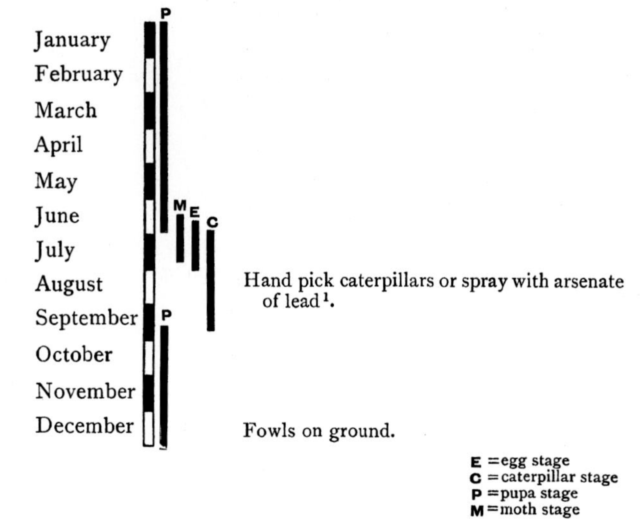


[1] See page 397.

## ERMINE MOTH (small)

Name *Hyponomeuta malinella* Class *Tortricidæ*
Order *Lepidoptera*

### 1. General

**Description**

**Larva (Caterpillar)**

| | |
|---|---|
| APPEARANCE. | About $\frac{3}{4}$ inch when mature, naked. |
| COLOUR. | Yellowish at first, grey with black spots later. |
| LOCATION. | Winter—on twigs in minute nests.<br>Spring—on buds and leaves.<br>Summer—in nests on leaves. |
| APPEARS IN | September and October. |
| DURATION. | Throughout winter till end of June. |

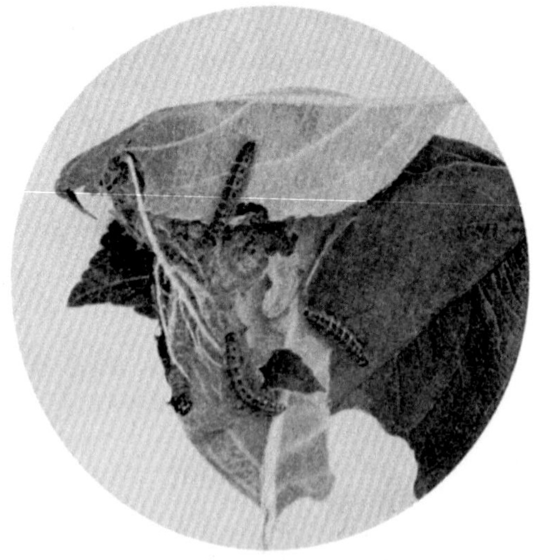

Fig. 47. Caterpillars of Little Ermine Moth. Natural size.

### Pupa (Chrysalis)

| | |
|---|---|
| APPEARANCE. | In greyish-white spindle-shaped cocoons side by side. |
| COLOUR. | Brown. |
| LOCATION. | In nests amongst leaves. |
| APPEARS IN | End of June. |
| DURATION. | About 2 weeks. |

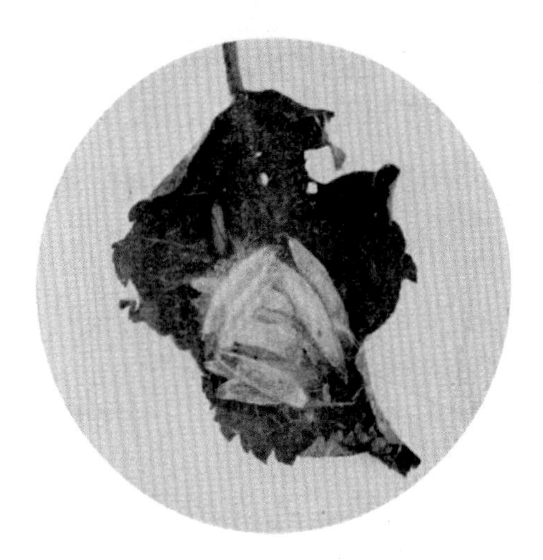

Fig. 48. Cocoons of Little Ermine Moth containing the pupæ. Natural size.

### Adult Insect (Moth)

| | |
|---|---|
| SIZE ACROSS WINGS. | About $\frac{1}{2}$ inch. |
| COLOUR (WINGS). | Fore—snowy white with rows of black spots.<br>Hind—pale grey. |
| ,, (BODY). | Chest—snowy white with black spots.<br>Abdomen—white. |
| APPEARS IN | July and August. |

### Ova (Eggs)

| | |
|---|---|
| Colour. | At first yellow; then brown to resemble the bark. |
| Arrangement. | Circular patches of 50–80, overlapping like the tiles of a house. |
| Location. | The small twigs. |
| Appear in | July and August. |
| Hatching period. | About 2 months. |

## Distribution

Widespread over Great Britain.

Fig. 49. Little Ermine Moths. Natural size.

## Life-history

The pretty moths appear in July and August and lay the eggs, overlapping each other, in circular patches. These are then covered over with a glutinous substance, which gradually assumes the colour of the branch, thus effectively screening the eggs from notice.

In the autumn the eggs hatch, and the minute caterpillars live together through the winter under a roof of egg-shells and debris.

On the arrival of spring, they grow active and enter the expanding buds, afterwards mining into the leaves. Later they feed upon the surfaces of the leaves.

As the season progresses they become associated together once more in "nests" of grey silk, spun between a number of leaves, and afterwards enlarge these so as to include whole boughs.

The mature caterpillars pupate about the end of June, spinning spindle-shaped cocoons of silk close together in the nest. The chrysalis stage lasts only about a fortnight.

## 2. Economic

### Trees Attacked

Apples.

### Frequency of Pest

Common.

### Nature of Attack

The blossom and leaves: the latter are usually skeletonised (see fig. 50).

### Degree of Damage

Often serious, though usually confined to a few parts of the tree.

### Natural Enemies

Starlings, and a species of parasitic fly (ichneumon).

### Remedies

1. Spraying thoroughly with concentrated LIME-SULPHUR[1] or CAUSTIC ALKALI EMULSION[2] *during the winter* will kill many of the hibernating young caterpillars, by the destruction of the gummy material which shelters them.
2. ARSENATE OF LEAD[3] forced into the nests with a coarse spraying nozzle *in summer* is effective but difficult and expensive.
3. Cutting out nests when these appear and burning them (or placing in quicklime) is the best method of dealing with an attack if time and labour permit.

[1] See page 612. [2] See page 420. [3] See page 397.

Fig. 50. Showing damage caused by Little Ermine Caterpillars.

**Calendar of Treatment**

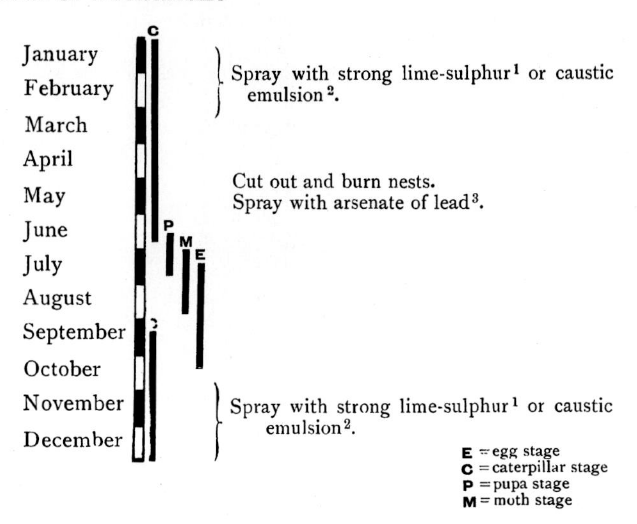


## FIGURE OF EIGHT MOTH (bluehead)

Name *Diloba cæruleocephala* Class *Noctuæ* Order *Lepidoptera*

### 1. General

**Description**

**Larva (Caterpillar)**

| | |
|---|---|
| APPEARANCE. | About 1 inch long: well marked. |
| COLOUR. | Head *blue* with two black spots. Body variable: yellow to grey. |
| LOCATION. | On foliage. |
| APPEARS IN | Late March or April. |
| DURATION. | Till end of June. |

[1] See page 612. [2] See page 420. [3] See page 397.

**Pupa (Chrysalis)**

| | |
|---|---|
| APPEARANCE. | In grey cocoons. |
| COLOUR. | Brown. |
| LOCATION. | On boughs of trees, or near by. |
| APPEARS IN | June. |
| DURATION. | About 3 months. |

Fig. 51. Diagram of Caterpillar of Figure of 8 Moth. Natural size.

**Adult Insect (Moth)**

| | |
|---|---|
| SIZE ACROSS WINGS. | 1 to $1\frac{1}{4}$ inches (variable). |
| COLOUR (WINGS). | Fore :—greyish brown, with pale spots in form of an "8" on each wing.<br>Hind :—greyish brown (lighter). |
| APPEARS IN | September and October. |
| DURATION. | 2 to 3 months. |

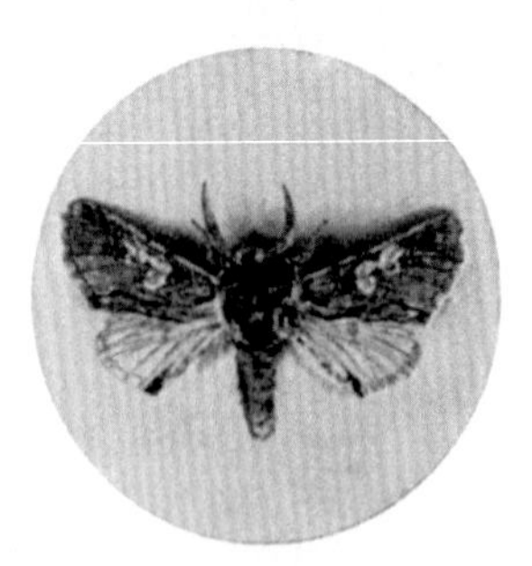

Fig. 52. Figure of 8 Moth. Natural size.

**Ova (Eggs)**

| | |
|---|---|
| APPEARANCE. | Shell-like with star-shaped ribs: conspicuous. |

| | |
|---|---|
| COLOUR. | Grey or greyish brown. |
| ARRANGEMENT. | Singly, or in groups of 5 to 10. |
| LOCATION. | On shoots of trees. |
| APPEAR IN | September to November. |
| DURATION. | Throughout winter till following spring. |

**Distribution**

Widespread in England.

**Life-history**

The moths appear in late autumn, flying at dusk or dark along the hedgerows. The eggs are very conspicuous on the boughs of the trees, being of a peculiar umbrella-top shape. They hatch out early in the following spring when the buds are swelling. The caterpillars live on the foliage of the tree and are mature about the end of June, pupating in grey cocoons on the tree, or near by.

### 2. Economic

**Trees Attacked**

Apples, plums, cherries; also the blackthorn, sloe, hawthorn.

**Frequency of Pest**

Not common.

**Nature of Attack**

The caterpillars devour the foliage.

**Duration of Attack**

Throughout spring and early summer.

**Degree of Damage**

Not usually severe.

**Preventive Measure**

Light-traps used at night in the late autumn where the attack is expected catch a good many moths and so prevent egg-laying.

**Remedy**

ARSENATE OF LEAD[1] is the correct spray, used in spring or summer when the attack is serious.

[1] See page 397.

**Calendar of Treatment**

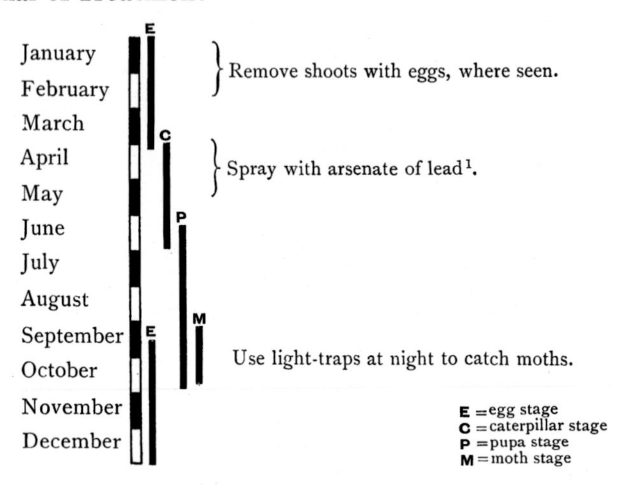


## GOAT MOTH

Name *Cossus ligniperda* Class *Cossidæ* Order *Lepidoptera*

### 1. General

**Description**

**Larva (Caterpillar)**

| | |
|---|---|
| APPEARANCE. | Naked and shiny; grows to 3—3½ inches long. |
| COLOUR. | At first pink, later yellowish and mahogany-red on back with 2 brown spots on first segment. |
| LOCATION. | Inside the woody tissue of trees. |
| APPEARS IN | July. |
| DURATION. | 3 years. |
| REMARKS. | The caterpillars frequently leave the trees, and attack the roots of other plants. |

[1] See page 397.

**Pupa (Chrysalis)**

| | |
|---|---|
| APPEARANCE. | Large, with rings of sharp spines on abdomen. |
| COLOUR. | Reddish brown. |
| LOCATION. | Near entrance of tunnel inside tree-trunk. |
| APPEARS IN | End of April and May. |
| DURATION. | 6—8 weeks |

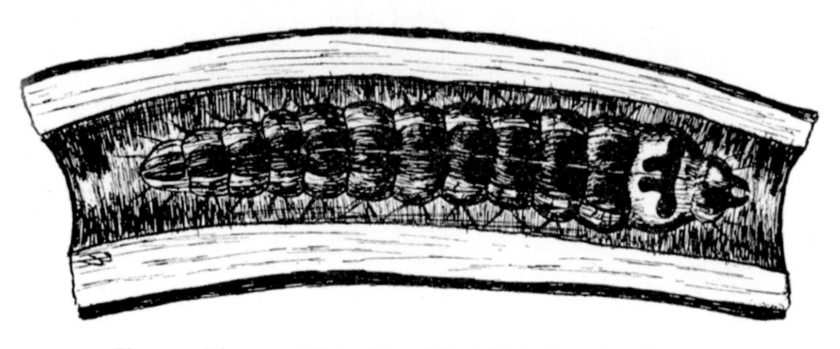

Fig. 53. Diagram of Caterpillar of Goat Moth (branch split open to show larva). Natural size.

**Adult Insect (Moth)**

| | |
|---|---|
| SIZE ACROSS WINGS. | Female 3—3½ inches : male 2¾—3 inches. |
| COLOUR (WINGS). | Greyish brown, mottled. |
| ,, (BODY). | Large, hairy, with dark grey bands. |
| REMARKS. | Its protective colouring when at rest on the trunks of trees effectively screens it from notice. |
| APPEARS IN | June and July. |

**Ova (Eggs)**

| | |
|---|---|
| APPEARANCE. | Round above, ribbed. |
| COLOUR. | Brownish. |
| ARRANGEMENT. | In groups of 10—50. |
| LOCATION. | In crevices of bark of trees. |
| APPEAR IN | June and July. |
| HATCHING PERIOD. | 10 days. |

## Distribution

Widespread : in England, chiefly in the south and east.

Fig. 54. Goat Moths: female (above), male (below). Natural size.

## Life-history

The moths emerge from their pupal skins, which are often seen at the mouth of the holes in the trunk of the tree attacked. They are sluggish, and usually lay their eggs near by.

The eggs are placed, by means of a strong horny ovipositor (egg-depositor), far into the crevices of the bark in numbers of 10 to 50 at a time, each moth being capable of laying some hundreds of eggs. The moths readily escape observation by their protective colouring[1], when resting with wings folded on the bark of the trees. The eggs hatch out in a few days, and the young caterpillars at once commence to burrow into the tree. In it they live for 3 YEARS before they mature, tunnelling into the wood. At the end of this time, unless they leave the tree before (see above), they bore an outlet from the tree, spin a cocoon near the exit hole, and change into the pupal state. The chrysalis forces its way to the exit by means of its spines (v.s.) and the moth in due time breaks through the skin, leaving this at the exit hole in the tree.

## 2. Economic

**Trees Attacked**

Apples and pears.

Forest trees are more commonly attacked.

**Frequency of Pest**

Not very common on fruit.

**Nature of Attack**

The wood is tunnelled through by the caterpillars, and ultimately becomes honeycombed, and dies.

**Duration of Attack**

Several years.

**Degree of Damage**

Attacked trees seldom survive.

**Preventive Measure**

Dress trees in the neighbourhood of those attacked with a mixture of clay, lime, and soft soap[2] to a height of 8 feet, or as far as the trunk extends.

**Natural Enemies**

Bats, goat suckers, and owls eat the moths, woodpeckers devour the caterpillars and pupæ, and tits the eggs.

The pupæ are parasitised by an ichneumon fly.

[1] I.e. the colour and markings of the wings and body resemble their surroundings very closely.

[2] See page 452.

**Remedies**

1. When the holes are found, place a lump of CYANIDE OF POTASH[1] (*caution—deadly poison*) in the holes, and seal these up with clay. The fumes of the cyanide destroy the caterpillar.
2. Sulphur or Nicotine fumes have been also used, blowing these into the holes by means of bellows, but this is neither so easy nor so effective as 1.
3. Cut down and destroy dead trees before the caterpillars escape.

**Calendar of Treatment**

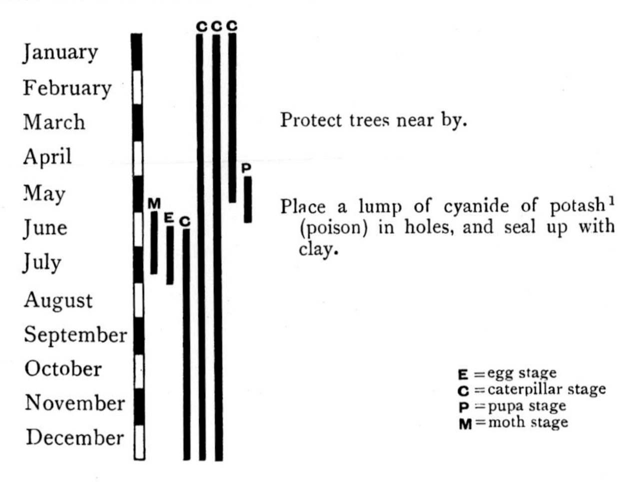


## GOLD-TAIL MOTH

Name *Porthesia similis* Family *Liparidæ*, "*Tussock Moths*" Order *Lepidoptera*

### 1. General

**Description**

**Larva (Caterpillar)**

LENGTH ETC. Winter stage, about $\frac{1}{3}$ inch (see fig. 60); full grown, $1\frac{1}{4}$ inches (see fig. 55), with tufts of brown hair—striking appearance.

[1] See page 423.

| | |
|---|---|
| COLOUR. | Deep brownish-black with bright red stripe on middle of back and a red line on each side, spotted with white. |
| LOCATION. | Winter—in crevices of bark in small grey cocoons. Summer—on leaves. |
| APPEARS IN | August and September; from caterpillar cocoons, end of March. |
| DURATION. | Throughout winter till following June. |
| REMARKS. | The young caterpillars hibernate in cocoons (see fig. 61). |

**Pupa (Chrysalis)**

| | |
|---|---|
| COLOUR. | Deep brown. |
| LOCATION. | On, or between leaves, or palings etc. near by. |
| APPEARS IN | July. |
| DURATION. | About a month. |

Fig. 55. Gold-tail Caterpillar nearly full grown. Natural size.

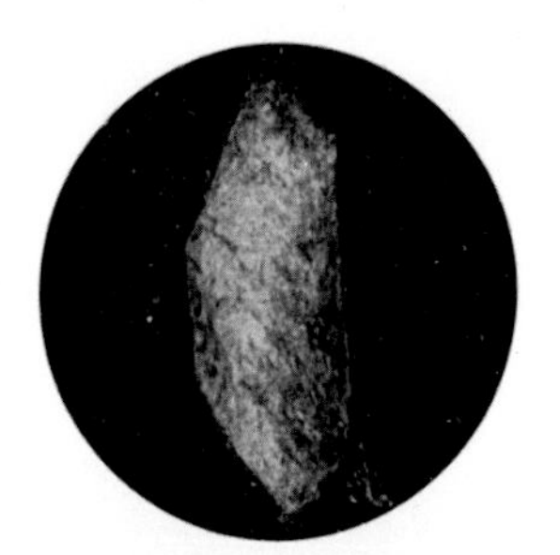

Fig. 56. The Gold-tail cocoon. Natural size.

**Adult Insect (Moth)**

| | |
|---|---|
| SIZE ACROSS WINGS. | Male 1—1½ inches; female slightly larger. |
| COLOUR (WINGS). | Female, pure satiny white; male has spot on each forewing. |
| „ (BODY). | Thorax and abdomen pure white, with golden-yellow tail-tuft. |
| APPEARS IN | August. |

**Ova (Eggs)**

| | |
|---|---|
| APPEARANCE. | In nests of golden hair from tails of moths. |
| LOCATION. | On leaves and branches. |
| APPEAR IN | August. |
| HATCHING PERIOD. | 7—10 days. |

Fig. 57. Cocoon opened to show pupa. Natural size.

Fig. 58. Gold-tail Moths: female (above) and male (below). Natural size.

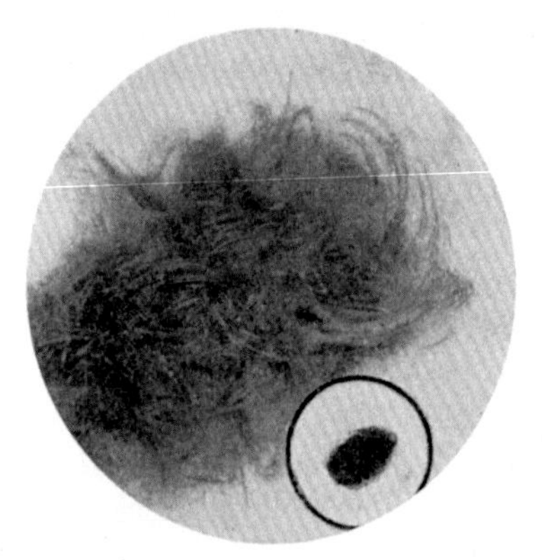

Fig. 59. Eggs of Gold-tail Moth enveloped in tail-hairs. Magnified (× 5). Inset, natural size.

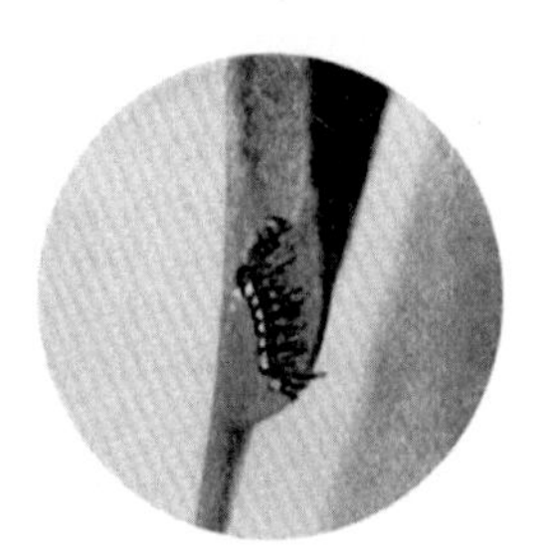

Fig. 60. Young Gold-tail Caterpillar before hibernating in autumn. Natural size.

### Distribution

General throughout England.

### Life-history

The moths are produced during August, and are found, with pure white wings folded, on palings etc. during the daytime. The eggs hatch out and the young caterpillars eat the leaves of the host-tree and then moult and winter in small grey cocoons in the crevices of the bark, under moss, etc. As soon as the buds swell in the following spring, the caterpillars emerge and commence to feed upon the young leaves. They grow to a large size and pupate about the end of June or early July.

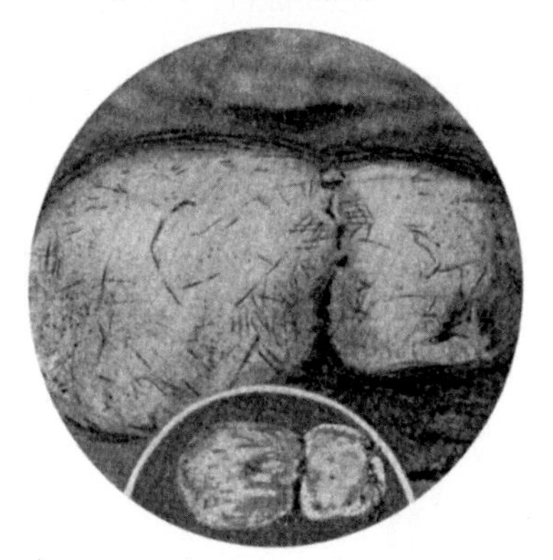

Fig. 61. Winter Caterpillar cocoons of Gold-tail Moth, magnified ($\times 2\frac{1}{2}$). Inset, natural size.

## 2. Economic

### Trees Attacked

Apples, plums, pears, cherries, also the hawthorn, hazel, cob, oak, chestnut, rose, etc.

### Frequency of Pest

Occurs in small quantities each year, and is often harmful locally.

### Nature of Attack

The young leaves, and occasionally fruitlets attacked.

### Degree of Damage

Variable; probably worst early in the spring, when the caterpillars, lready fairly large, are very voracious.

## Natural Enemies

Cuckoos are reported to feed upon them. Parasitic flies have also been recorded.

## Remedies

1. If attack is serious spray in spring, when caterpillars are observed on the leaves, with ARSENATE OF LEAD[1].
2. Place bands of sacking around trees in autumn to entice the young caterpillars to shelter inside; these are then destroyed.
3. Spraying with strong LIME-SULPHUR[2] just before buds burst will probably kill a proportion of the young caterpillars, and CAUSTIC EMULSION used in the winter is also effective.

## Calendar of Treatment

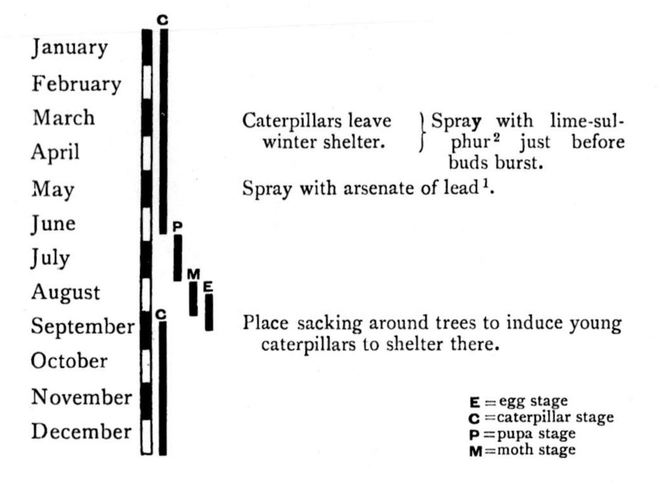


[1] See page 397. [2] See page 612.

## HAWK (EYED) MOTH

Name *Smerinthus ocellatus* Class *Sphingidæ*
Order *Lepidoptera*

### 1. General

**Description**

**Larva (Caterpillar)**

| | |
|---|---|
| SIZE. | 3 inches long when mature. |
| COLOUR. | Green to bluish green, dotted with white. |
| LOCATION. | On leaves. |
| APPEARS IN | June to September. |
| DURATION. | 6 weeks. |
| REMARKS. | The caterpillar enters the ground to pupate. |

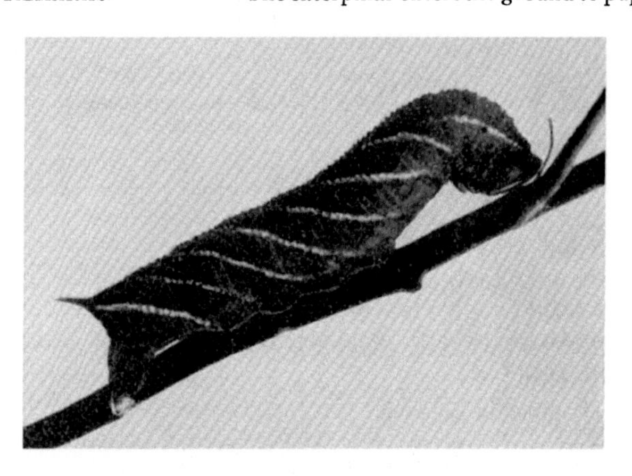

Fig. 62. Caterpillar of Eyed Hawk Moth. Almost full fed. Natural size.

**Pupa (Chrysalis)**

| | |
|---|---|
| SIZE. | About $1\frac{3}{4}$ inches long. |
| COLOUR. | Deep brown. |
| LOCATION. | In a cavity in soil, about 2 inches below surface of ground. |
| APPEARS IN | Autumn. |
| DURATION. | Throughout winter till following summer. |

### Adult Insect (Moth)

| | |
|---|---|
| SIZE ACROSS WINGS. | $2\frac{1}{2}$ to $3\frac{3}{4}$ inches. |
| COLOUR (WINGS). | Rich grey-brown (see index sheet). |
| APPEARS IN | End of May to middle July. |

Fig. 63. Pupa of Eyed Hawk Moth. Natural size.

### Ova (Eggs)

| | |
|---|---|
| APPEARANCE. | Large, oval; upper surface shrunken in centre. |
| COLOUR. | Pale greenish yellow. |
| ARRANGEMENT. | Singly. |
| LOCATION. | On leaves. |
| APPEAR IN | End May to middle July. |

### Distribution

Widespread in England.

### Life-history

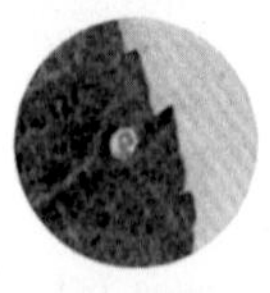

Fig. 64. Egg of Eyed Hawk Moth. Natural size.

This large and handsome moth appears in June and July and lays its eggs singly upon the leaves. The young caterpillars feed upon the leaves and are full grown in about 6 weeks, being then upwards of 3 inches long and of a striking appearance. They enter the soil and there pupate, the chrysalis remaining in the ground till the following summer.

Fig. 65. Eyed Hawk Moths: female (above), male (below). Natural size.

## 2. Economic

**Trees Attacked**

Apples, peaches, almonds, sloes, also the willow, sallow, and poplar.

**Frequency of Pest**

Not very common.

**Nature of Attack**

The large caterpillars rapidly devour the foliage.

**Degree of Damage**

Great, if in sufficient numbers.

## Remedies

1. Hand picking: the caterpillars are so large and easily seen that this is readily done.
2. If the severity of the attack warrants it, spray with ARSENATE OF LEAD[1] early in the summer.

## Calendar of Treatment

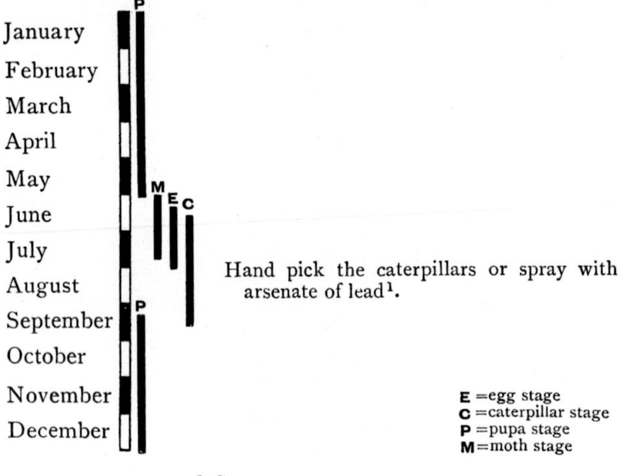


[1] See page 397.

## LACKEY MOTH

Name *Clissiocampa neustria* Class *Bombycidæ*
Order *Lepidoptera*

### 1. General

### Description

#### Larva (Caterpillar)

| | |
|---|---|
| APPEARANCE. | About $1\frac{1}{2}$ inches long, when full grown: somewhat hairy. |
| COLOUR. | Dark at first, then brilliantly striped from head to tail end: two black spots on head. |

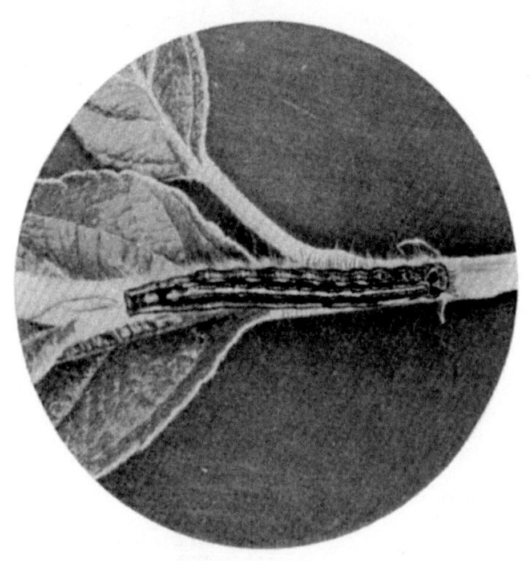

Fig. 66. Lackey Caterpillar, full grown. Natural size.

| | |
|---|---|
| LOCATION. | In "nests" enclosing leaves, or, when mature, on fork of tree. |
| APPEARS IN | End of April. |
| DURATION. | Till middle or end of July. |
| REMARKS. | Somewhat timid, readily fall to ground on shaking the tree. Hundreds may live together in the "nests." |

**Pupa (Chrysalis)**

| | |
|---|---|
| APPEARANCE. | In loosely spun cocoons. |
| COLOUR. | Dark brown. |
| LOCATION. | Among leaves, or on fences etc. |
| APPEARS IN | July and August. |
| DURATION. | 2—3 weeks. |

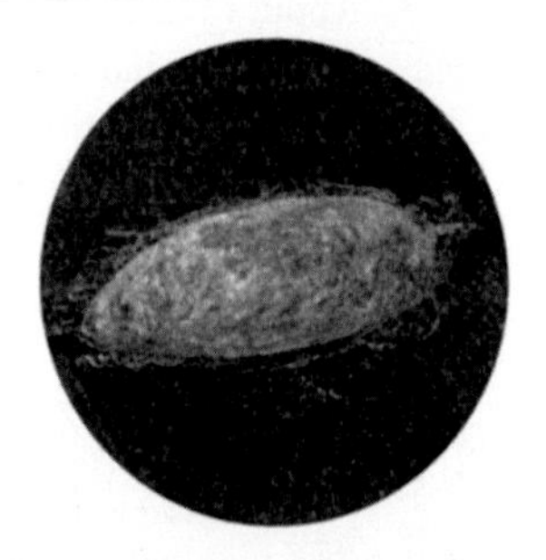

Fig. 67. Cocoon of Lackey Moth containing the pupa.
Natural size.

**Adult Insect (Moth)**

| | |
|---|---|
| SIZE ACROSS WINGS. | Male, about 1 inch: female larger (about $1\frac{1}{2}$ inches). |
| COLOUR (WINGS). | Variable: fore-wings brown to red with a bar across. |
| APPEARS IN | July and August. |

**Ova (Eggs)**

| | |
|---|---|
| APPEARANCE. | Easily visible, fairly large. |
| COLOUR. | Grey, with dark hollow in centre. |

| | |
|---|---|
| ARRANGEMENT. | 50 to 200 arranged in bands. |
| LOCATION. | Around the young branches. |
| APPEAR IN | July and August. |
| DURATION. | Throughout the winter till the end of April. |

Fig. 68. Lackey Moths: female (above), male (below). Natural size.

**Distribution**

In England, chiefly in South, West and Midlands; most abundant in South.

**Life-history**

The moths have moderate powers of flight. The eggs are deposited in bands on the young branches, and remain throughout the winter. On hatching, in the spring, the young caterpillars form a WEB of fine silk enclosing a few leaves, and beneath this they continue to feed for some time. As they grow, this "nest" is enlarged until it may reach over a foot in length.

Later on the caterpillars spread over the whole tree, but return to the nest in dull weather or at night. On warm days they group themselves in parallel rows on the outsides of the

Fig. 69. Egg-band of Lackey Moth on twig of apple. Enlarged (× 3). Inset, natural size.

nests or among the branches. After the final moult they are usually found in the fork of the tree.

The large nests are called TENTS.

The cocoons are spun among the leaves, and on fences, and the moth hatches out in 2—3 weeks.

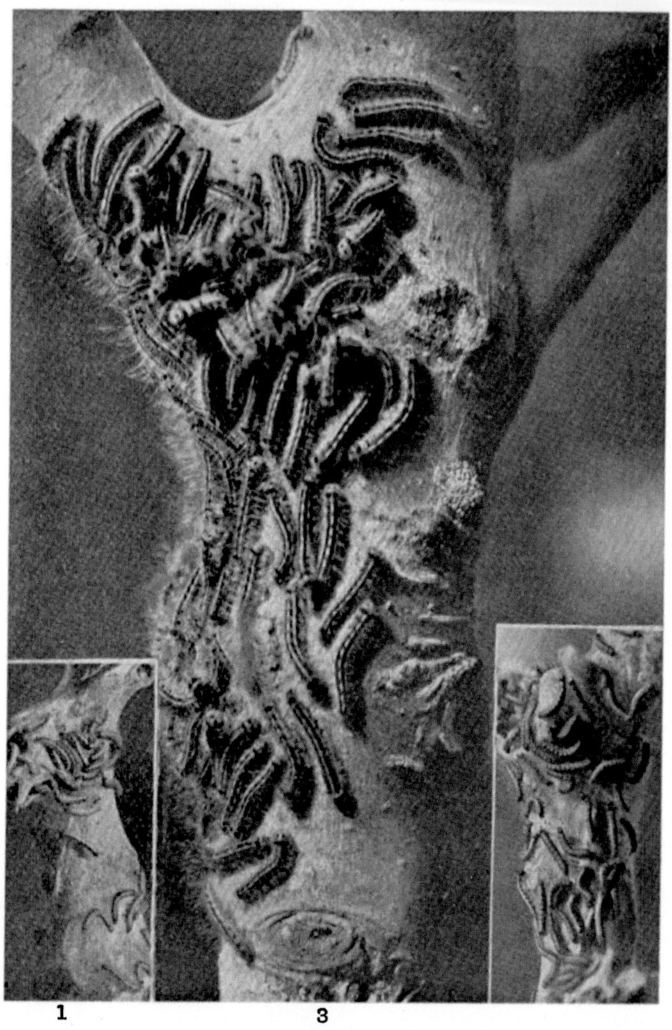


Fig. 70. Caterpillars of Lackey Moth. 1, after first moult on small branch. 2, after second moult on larger branch. 3, before last moult: manufacturing the tent on fork of tree. All slightly reduced in size.

## 2. Economic

**Trees Attacked**

Apples, pears, also the oak, elm, hawthorn, rose.

**Frequency of Pest**

Fairly common, and periodically occurs as a "plague" in certain localities.

**Nature of Attack**

The leaves and small branches are completely devoured.

**Duration of Attack**

Usually from end of April till middle of July.

**Degree of Damage**

Severe, and in certain cases enormous.

**Preventive Measure**

Remove the branches which have EGG BANDS, and are easily seen, during the winter. (Winter washes are useless.)

**Natural Enemies**

Few.

The cuckoo has been observed eating the caterpillars.

**Remedies**[1]

[The first two are especially of service in large plantations.]

1. Destroy the caterpillars by rubbing the nests against branches with sacking tied on the end of long poles.

2. Disperse nests by shooting with a muzzle-loading gun and small charge of gunpowder with paper wad.

   This specially applies when the caterpillars are in the larger branches or fork of the trees.

3. Cut out nests whole and burn them, or place in a bucket of fresh lime and water.

   Be careful caterpillars do not drop on the ground (best to place sacks underneath) and so escape meanwhile.

[1] If a thorough spraying with arsenate of lead is given early in the season for other caterpillars (Winter Moth, etc.) the lackeys are usually killed while young. Where bad attacks occur locally it is even worth while to spray for the lackey alone in early May.

4. STICKY-BAND[1] the trees attacked, using plenty of material; then shake the caterpillars off on to the ground. On attempting to reascend the tree, they are caught on the bands. (Care must be taken that none crawl to other trees.)

**Calendar of Treatment**

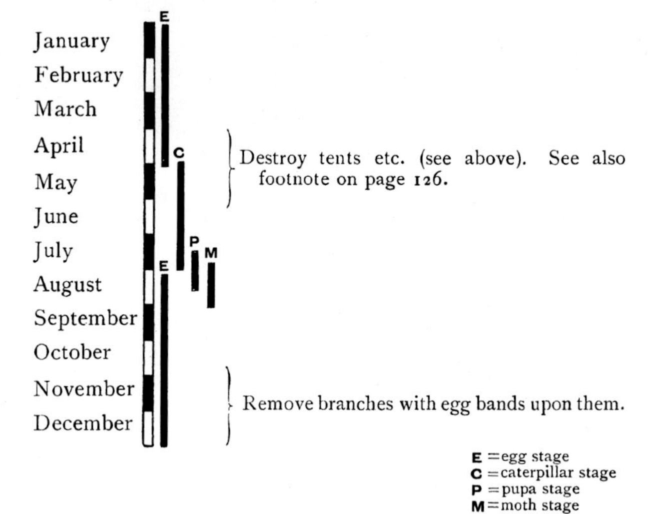


## LAPPET MOTH

Name *Lasiocampa quercifolia* Family *Lasiocampidæ*
Order *Lepidoptera*

### 1. General

**Description**

**Larva (Caterpillar)**

| | |
|---|---|
| APPEARANCE. | Very large (over 4 inches, full grown): covered with long grey hairs, and with fleshy pad-like appendages. |
| COLOUR. | Grey or grey-brown with 2 blue or purple bands. |

[1] See page 404.

| | |
|---|---|
| LOCATION. | On stems or leaves. |
| APPEARS IN | Late summer and autumn. |
| DURATION. | Throughout the winter till following May. |
| REMARKS. | The largest caterpillar pest attacking fruit. |

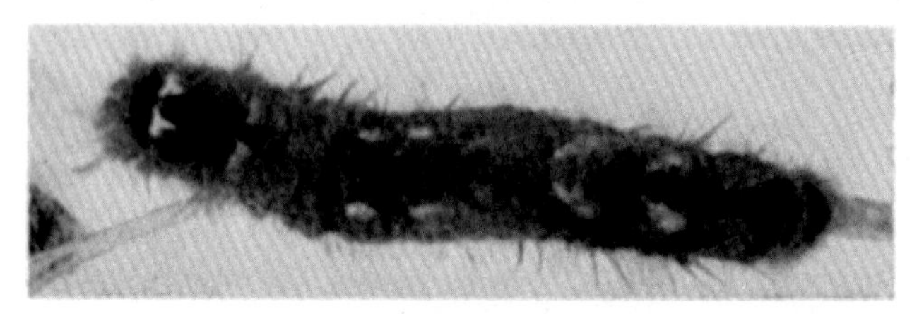

Fig. 71. Lappet Caterpillar, nearly full grown. Natural size.

**Pupa (Chrysalis)**

| | |
|---|---|
| APPEARANCE. | In large, oval cocoon. |
| COLOUR. | Dark brown. |
| LOCATION. | On twigs of trees, or under rubbish on ground. |
| APPEARS IN | May. |
| DURATION. | Till end of June or July. |
| REMARKS. | The chrysalis is a well-known object of curiosity for its large size and the rapid movements of its "tail." |

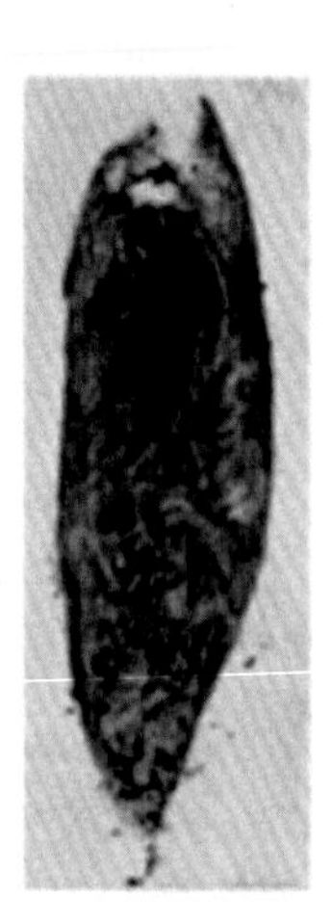

Fig. 72. Lappet cocoon, containing pupa. Natural size.

**Adult Insect (Moth)**

| | |
|---|---|
| SIZE ACROSS WINGS. | Male about $2\frac{1}{4}$ inches, female up to $3\frac{1}{4}$ inches. |
| COLOUR (WINGS). | Rich brown; edges scalloped. |
| ,, (BODY). | Very furry. |
| REMARKS. | When at rest, the hind-wings project below the others in a characteristic manner (see fig. 75). |
| APPEARS IN | End of June and July. |

**Ova (Eggs)**

| | |
|---|---|
| APPEARANCE. | Very marked, large (see fig. 74). |
| COLOUR. | Grey. |
| LOCATION. | On leaves. |
| APPEAR IN | End of June and July. |
| DURATION. | Till late summer or autumn. |

Fig. 73. Lappet Moths: male (above), female (below). Natural size.

**Life-history**

The large brown moths appear in middle summer, and lay the well-marked eggs. These hatch out in late summer and autumn and in this state exist all through the winter, extended on the twigs of trees.

They mature in summer and then spin a long oval cocoon among the twigs of the trees, or under rubbish, the large, dark brown, active pupa often attracting attention.

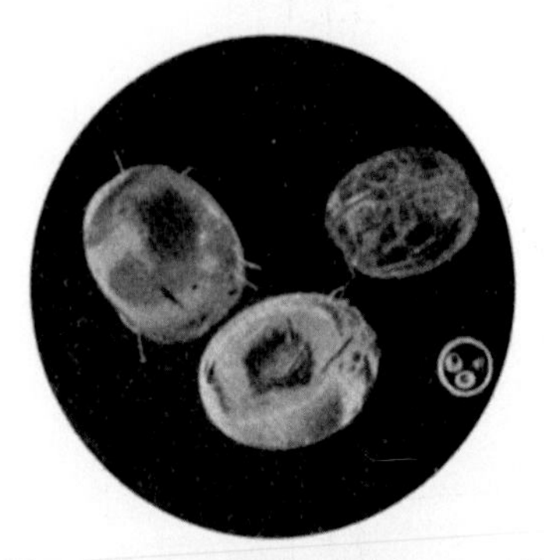

Fig. 74. Eggs of Lappet Moth. Highly magnified (× 10). Inset, natural size.

Fig. 75. Lappet Moth at rest. Natural size.

## 2. Economic

### Trees Attacked

Apple.

### Frequency of Pest

Not common.

### Degree of Damage

Only serious if the caterpillar is present in large numbers.

### Remedies

1. Hand picking of the large caterpillars is comparatively easy.
2. Spray with ARSENATE OF LEAD[1] if the number of caterpillars warrants the expenditure.

### Calendar of Treatment

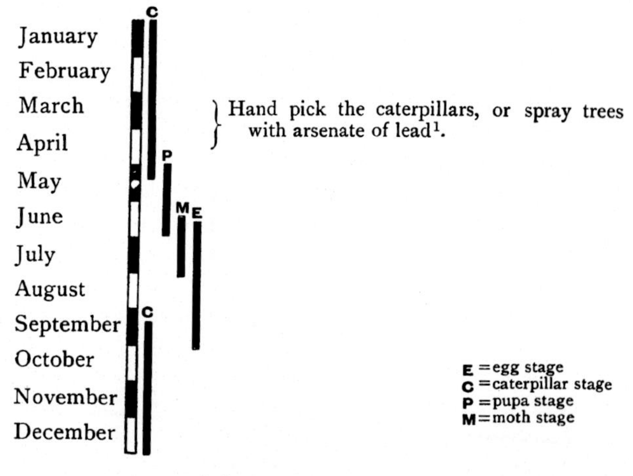


[1] See page 397.

## LEAF MINER MOTH (apple)

Name *Lyonetia clerckella* Class *Tineidæ* Order *Lepidoptera*

### 1. General

**Description**

**Larva (Caterpillar)**

| | |
|---|---|
| SIZE. | Less than $\frac{1}{4}$ inch when full grown. |
| COLOUR. | Pale to deep green, with dark head. |
| LOCATION. | Inside leaves. |
| APPEARS IN | (1) April and May, (2) August, (3) October. |
| DURATION. | 4—5 weeks. |
| REMARKS. | The caterpillars leave the interior of the leaves when mature. |

**Pupa (Chrysalis)**

| | |
|---|---|
| APPEARANCE. | In cocoons about $\frac{1}{7}$ inch long; with fine, silky threads. |
| COLOUR. | Pale green. |
| LOCATION. | On leaves. |
| APPEARS IN | (1) June, (2) September, (3) November. |
| DURATION. | 10—20 days. |

**Adult Insect (Moth)**

| | |
|---|---|
| SIZE ACROSS WINGS. | About $\frac{1}{3}$ inch. |
| COLOUR (WINGS). | Brownish-white with long fringe of hairs. |
| APPEARS IN | June and July, and there are often two other broods. |
| DURATION. | Throughout winter. |

Fig. 76. Apple Leaf Miner Moth. Natural size.

**Ova (Eggs)**

| | |
|---|---|
| ARRANGEMENT. | Single. |
| LOCATION. | On leaf. |
| APPEAR IN | June and July etc. |

**Life-history**

There may be as many as three broods in a year. The moths hibernate over the winter, depositing a single egg on the leaf in the following spring.

The minute caterpillar, on hatching, at once enters the leaf and then forms a twisted burrow, feeding upon the soft interior tissue of the leaf.

When mature, the caterpillar (being then only about $\frac{1}{4}$ inch long) comes on to the surface of the leaf and spins a delicate cocoon, from which the moth emerges after 2—3 weeks.

## 2. Economic

**Trees Attacked**

Apples, cherries.

**Frequency of Pest**

Common.

**Nature of Attack**

The attacked leaves often shrivel or wither and damage the tree by checking the sap.

**Natural Enemies**

Is often attacked by a very minute fly (a Chalcid).

**Remedies**

Since the caterpillar is only exposed to attack on the surface of the leaf for a very short period there is little opportunity for any treatment at all.

If the pest becomes very bad, resort may be had to hand picking of the infested leaves.

**Calendar of Treatment** (3 broods)

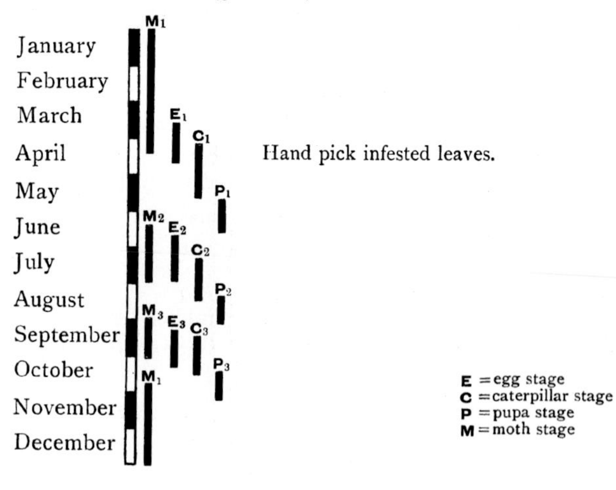

## MAGPIE MOTH

Name *Abraxas grossulariata* Class *Geometridæ*
Order *Lepidoptera*

### 1. General

**Description**

**Larva (Caterpillar)**

| | |
|---|---|
| APPEARANCE. | Conspicuous: about 1½ inches long. |
| COLOUR. | White, spotted with black, and orange-yellow. |
| LOCATION. | (1) on leaves, in spring, (2) under dead leaves and rubbish in winter. |
| PROGRESSION. | By "*looping.*" |
| APPEARS IN | July and August. |
| DURATION. | Throughout winter till end of following June. |
| REMARKS. | The young caterpillars hibernate during the winter and emerge to feed upon the young leaves in the spring. |

**Pupa (Chrysalis)**

| | |
|---|---|
| APPEARANCE. | Marked, glossy. |
| COLOUR. | Black, with 3 golden-yellow rings round body. |
| LOCATION. | Suspended from the leaves. |
| APPEARS IN | End of June and July. |
| DURATION. | 4 to 5 weeks. |

**Adult Insect (Moth)**

| | |
|---|---|
| SIZE ACROSS WINGS | 1¼ to 1¾ inches: variable. |
| COLOUR (WINGS) | Creamy white spotted with black, but *very variable.* |
| ,, (BODY) | Yellow and black. |
| APPEARS IN | July and August. |

Fig. 77. Magpie Caterpillars and Pupa on currant leaves.

**Ova (Eggs)**

| | |
|---|---|
| APPEARANCE. | Oval. |
| COLOUR. | Cream. |
| ARRANGEMENT. | Singly or in heaps. |
| LOCATION. | On leaves. |
| APPEAR IN | July and August. |
| HATCHING PERIOD. | 6—15 days. |

Fig. 78. Magpie Moth. Natural size.

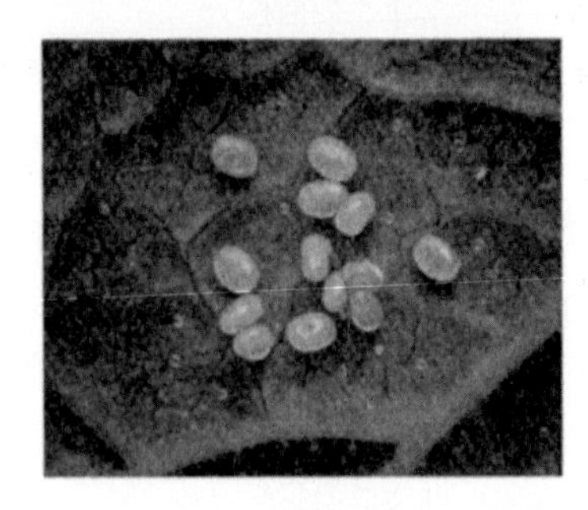

Fig. 79. Eggs of the Magpie Moth. Magnified.

**Distribution**

Widely in Britain.

**Life-history**

The gaily-coloured moths appear in middle to late summer at dusk or in the daytime, being sluggish fliers.

The eggs are laid in groups, or singly, on the leaves, and hatch out in a few days into small, nearly black, LOOPER caterpillars, which feed upon the leaves and then seek shelter for the winter among dead leaves, under stones, etc.

In spring they leave their hiding places and commence to feed upon the young leaves, being mature about the end of June.

They then suspend themselves from the leaves and pupate in this position, giving rise to the adult moths in a few weeks.

## 2. Economic

**Trees Attacked**

Currants, gooseberries, also apricots, nuts, and many woodland trees.

**Susceptible Varieties**

Chiefly black currants.

**Frequency of Pest**

Very common in South of England.

**Nature of Attack**

The leaves and young shoots are devoured and, in bad cases, only the main branches left.

**Degree of Damage**

Frequently serious, especially in gardens and allotments.

**Preventive Measure**

Placing STICKY-BANDS[1] round the stems of the bushes in early spring prevents the return of the caterpillars from their winter quarters on the ground, etc.

**Remedies**

1. Spray the affected bushes in spring or, better still, in autumn (after an attack) with ARSENATE OF LEAD[2].
2. Hoe the ground under bushes in late winter and mix in lime and soot, removing and burning dead leaves. This mixture destroys the young caterpillars.

[1] See page 404. [2] See page 397.

**Calendar of Treatment**

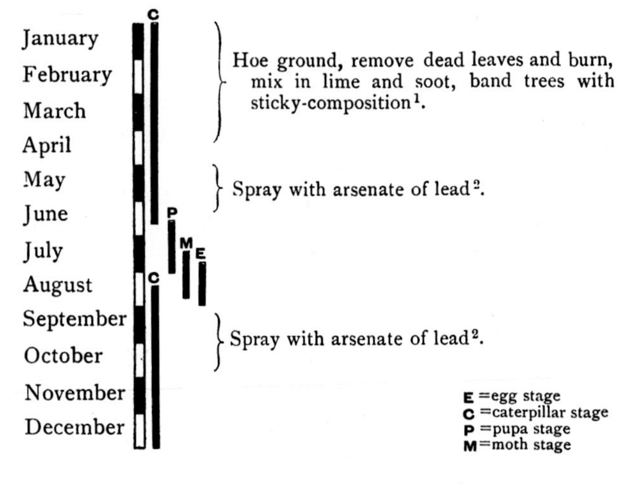


[1] See page 404. [2] See page 397.

## MARCH MOTH

Name *Anisopteryx æscularia* Class *Geometridæ*
Order *Lepidoptera*

### 1. General

**Description**

**Larva (Caterpillar)**

| | |
|---|---|
| APPEARANCE. | Smooth skinned, slender, cylindrical, about 1 inch in length (mature). |
| COLOUR. | Bright to yellowish green. |
| LOCATION. | On leaves. |
| PROGRESSION. | By "*looping*." |
| APPEARS IN | April. |
| DURATION. | About 3 months. |
| REMARKS. | Resembles the caterpillar of the "Winter Moth" but is much thinner and more slender. |

Fig. 80. Diagram of Caterpillars of March Moth. Natural size.

**Pupa (Chrysalis)**

| | |
|---|---|
| APPEARANCE. COLOUR. | Brown with yellow silk cocoon. |
| LOCATION. | In the ground. |
| APPEARS IN | June. |
| DURATION. | Till following February or later. |

**Adult Insect (Moth)**

| | Male | Female |
|---|---|---|
| SIZE ACROSS WINGS. | $1\frac{1}{4}$ to $1\frac{1}{2}$ inches. | Wingless. |
| COLOUR (WINGS). | Greyish brown with wavy lines. | — |
| „ (BODY). | Grey-brown. | Grey-brown. |
| REMARKS. | The wingless female has a tail tuft of grey hairs. | |
| APPEARS IN | Mid-February till end of April (variable). | |

Fig. 81. March Moth. Wingless female. Magnified (× $3\frac{1}{2}$). Inset, natural size.

Fig. 82. March Moth. Male. Natural size.

**Ova (Eggs)**

| | |
|---|---|
| APPEARANCE. | Small, glossy. |
| COLOUR. | Black. |
| ARRANGEMENT. | In bands of parallel rows, with hairs from tail tuft of female. |
| LOCATION. | Around the small branches. |
| APPEAR IN | February till April. |
| HATCH IN | April. |

**Distribution**

Fairly widespread in Great Britain.

**Life-history**

Almost identical with the WINTER MOTH, but the females ascend the trees *much later*. The wingless females leave their pupa cases in the ground and crawl to the nearest tree trunk, which they ascend in February or up to April. On their way up the trees they are fertilised by the males. The eggs are laid in parallel rows on the branches embedded in hairs from the tail tuft of the moth. These hatch out in April and the young caterpillars at once devour the young leaves. They mature in June, and then fall to the ground; and pupate in the soil in yellow silk cocoons.

**Remarks**

These caterpillars, like those of the Winter Moth, Mottled Umber, Magpie, Peppered Moth and some other pests, move by "looping," that is, they arch their bodies, fixing their tail-claspers near the head, and by this support extend the whole body straight, re-looping the hinder part as before. This gives them the name of "earth-measurers" (*Geometridæ*), and they are all noted for being of very active habits, often moving from tree to tree in search of food.

## 2. Economic

**Trees Attacked**

Apples, plums, pears. Its normal food is the blackthorn, but it also infests several forest trees.

**Frequency of Pest**

Not frequent as a destructive pest.

**Nature of Attack**

Foliage devoured by caterpillars.

**Duration of Attack**

Late spring and early summer.

**Degree of Damage**

Considerable when in sufficient numbers.

**Preventive Measures**

1. Banding the trees with sticky material (see under page 180, Winter Moth, and page 404). The bands must be kept in operation till the end of April to catch the late-comers.

2. Destroying branches seen to carry the egg-bands.
3. Hoeing the ground under the trees during the autumn and winter months.

## Remedies

1. Spraying the affected trees during the spring and summer with ARSENATE OF LEAD[1].
2. Some insecticides, such as strong NICOTINE[2], or NICOTINE SUBSTITUTES used with soap, will kill the young caterpillars.

## Calendar of Treatment

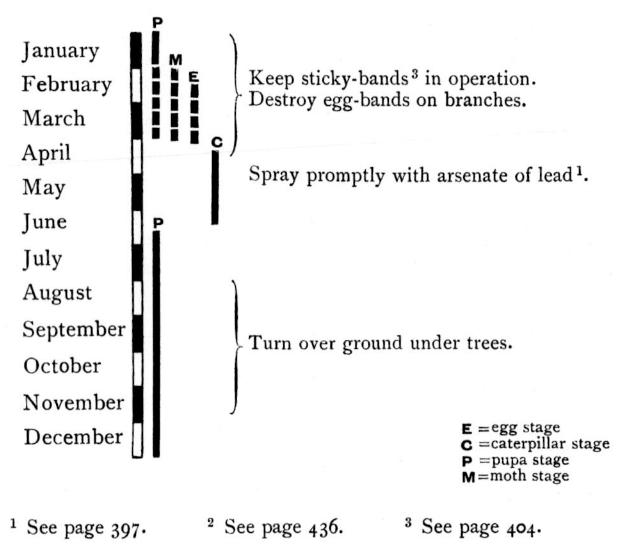


[1] See page 397. [2] See page 436. [3] See page 404.

## MOTTLED UMBER MOTH

Name *Hybernia defoliaria* Class *Geometridæ*
Order *Lepidoptera*

### 1. General

**Description**

**Larva (Caterpillar)**

| | |
|---|---|
| APPEARANCE. | Slender, smooth skinned, about 1½ inches in length (mature). |
| COLOUR. | Chestnut brown and creamy yellow. |
| LOCATION. | On leaves. |
| PROGRESSION. | By "*looping*." |
| APPEARS IN | April. |
| DURATION. | Till end of June. |
| REMARKS. | On maturity they fall to the ground and enter the soil to pupate. |

Fig. 83. Mottled Umber Caterpillars. Natural size.

**Pupa (Chrysalis)**

| | |
|---|---|
| COLOUR. | Brown. |
| LOCATION. | In soil. |
| APPEARS IN | End of June. |
| DURATION. | 4 to 7 months. |

**Adult Insect (Moth)**

| | Male | Female |
|---|---|---|
| SIZE ACROSS WINGS. | About $1\frac{3}{4}$ inches. | Wingless. |
| COLOUR (WINGS). | Yellowish brown. | — |
| ,, (BODY). | Yellowish grey. | Mottled brown. |
| REMARKS. | The wingless females ascend the tree trunks. | |
| APPEARS IN | End October—February. | |

**Ova (Eggs)**

| | |
|---|---|
| APPEARANCE. | Oval. |
| COLOUR. | Deep straw colour. |
| ARRANGEMENT. | Not systematic. |
| LOCATION. | Buds, twigs, bark cavities, or pruned surfaces. |
| APPEAR IN | October to February. |
| HATCH IN | April. |

Fig. 84. Pupa of Mottled Umber Moth. Natural size.

**Distribution**

Chiefly in Midlands and South of England.

Fig. 85. Mottled Umber Moth, female. Magnified ($\times 2\frac{1}{2}$). Inset, natural size.

## Life-history

The moths appear from the soil at very variable times, from the end of October until February, according to the season.

The female ascends the trunk (being without wings) and is there fertilised by the winged male moth. Eggs are deposited on the

Fig. 86. Three varieties of male Mottled Umber Moths. Natural size.

twigs, buds etc. The caterpillars hatch out in April and feed upon the foliage until June when they drop to the ground and pupate under the soil.

## Remarks

The caterpillar is a "looper and very active (see page 56).

## 2. Economic

**Trees Attacked**

Apples, plums, cherries, pears, cobs, filberts; also forest trees.

**Frequency of Pest**

In the case of orchards near oak, or forest trees, bad attacks often occur locally.

**Nature of Attack**

The caterpillars devour the leaves, and often attack the fruit, especially growing cherries.

**Duration of Attack**

April—July.

**Degree of Damage**

Often serious defoliation, and damage to fruits.

**Preventive Measures**

1. Keep the trees banded with sticky material (see page 404) from October onwards to catch the ascending females and prevent egg laying.
2. Turn over ground under trees in late summer and early autumn.

**Remedies**

1. Spray with ARSENATE OF LEAD[1].
2. Soap sprays with NICOTINE[2] or NICOTINE SUBSTITUTES in sufficient quantities will kill the young caterpillars.

**Calendar of Treatment**

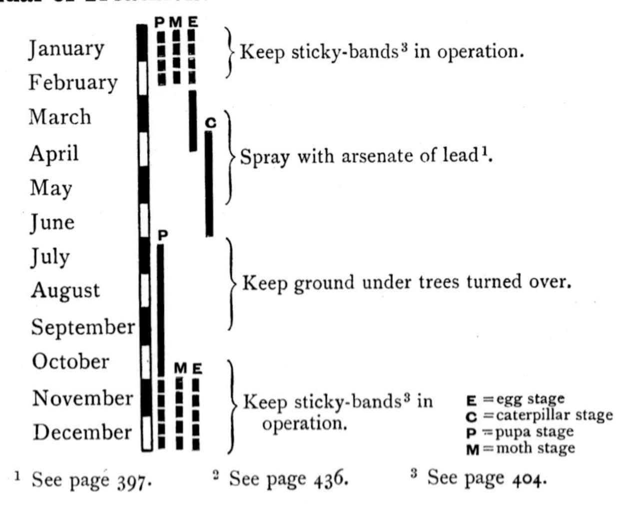


[1] See page 397. [2] See page 436. [3] See page 404.

## PEPPERED MOTH

Name *Amphidasys betularia* Class *Geometridæ*
Order *Lepidoptera*

### 1. General

#### Description

**Larva (Caterpillar)**

| | |
|---|---|
| APPEARANCE. | About 2 inches when mature—resembles a small twig. |
| COLOUR. | Green to brown (protectively coloured). |
| LOCATION. | On shoots. |
| PROGRESSION. | By "*looping*." |
| APPEARS IN | Late July and August. |
| DURATION. | Till late October. |
| REMARKS. | The caterpillars are very difficult to find, as they mimic, very wonderfully, the twigs of trees. |

**Pupa (Chrysalis)**

| | |
|---|---|
| COLOUR. | Brown. |
| LOCATION. | In the ground. |
| APPEARS IN | Late October—November. |
| DURATION. | Till following June (throughout winter). |

**Adult Insect (Moth)**

| | |
|---|---|
| SIZE ACROSS WINGS. | $1\frac{1}{2}$ to $2\frac{1}{4}$ inches (female slightly larger). |
| COLOUR (WINGS). | White, speckled with black. |
| APPEARS IN | Late May—July. |

**Ova (Eggs)**

| | |
|---|---|
| ARRANGEMENT. | Singly. |
| LOCATION. | On leaves. |
| APPEAR IN | Late May—July. |
| HATCHING PERIOD. | 3—4 weeks. |

#### Distribution

Widespread in England.

#### Life-history

The moths commonly appear in June and July, and both the sexes are winged.

The caterpillars hatch out in August, and do not mature till late in the autumn, by which time they can strip the trees and the damage often pass unnoticed, since the fruit is gathered and the leaves falling.

Fig. 87. Caterpillars of Peppered Moths in characteristic attitudes. Natural size.

The caterpillars also escape notice by their wonderful resemblance to twigs.

Pupation takes place in the soil in late autumn and the winter is spent in this state in the ground.

Fig. 88. Pupæ and adults of Peppered Moth. Natural size.

## 2. Economic

### Trees Attacked

Apples, cherries.

### Frequency of Pest

Not uncommon.

### Nature of Attack

The caterpillars eat the leaves of the trees in the autumn and so weaken the plant (see page 16). They are difficult to detect on account of their resemblance to dead twigs.

### Degree of Damage

Often more serious than suspected: see page 16.

### Preventive Measure

Turning the ground over under the trees in the winter and spring exposes the pupæ to the attacks of birds and fowls.

### Remedies

1. Spray in the autumn with ARSENATE OF LEAD[1].
2. In small plantations it may be worth while to hand pick the trees.

This is a little difficult till experience has been gained of the difference between dry twigs and the caterpillar, but it is easier to see than in the spring and summer owing to the absence of many of the leaves from the trees.

### Calendar of Treatment

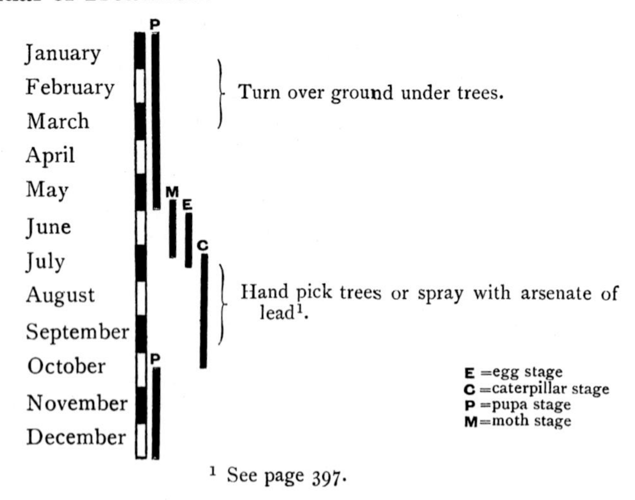


[1] See page 397.

## PITH MOTH

Name *Blastodacna hellerella* Class *Tineidæ*
Order *Lepidoptera*

### 1. General

#### Description

**Larva (Caterpillar).** See fig. 92.

| | |
|---|---|
| SIZE. | About $\frac{1}{3}$ inch full grown. |
| COLOUR. | Dull reddish brown with dark brown head. |
| LOCATION. | (1) leaves, (2) inside bud, (3) under rind of shoot, (4) tunnelling in shoot. |
| APPEARS IN | Late summer. |
| DURATION. | Throughout winter till end of following June. |

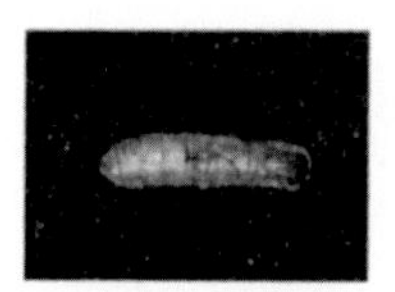

Fig. 89. Pupa of Pith Moth. Magnified.

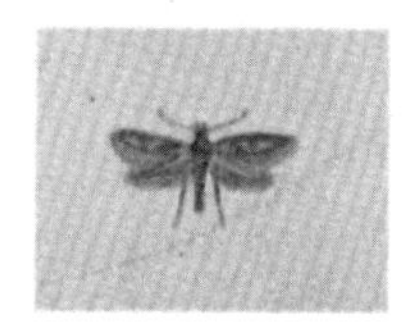

Fig. 90. Pith Moth. Natural size.

**Pupa (Chrysalis)**

| | |
|---|---|
| SIZE. | About $\frac{1}{4}$ inch long. |
| COLOUR. | Yellowish: head and tip mahogany red. |
| LOCATION. | Near end of shoot (inside). |
| APPEARS IN | June. |
| DURATION. | 2—3 weeks. |

**Adult Insect (Moth)**

| | |
|---|---|
| SIZE ACROSS WINGS. | Up to $\frac{1}{2}$ inch. |
| COLOUR (WINGS). | Fore-wings, black or mottled brown. Hind-wings, grey; densely fringed. |
| REMARKS. | Extremely active. |
| APPEARS IN | July and August. |

### Distribution

Widespread in Britain.

### Life-history

The small, very active moths appear in July and August, and lay eggs which have not yet been discovered. These hatch out into minute caterpillars which appear to feed first on the leaves, and when larger enter the buds, and afterwards bore into the shoots near the bud and remain in this position during the winter.

Fig. 91. Withered truss showing hole (×) made by the caterpillar which is the sign of the Pith Moth, and distinguishes it from injuries caused by other diseases. Natural size.

Their presence may be detected by a small round hole near the bud, and a brown blister where they have been working. In spring they bore up into the shoot and pupate in this position, when mature, in June.

## 2. Economic

### Trees Attacked

Apples, pears, especially nursery stock and bush fruit.

### Susceptible Varieties

Worcesters.

### Nature of Attack

The leaves and blossoms flag and then turn brown and die : the dead parts may remain on the trees for some time before falling off. Distinguished from BUD MOTH (page 65) by the absence of leaves and blossom spun together, and by the whole mass dying back. In appearance very similar to CANKER.

Fig. 92. Shoot cut open to show Pith Moth Caterpillar in position. Natural size.

### Degree of Damage

Sometimes very serious.

### Remedies

1. In case of small trees, hand pick the dead shoots off before the moths emerge in June.
2. In winter, prune off the attacked shoots.
3. Spray heavily with ARSENATE OF LEAD[1] as soon as the fruit is gathered.

[1] See page 397.

**Calendar of Treatment**

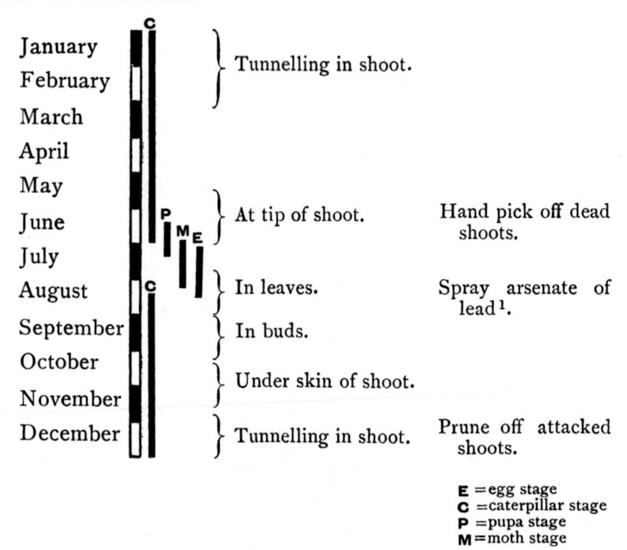


## PLUM FRUIT MOTH (Red plum maggot)

Name *Opadia funebrana* Class *Tineidæ* Order *Lepidoptera*

### 1. General

**Description**

**Larva (Caterpillar)**

| | |
|---|---|
| SIZE. | About $\frac{5}{8}$ inch (mature). |
| COLOUR. | Red, sides often yellow; head shiny brown, black spots on 7, 8, 9, 11 and 13 segments. |
| LOCATION. | Inside fruit (plums). |
| APPEARS IN | July. |
| DURATION. | Till following summer. |
| REMARKS. | The grubs leave the fruits about the end of August and shelter under the bark or in rubbish, spinning a cocoon of grey silk. |

[1] See page 397.

**Pupa (Chrysalis)**

| | |
|---|---|
| APPEARANCE. | In the caterpillar cocoons. |
| COLOUR. | Yellowish, with dark brown "tail." |
| LOCATION. | In crevices of bark or other sheltered spots. |

**Adult Insect (Moth)**

| | |
|---|---|
| SIZE ACROSS WINGS. | About $\frac{1}{2}$ inch. |
| COLOUR (WINGS). | Purplish grey, clouded. |
| APPEARS IN | June and July. |

**Ova (Eggs)**

| | |
|---|---|
| ARRANGEMENT. | Singly. |
| LOCATION. | At base of stalk. |
| APPEAR IN | June and July. |
| HATCHING PERIOD. | About 10 days. |

Fig. 93. Diagram of fruit opened showing Caterpillar of Plum Fruit Moth. Natural size.

### Distribution

Fairly widespread.

### Life-history

The moths appear in middle summer and lay a single egg on the fruit-stalk. The caterpillar, on hatching, enters the fruit and lives inside until mature, at about the end of August, when it leaves the plum and finds shelter under the bark or beneath rubbish etc.; spinning a cocoon of grey silk and wintering in this condition. It pupates in the following summer.

## 2. Economic

### Trees Attacked

Plums and damsons.

### Frequency of Pest

Fairly common in West of England.

## Nature of Attack

The fruit is entered by the maggot which frequently causes it to fall before it is quite ripe. There is also much bitter "frass" formed in the interior of the plum which makes it unpalatable.

## Remedies

1. Where practicable prepare TRAPS of sacking tied round the trunks of the trees (as for Codling Moth, see page 92) to catch the wintering caterpillars. STICKY-BANDS[1] are also good.
2. Apply strong LIME-SULPHUR[2], or other winter wash, during the dormant season.
3. Shake the tree in the summer, and destroy the prematurely ripe plums which fall. These contain the maggots (future trouble is thus prevented).

## Calendar of Treatment

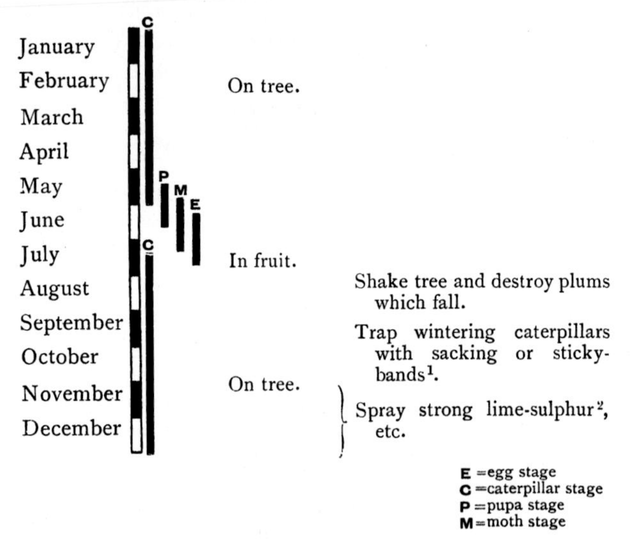


[1] See page 404. [2] See page 612.

## RASPBERRY MOTH (Red raspberry grub)

Name *Tinea rubiella* Class *Tineidæ* Order *Lepidoptera*

### 1. General

#### Description

**Larva (Caterpillar)**

| | |
|---|---|
| SIZE. | About $\frac{1}{4}$ inch when full grown. |
| COLOUR. | At first pale, then pink and finally reddish: head black or brown. |
| LOCATION. | (1) in flowers, (2) in shelter in cocoons, (3) in buds, (4) in cane. |
| APPEARS IN | June (from eggs), middle to end of March (from winter quarters). |
| DURATION. | Throughout winter till following May. |

Fig. 94. Diagram of Caterpillars of Raspberry Moth. Natural size.

**Pupa (Chrysalis)**

| | |
|---|---|
| SIZE. | About $\frac{1}{4}$ inch long. |
| COLOUR. | Yellowish red. |
| LOCATION. | In cavity in cane or shoot. |
| APPEARS IN | May. |
| DURATION. | 15 to 30 days. |

**Adult Insect (Moth)**

| | |
|---|---|
| SIZE ACROSS WINGS. | Up to $\frac{1}{2}$ inch. |
| COLOUR (WINGS). | Front—brown with yellow spots.<br>Back—brown. |
| ,, (BODY). | Brown. |
| APPEARS IN | End of May and June. |

**Ova (Eggs)**

| | |
|---|---|
| LOCATION. | On the flowers. |
| APPEAR IN | End of May and June. |
| DURATION. | 5 to 7 days. |

**Distribution**

Widely distributed in Britain.

**Life-history**

The small moths appear at the end of May and June, and are very active, flying both by day and night: they may be seen over the canes or on the flowers at this period.

The eggs, which are laid, usually, in the flowers, hatch out, and the minute whitish caterpillars penetrate to the base of the bloom where they remain (not being harmful at this stage) until the fruit is almost ripe, when they drop to the ground and seek a secure shelter for the winter[1], where they spin a small cocoon of grey silk.

In late March they issue from their hiding places and enter the base of the buds. Here the caterpillar becomes pink and, later on, red. It feeds on the bud and then penetrates into the pith of the shoot. The attacked buds and shoots all die before long.

**Remarks**

When the caterpillar is mature, in May, it makes a hollow chamber in the cane or shoot and here pupates.

## 2. Economic

**Fruit-plants Attacked**

Raspberry, also blackberry and probably loganberry.

**Frequency of Pest**

Common.

**Nature of Attack**

The buds and shoots are attacked by the caterpillars with the result that the buds either do not open, or they expand and

[1] Either on the canes under bark etc. or under stones, debris etc.

die and finally all the leaves and blossoms perish on the attacked shoots.

## Degree of Damage

Very serious: entire crops may be ruined.

## Preventive Measures

1. Remove and burn all old stakes in the winter.
2. Cut back all old canes close to the ground in the autumn and burn them at once.
3. Hoe the ground around canes deeply in the winter, and mix in soot, lime and ashes.
4. In small house-gardens it is a good plan to apply BANDING COMPOSITION[1] to the tops of the canes under shoots so as to catch the ascending caterpillars. This should be done in *early March*.

## Remedy

Apply a spray of ARSENATE OF LEAD[2] late in March, and a further spray a few days later on. This will kill the caterpillars, who will take a dose of the poison on eating their way into the buds.

## Calendar of Treatment

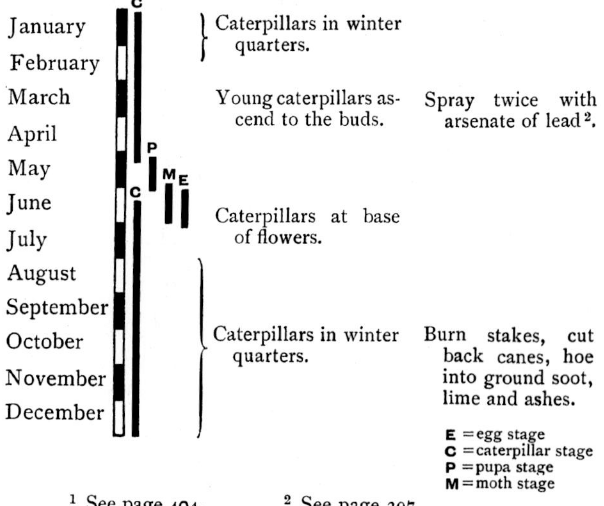


[1] See page 404. [2] See page 397.

## SWIFT MOTH

Name *Hepialus lupulinus* Family *Hepalidæ*
Order *Lepidoptera*

### 1. General

**Description**

**Larva (Caterpillar)**

| | |
|---|---|
| SIZE. | About $1\frac{1}{2}$ inches when full grown. |
| COLOUR. | Dull white, with shiny brown head. |
| LOCATION. | In the ground. |
| APPEARS IN | June and July. |
| DURATION. | Throughout winter to following April. |

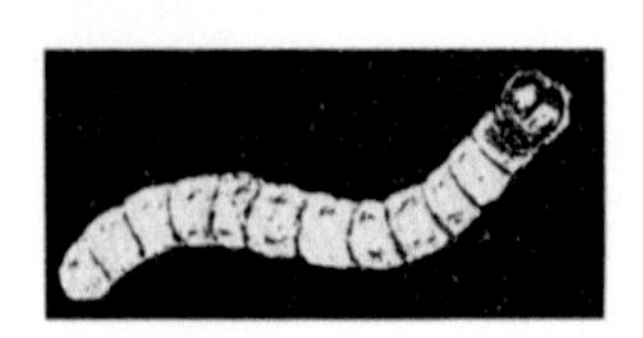

Fig. 95. Diagram of Caterpillar of Swift Moth. Natural size.

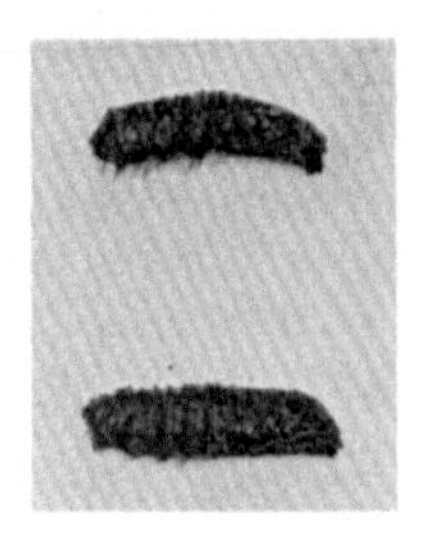

Fig. 96. Pupæ of Swift Moth. Natural size.

**Pupa (Chrysalis)**

| | |
|---|---|
| APPEARANCE. | In cocoon, or naked: about $\frac{2}{3}$ inch long with 5 rows of spines on abdomen. |
| COLOUR. | Brown. |
| LOCATION. | In soil. |
| APPEARS IN | April. |
| DURATION. | 2 to 3 weeks. |

**Adult Insect (Moth)**

| | |
|---|---|
| SIZE ACROSS WINGS. | About 1 to $1\frac{1}{4}$ inches. |
| COLOUR (WINGS). | Pale brown with white streak (variable). |
| APPEARS IN | Middle May. |
| REMARKS. | The male moth has a curious, jerky habit of flight. |

**Ova (Eggs)**

ARRANGEMENT. Dropped by the females during flight.
LOCATION. On ground.
HATCHING PERIOD. About 10 days.

**Distribution**

Widespread.

Fig. 97. Swift Moths. Natural size.

**Life-history**

The moths emerge from the ground from the middle of May onwards for a few weeks, and appear in flight at dusk. The male moth flies with a swift jerky movement (hence name). The eggs are *dropped by the female in flight* and fall to the ground, hatching out in about a week. The young caterpillar then enters the soil and commences feeding upon the roots of the plants attacked, till the following spring, when it pupates in the soil, either naked or spinning a cocoon of grey silk.

## 2. Economic

### Plants Attacked

Strawberries. Also many other roots of plants.

### Frequency of Pest

Fairly common.

### Nature of Attack

The roots are attacked, first the young rootlets, then the main roots even up to the crown of the plant (in which case it is entirely destroyed).

### Degree of Damage

Often serious, especially during mild winters.

### Natural Enemies

A species of bug (*Anthocoris*[1]). Also a fungus (*Cordyceps entomorrhiza*) often attacks it, filling its body with the mould substance.

### Remedies

1. Hoe the beds constantly.
2. Work into the soil a SOIL INSECTICIDE[2], best in early autumn.
3. Dress with one or more of the following : soot, kainit, wood ashes, gas lime.

### Calendar of Treatment

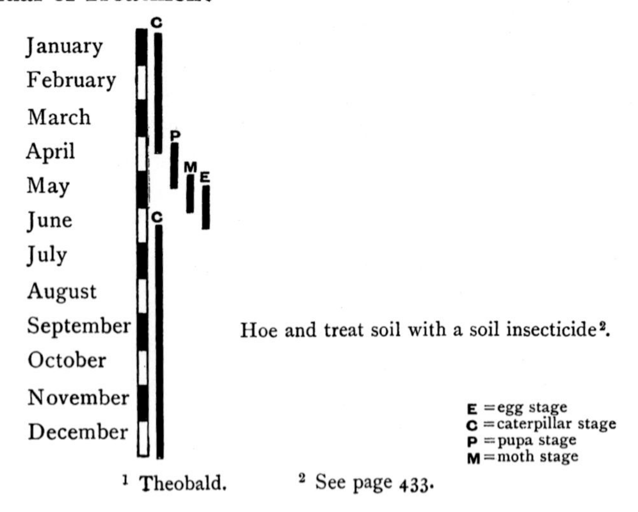


[1] Theobald. [2] See page 433.

## TORTOISESHELL BUTTERFLY (large)

Name *Vanessa polychloros* Order *Lepidoptera*

### 1. General

### Description

#### Larva (Caterpillar)

| | |
|---|---|
| SIZE. | About 2 inches when full grown. |
| COLOUR. | Black or brownish black, with yellow side lines. |
| LOCATION. | On leaves. |
| APPEARS IN | May. |
| DURATION. | About 6 weeks. |
| REMARKS. | The caterpillars first live in colonies, and afterwards disperse over the leaves. |

Fig. 98. Caterpillar and Pupa of Large Tortoiseshell Butterfly. Natural size.

#### Pupa (Chrysalis)

| | |
|---|---|
| APPEARANCE. | Suspended head downwards: has two rows of blunt spines. |
| COLOUR. | Grey ringed with black: yellow at end. |

| | |
|---|---|
| LOCATION. | Attached to twigs and leaves. |
| APPEARS IN | Mid June. |
| DURATION. | 3 to 4 weeks (variable). |

**Adult Insect (Butterfly)**

| | |
|---|---|
| SIZE ACROSS WINGS. | $2\frac{1}{2}$ to 3 inches. |
| COLOUR (WINGS). | Rich orange-brown etc. (see fig.). |
| " (BODY). | Brown, hairy. |
| APPEARS IN | July. |
| DURATION. | Winters in sheltered places, and lays eggs in following May. |

Fig. 99. Large Tortoiseshell Butterfly. Natural size.

**Ova (Eggs)**

| | |
|---|---|
| ARRANGEMENT. | In rings around twigs. |
| LOCATION. | On twigs. |
| APPEAR IN | May. |

**Distribution**

Widespread.

**Life-history**

This beautiful butterfly appears on the wing in July, and winters in sheltered places, laying its eggs in rings around the branches in the following May.

The young caterpillars first live in colonies, but afterwards disperse over the trees. They pupate in mid June, suspending themselves head downwards from the branches for this purpose.

## 2. Economic

**Trees Attacked**

Cherries, apples. More often the elm, aspen, sallow, etc.

**Frequency of Pest**

Not frequent.

**Nature of Attack**

Foliage eaten.

**Degree of Damage**

Occasionally serious.

**Preventive Measure**

Destroy the egg-rings on branches when they appear if this is possible.

**Remedies**

1. Collect and destroy the young colonies of caterpillars.
2. Spray the tree with ARSENATE OF LEAD[1].

**Calendar of Treatment**

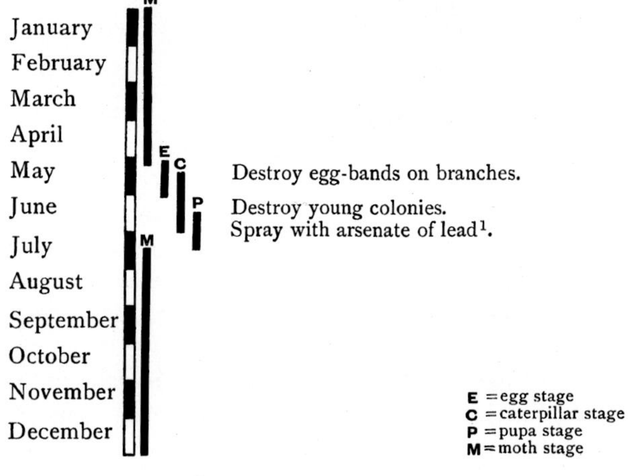


[1] See page 397.

## TORTRIX MOTHS

A large number of these small moths attack fruit trees, and cause a great amount of damage every year.

The BUD MOTH belongs to this group, also the CODLING, but these are specially described on account of their serious attacks upon the young blossoms and fruit respectively.

The **caterpillars** can be recognised by their trick of *wriggling backwards* when touched, and their great activity.

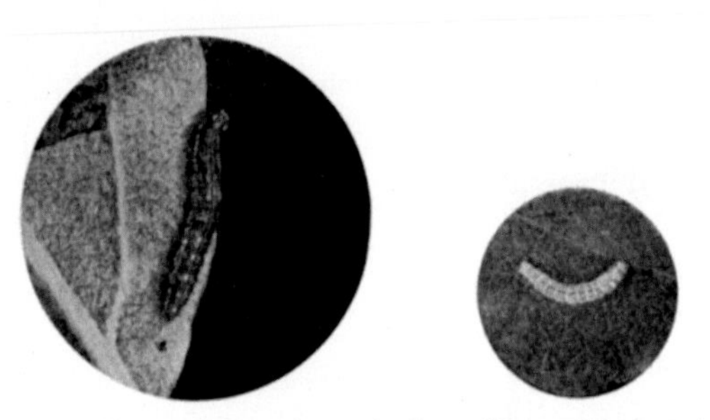

Fig. 100. Two varieties of Tortrix Caterpillars. Natural size.

Most of them feed also upon the blossom, but, except in the case of the Bud Moth, the leaves are chiefly attacked. They almost invariably *unite leaves (and blossom) together* by silken threads and feed between them, and the pupal stage is passed in the pockets so formed.

The following calendar shews roughly the dates of the different stages, although variations sometimes occur, some caterpillars passing the winter in dead leaves, etc.

Treatment follows the general lines for other caterpillars, but the spraying should be done while the caterpillars are very young and before they spin the leaves together.

**Calendar**

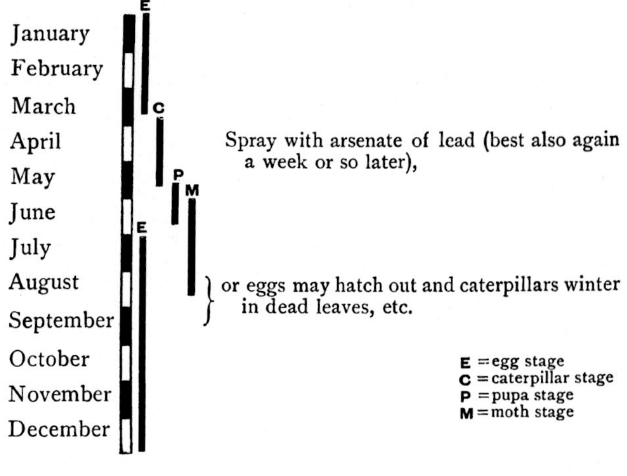

Fig. 101. Various Tortrix Moths. 1, Allied Bud. 2, Common Tortrix. 3, *Spilonota ribeana*. 4, Allied Tortrix (*Heparana*). Natural size.

The following are the scientific names of the chief Tortrix Moths recorded as pests on fruit:

| Scientific Name | Common Name | Fruit attacked |
| --- | --- | --- |
| Tortrix ribeana | The common Tortrix | Apples, plums |
| Tortrix heparana | Allied Tortrix | ,, |
| Tortrix rosana | —— | Apples, plums, cherries, pears, etc. |
| Sideria achatana | —— | Apples |
| Penthina variegana | Allied Bud Moth | Apples, pears |
| Penthina pruniata | —— | Plums, apples, cherries, nuts |
| Tortrix podana | —— | Apples, currants |
| Tortrix relinquana | —— | Vines |
| Tortrix viridana | Green oak Tortrix | Chestnuts |
| Carpocapsa splendidana | Nut fruit Tortrix | ,, |

Fig. 102. Green Pug Moths. Occasionally pests on fruit. Natural size.

## TORTRIX MOTH (common Tortrix)

Name *Tortrix ribeana* Class *Tortricidæ* Order *Lepidoptera*

### 1. General

**Description**

**Larva (Caterpillar)**

| | |
|---|---|
| APPEARANCE. | Slightly hairy. |
| COLOUR. | Dark olive-green with black spots. Head, dark brown. |
| LOCATION. | On or inside spun leaves. |
| APPEARS IN | April and May. |
| DURATION. | Till June. |

Fig. 103. Caterpillar of Common Tortrix. Magnified (× 3). Inset, natural size.

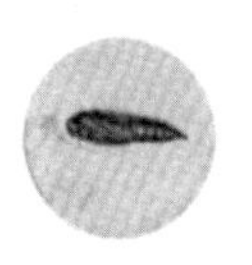
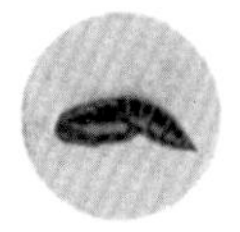

Fig. 104. Common Tortrix Moth pupæ. Natural size.

**Pupa (Chrysalis)**

| | |
|---|---|
| APPEARANCE. | Small, slender. |
| COLOUR. | Reddish brown. |
| LOCATION. | Between leaves spun together. |
| APPEARS IN | June. |
| DURATION. | 2 to 3 weeks. |

**Adult Insect (Moth)**

| | |
|---|---|
| SIZE ACROSS WINGS. | About $\frac{2}{3}$ inch. |
| COLOUR (WINGS). | Front—pale yellowish brown, with darker markings; hind—slaty grey. |
| APPEARS IN | End of June and July. |

**Ova (Eggs)**

| | |
|---|---|
| APPEARANCE. | Round, transparent. |
| COLOUR. | Yellowish. |
| ARRANGEMENT. | In batches. |
| LOCATION. | On leaves or branches. |
| APPEAR IN | End of June to July. |
| DURATION. | Throughout the winter till following spring. |

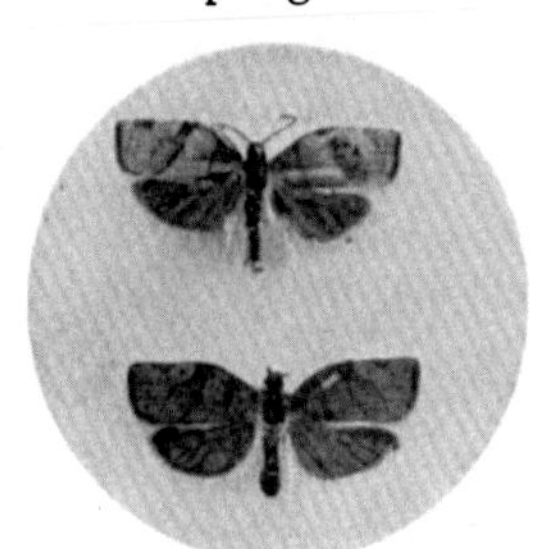

Fig. 105. Common Tortrix Moths. Natural size.

**Distribution**

Widespread.

**Life-history**

The winter is usually passed in the egg stage, but some eggs may hatch out in the late summer and the caterpillar then winters spun up in leaves or under rubbish.

The eggs hatch out fairly late in the spring and the caterpillars first feed upon the surface of the leaf, and then spin leaves together, and live in the pocket so formed. They mature in June and the pupa is formed in the leaf pocket, which is then usually withered. The moths hatch out in late June and July and lay the eggs upon the leaves or shoots.

## 2. Economic

### Trees Attacked

Apples, plums and many other trees and bushes.

### Frequency of Pest

Very common.

### Nature of Attack

The leaves and sometimes the blossoms are attacked and wither. Leaves are spun together.

### Degree of Damage

Dependent on number present: may be very serious.

### Remedy

Spraying with ARSENATE OF LEAD[1], while the young caterpillars are feeding openly on the leaves before these are spun together, is the best remedy.

### Calendar of Treatment

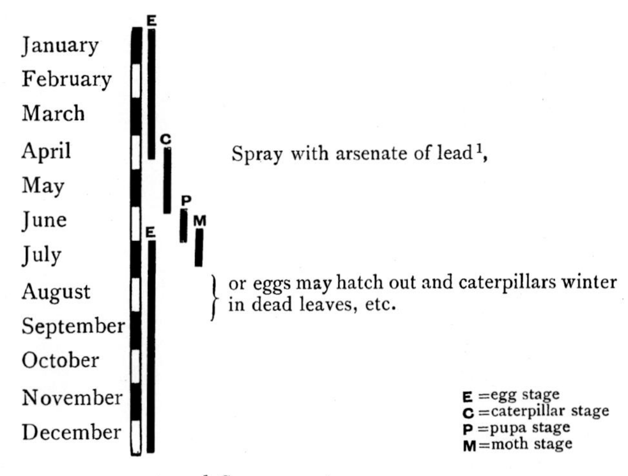


[1] See page 397.

## VAPOURER MOTH

Name *Orygia antiqua* Family *Liparidæ* Order *Lepidoptera*
(*Tussock Moths*)

### 1. General

**Description**

**Larva (Caterpillar)**

| | |
|---|---|
| APPEARANCE. COLOUR. | Remarkable: has four large tufts of creamy yellow and long tuft of dark hairs in front. Body dark grey spotted with red. |
| LOCATION. | On leaves. |
| APPEARS IN | Late April to May. |
| DURATION. | Till June or July. |

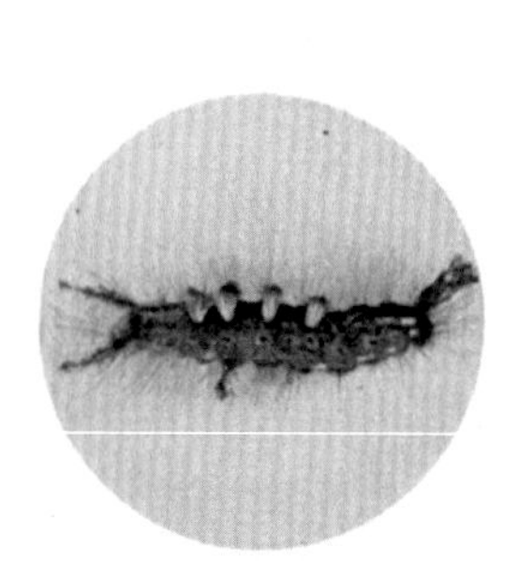
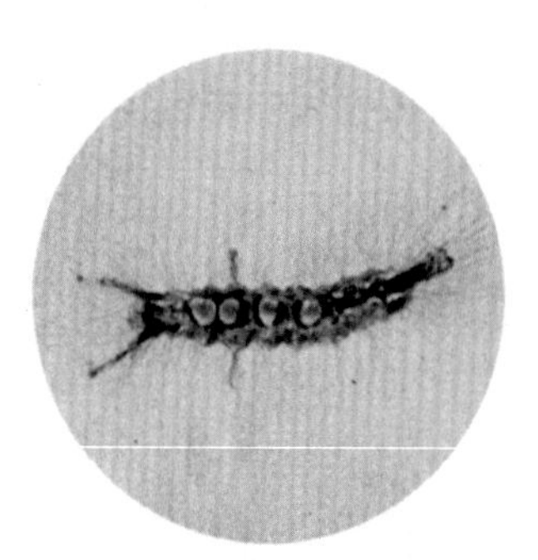

Fig. 106. Two views of Vapourer Caterpillars. Natural size.

**Pupa (Chrysalis)**

| | |
|---|---|
| APPEARANCE. | In cocoons of pale silk. |
| COLOUR. | Brown. |
| LOCATION. | As for eggs. |
| APPEARS IN | June. |
| DURATION. | 2 to 3 weeks. |

Fig. 107. Cocoon of Vapourer Moth. Natural size.

**Adult Insect (Moth)**

| | Male | Female |
|---|---|---|
| SIZE ACROSS WINGS. | 1 to $1\frac{1}{4}$ inches. | Almost wingless. |
| COLOUR (WINGS). | Chestnut brown, with white spot on each forewing. | ,, |
| ,, (BODY). | Brown. | Grey. |

REMARKS. The female is wingless like the *loopers*, but does not ascend the trunks of trees, *not moving from the pupa case.*

APPEARS IN Early July.

Fig. 108. Vapourer Moths: males (above), female (below). Natural size.

**Ova (Eggs)**

| | |
|---|---|
| APPEARANCE. | Conspicuous, with dark rings and central spot. |
| COLOUR. | Reddish brown to grey. |
| ARRANGEMENT. | Close, single-layered mass *on cocoon.* |
| LOCATION. | On the cocoons on twigs, stems, etc. or on fences near trees. |
| APPEAR IN | July. |
| HATCHING PERIOD. | Throughout winter till following May. |

**Distribution**

Widespread over Europe.

**Life-history**

The adults emerge from the cocoons from early July to September. The female is wingless and remains on the cocoon, on which, after fertilisation by the winged male, she deposits her eggs.

The male flies by day with a rapid motion (especially on sunny days). The eggs remain as conspicuous objects on the cocoon, and hatch out from the end of April to the beginning of June.

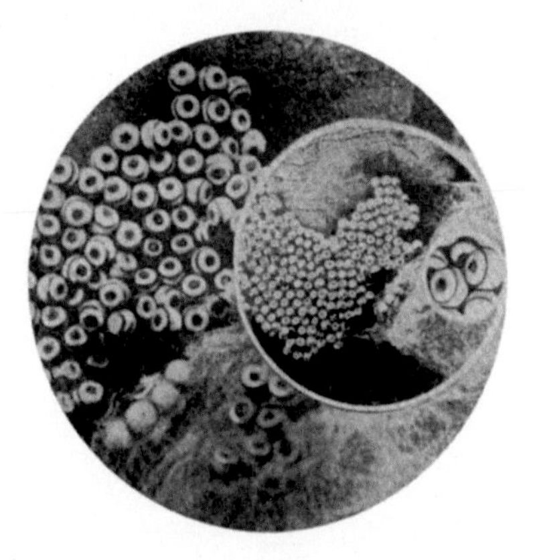

Fig. 109. Eggs of Vapourer Moth. Magnified (× 3). Inset, natural size and further enlarged (× 5).

The caterpillars are very curious and pretty, and feed on the leaves of the trees, pupating in a silken cocoon mixed with hairs from the tufts.

**Remarks**

Common in towns (e.g. London) as well as in the country.

## 2. Economic

**Trees Attacked**

Apples, pears, plums, damsons, walnuts, also the hawthorn, rose, sloe, acacia, elm, lime.

**Frequency of Pest**

Occurs locally as destructive pest at intervals.

**Nature of Attack**

Defoliation of the trees.

**Degree of Damage**

Not usually severe.

**Preventive Measure**

If practicable, the egg-masses, which are very noticeable, may be collected and destroyed in the winter.

**Remedy**

Spray with ARSENATE OF LEAD[1] as soon as the caterpillars become troublesome.

**Calendar of Treatment**

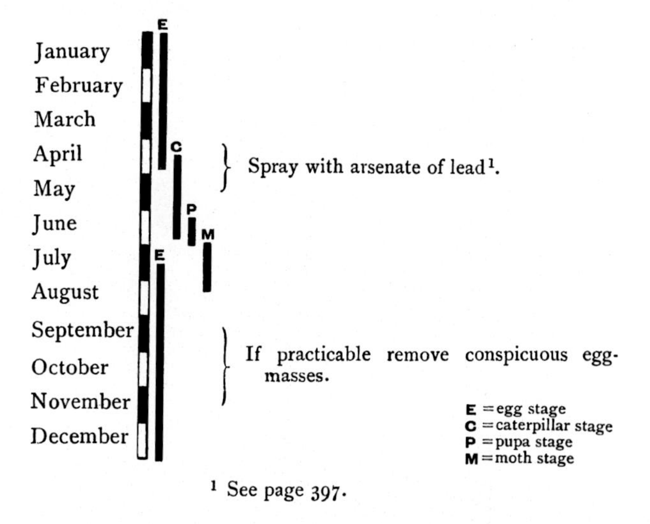


[1] See page 397.

## WINTER MOTH

Name *Cheimatobia brumata* Class *Geometridæ* (*loopers*)
Order *Lepidoptera*

### 1. General

#### Description

##### Larva (Caterpillar)

| | |
|---|---|
| APPEARANCE. | Smooth skinned; about 1¼ inches long when full fed. |
| COLOUR. | At first dark, then green with pale lines along body. |
| LOCATION. | On the leaf buds, blossom trusses, and, later, the leaves. |

Fig. 110. Caterpillars of Winter Moth. Natural size.

| | |
|---|---|
| PROGRESSION. | By *looping*. |
| APPEARS | Just before bursting of buds. |
| DURATION. | Till about middle of June. |
| REMARKS. | Similar to March Moth Caterpillar, but not so slender as latter. |

**Pupa (Chrysalis)**

| | |
|---|---|
| COLOUR. | Brown. |
| LOCATION. | 2 to 3 inches below ground in soil under trees. |
| APPEARS IN | June or July. |
| DURATION. | Till October—January. |

Fig. 111. Pupæ of Winter Moth in soil. Natural size.

**Adult Insect (Moth)**

| | Male | Female |
|---|---|---|
| SIZE ACROSS WINGS. | About $1\frac{1}{4}$ inches. | Almost wingless. |
| COLOUR (WINGS). | Greyish brown to brown. | |
| „ (BODY). | Greyish brown. | Greyish brown. |
| REMARKS. | The females crawl up the tree trunks. | |
| APPEARS IN | October to January (variable), males some days before females. | |

**Ova (Eggs)**

| | |
|---|---|
| APPEARANCE. | Small, oval, flat at ends, shells thick. |
| COLOUR. | At first, *pale yellowish green*; later *brick red.* |
| ARRANGEMENT. | Not systematic. |
| LOCATION. | Round buds, and in crevices of twigs and boughs. |
| APPEAR IN | October to January. |
| HATCH IN | Late March or April. |

Fig. 112. Winter Moth: female. Natural size.

Fig. 113. Winter Moth: female. Magnified (× 2½).

Fig. 114. Winter Moth: male. Natural size.

Fig. 115. Eggs of Winter Moth in position, and dead female moth. Magnified (× 2¾). Inset, natural size.

**Distribution**

Very widespread.

**Life-history**

The females leave the ground and crawl to the tree-trunks, which they ascend, being usually fertilised by the winged male moth during this time.

The eggs are then laid on the branches in crevices, a female laying as many as 300 eggs.

The eggs hatch out *just before the bursting of the leaf-buds* in spring and the small looper caterpillars attack the young leaves, and enter the flower trusses spinning these together. Later they feed openly on the leaves, and may attack the young fruit.

Pupation usually occurs during June, the caterpillar dropping to the ground and entering the soil. Before pupating, an oval cocoon of silk is usually spun, and a small cavity in the soil chosen.

**Remarks**

1. The males are seen to fly at dusk just before the females ascend the trees.
2. It is at present uncertain whether, in some few cases, the males do not fly with the females (*in copula*) to the branches. In any case, it is not a common practice, as is shown by the large number of females caught upon the sticky bands.

## 2. Economic

**Trees Attacked**

Apples, plums. Many forest trees attacked.

**Susceptible Varieties**

All varieties open to attack.

**Frequency of Pest**

Very common, probably the commonest and most destructive of all caterpillar pests.

**Nature of Attack**

The caterpillars eat the leaves continuously from the bursting of the buds till the middle of June.

Flowers and fruit may also be attacked.

**Duration of Attack**

During spring and early summer.

**Degree of Damage**

Unless promptly dealt with, the trees may be entirely defoliated.

**Preventive Measures**

Turn over soil under trees and keep fowls in orchard—many pupæ may thus be destroyed.

**Remedies**

1. Applying grease-proof paper-bands to the trunks of the trees and coating with BANDING COMPOSITION[1] to catch the ascending female moths and so prevent egg laying.
2. Spraying the trees with ARSENATE OF LEAD[2] when the attack first appears, or subsequently.

   The eggs are too resistant to be successfully dealt with by winter washes, despite many interested statements to the contrary.
3. Applying a thick LIME and SALT[3] cover-wash to the trees in late winter to seal the eggs and hinder their hatching.

**Remarks on Remedies**

The life-history of the Winter Moth offers peculiar opportunities to the grower for successful treatment of this pest.

The females, being wingless, are obliged to crawl up the trunks of the trees to gain the branches. If therefore the trunks are properly banded (see method of banding, page 406) and the bands kept in efficient operation during the whole time when the moths appear, all egg laying should be prevented[4] (see however Remarks, 2 on page 179). The bands

[1] See page 404. [2] See page 397. [3] See page 428.

[4] As many as 300 female Winter Moths have been caught on a single band!

should be in operation before the end of September and kept sticky till the spring (for March Moth).

**Calendar of Treatment**

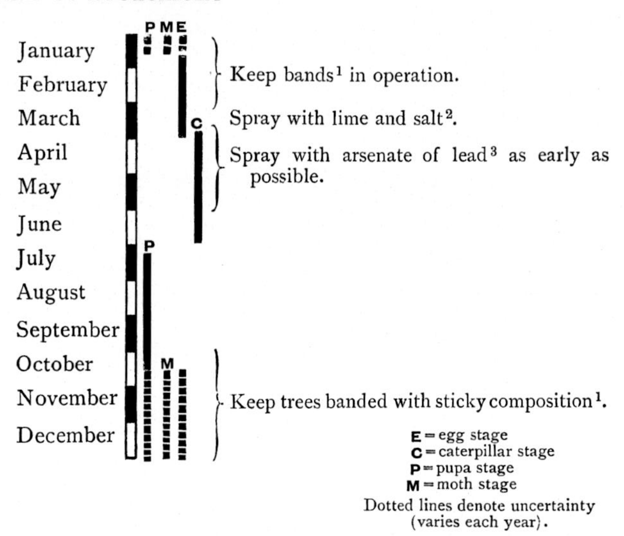


## WOOD LEOPARD MOTH

Name *Zeuzera pyrina* Family *Coccidæ* Order *Lepidoptera*

### 1. General

**Description**

**Larva (Caterpillar)**

| | |
|---|---|
| APPEARANCE. | Each segment with a black bristle. About 2 inches long. |
| COLOUR. | Yellowish white: head brown, with two black spots. |
| LOCATION. | In tunnels in the wood of trees. |
| APPEARS IN | June to August. |
| DURATION. | 10 months. |

[1] See page 404. [2] See page 428. [3] See page 397.

**Pupa (Chrysalis)**

| | |
|---|---|
| APPEARANCE. | In silken cocoon mixed with particles of wood. |
| COLOUR. | Bright brown. |
| LOCATION. | Under the bark. |
| APPEARS IN | May. |
| DURATION. | Few weeks. |

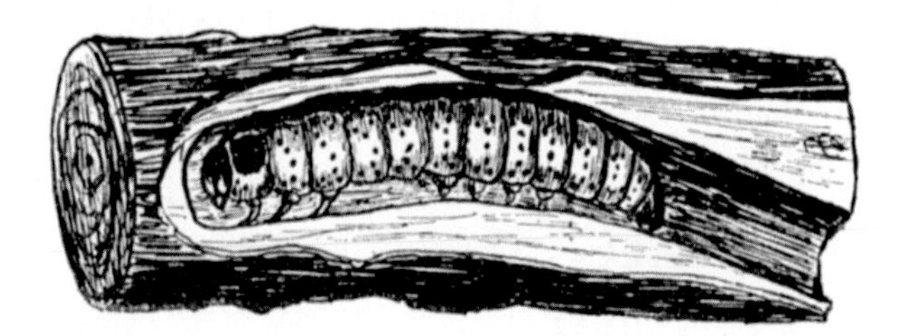

Fig. 116. Diagram of branch opened to show Caterpillar of Wood Leopard Moth. Natural size.

Fig. 117. Wood Leopard Moth: wings folded and open. Natural size.

**Adult Insect (Moth)**

| | |
|---|---|
| SIZE ACROSS WINGS. | About 2 inches (female larger than male). |
| COLOUR (WINGS). | White, with spots of steel-blue. |
| " (BODY). | Head covered with downy white; abdomen bluish black. |
| APPEARS IN | June and July. |

**Ova (Eggs)**

| | |
|---|---|
| COLOUR. | Dark or orange-yellow. |
| LOCATION. | Deep in the bark of trees. |
| APPEAR IN | June to August. |
| HATCHING PERIOD. | Few days. |

### Distribution

General over Europe.

### Life-history

The very striking moths may be seen resting near the ground on tree trunks.

The eggs are laid deep down in the bark of the trees by the powerful egg-depositor of the female. These eggs hatch out in a few days and the young caterpillars bore their way into the wood of the tree, remaining inside the branch and tunnelling through it till mature in about 10 months.

They then pupate near the surface, the brown pupa giving rise in a few weeks to the mature moth, which makes its escape from the tree.

## 2. Economic

### Trees Attacked

Apples, cherries, pears, chestnuts.

### Frequency of Pest

Not frequent.

### Nature of Attack

The caterpillars, by tunnelling through the wood of the branches, finally cause the death of the latter.

### Degree of Damage

Usually serious on attacked trees.

## Remedies

1. If holes are found, CARBON BISULPHIDE[1] vapour may be blown into the aperture by means of bellows, or
2. A piece of CYANIDE OF POTASH[2] (*caution, poison*) may be placed in the hole and then this plugged up with clay.
3. When branches are obviously dying and no holes can be seen, slit them up till the caterpillar is found and kill it.

## Calendar of Treatment

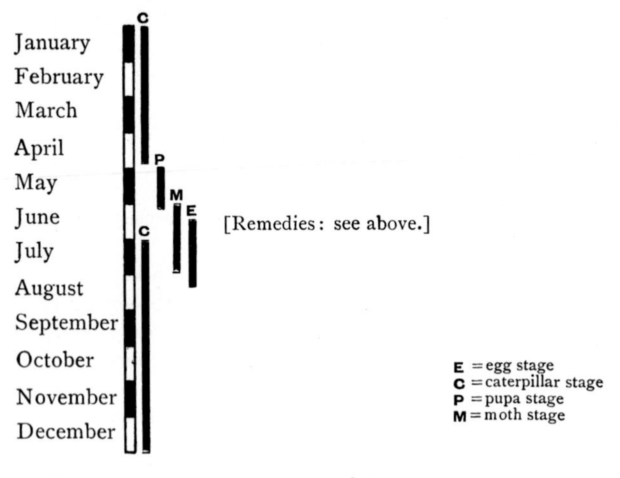


[1] See page 410. [2] See page 423.

# BEETLES

## (*COLEOPTERA*)

## and their larvæ (GRUBS)

# CHAPTER 13

## Beetles

(COLEOPTERA)

### 1. General

Most of the beetle pests of fruit trees are destructive in the adult or *beetle* stage. In the case however of two of the weevils, it is the grub only which does the damage, the adult beetle not feeding on the leaves to any serious extent, if at all. In the remainder of the beetle pests, both adults and grubs are injurious. They may therefore be classed as follows:

| GRUB ONLY INJURIOUS | ADULT BEETLE ONLY INJURIOUS | BOTH GRUBS AND BEETLES INJURIOUS |
|---|---|---|
| Apple-blossom weevil | Leaf weevils | Raspberry weevil |
| Nut weevil | Red-legged weevil | Fruit-bark beetle |
| | Twig-cutting weevil | Raspberry beetle |
| | Ground beetles | Shot-borer beetle |
| | | Cockchafer |
| | | Rosechafer |

Of all fruits perhaps raspberries are most seriously attacked by beetles of different kinds. The APPLE-BLOSSOM WEEVIL however is one of the worst pests of apples, and is apparently on the increase.

The methods of dealing with beetle attacks are much less satisfactory than in the case of caterpillars. Those weevils which lay their eggs inside the blossom cannot be killed with a stomach poison, such as arsenate of lead, as they do not appear to feed. A good coating of lime prevents the laying of the egg in the blossom and research is in progress on the lines of making the blossoms distasteful to the weevils by means of some chemical substance.

In the case of many beetles, the most generally successful method consists in catching the adults by jarring them on to tarred boards, but this is a laborious and tedious business.

A few which actually devour the leaves or blossoms may be successfully dealt with by spraying with arsenate of lead.

The two beetles which attack the trunks of trees, and feed internally —the Fruit-bark and Shot-borer beetles—present, so far, an unsolved problem. The only remedy is to cut down and destroy the trees and so prevent the spread of the pest.

The grubs of beetles attacking the roots of plants are dealt with by treating the ground with soil insecticides.

It should be remembered, in this connection, that all beetle grubs found in the soil are not injurious; many are on the contrary actually beneficial, living upon other insect life, especially snails, etc.

## 2. Scientific

Beetles have a strong armour-plate of chitinous material, which is a protection to insects which are often under the soil. The legs are strong, and the hind-wings very large and folded beneath the *elytra* or hardened fore-wings which act as cases or shields. The mouth is adapted for chewing.

All beetles undergo a complete transformation, having a true *pupal* resting stage.

The colour varies. Beetles living underground are dark in colour; others may be brown, and leaf-eating weevils are often green.

Larvæ which burrow in the ground, or in trees, may feed two or three years before reaching maturity. In contrast to this, the apple-blossom weevil goes through its whole life-history in a few weeks.

The beetle grubs may be footless, and resemble maggots, as in the case of many weevils, or they may have three pairs of legs. In chafers the legs are short, and the skin fleshy.

### *Classification.*

Order **COLEOPTERA**

Sub-order 1. LAMELLICORNIA.

Only one family are fruit pests, viz:

Scarabæidæ (chafers).

Cockchafer, *Melolontha vulgaris*.

Rose chafer, *Cetonia aurata*.

Sub-order 2. ADEPHAGA (predatory beetles).

Nearly all this group are beneficial, eating other insects. A few of the ground beetles sometimes attack plants.

Carabidæ (ground beetles).

Occasionally pests.

Sub-order 3. CLAVICORNIA.

Practically all these beetles are beneficial, being insect eaters. The most directly beneficial are the family

Coccinellidæ (lady-birds).

These and their larvæ devour aphides greedily (see page 469).

Sub-order 4. SERRICORNIA.

None of these beetles appear to attack fruit.

Sub-order 5. HETEROMERA.

Remarks as for (4).

Sub-order 6. PHYTOPHAGA (leaf-eating beetles).

Most of these are pests on various plants.

Raspberry beetle, *Byturus tomentosus*.

Sub-order 7. RHYNCHOPHORA (snout-beetles).

Head with snout and mouth parts at end.

Family *Curculionidæ* (weevils).

Antennæ, elbowed.

Apple-blossom weevil, *Anthonomus pomorum*.
Nut weevil, *Balaninus nucum*.
Raspberry weevil, *Otiorhyncus picipes*.
Red-legged weevil, *Otiorhyncus tenebricosus*.
Twig-cutting weevil, *Rhynchites cœruleus*.
Leaf weevils, *Phyllobius*, various species.

Family *Scolytidæ*.

Snout short: antennæ clubbed, not usually elbowed.

Fruit-bark beetle, *Scolytus rugulosus*.
Shot-borer beetle, *Xyleborus dispar*.

# CHAPTER 14

## Weevils

## APPLE-BLOSSOM WEEVIL

Name *Anthonomus pomorum* Class *Rhynchophora* (*Weevils*)
Order *Coleoptera* (*Beetles*)

### 1. General

#### Description

**Adult Insect (Weevil)**

| | |
|---|---|
| SIZE. | Small (about $\frac{1}{8}$ to $\frac{1}{6}$ inch long). |
| COLOUR. | Brownish black, with pale **V**-shaped mark on wing-cases. |
| APPEARS IN | Early April (old brood), May—June (new brood). |
| DURATION. | Throughout summer, autumn, winter and early spring. |
| REMARKS. | The snout long, slender and curved, with feelers (antennæ) near the end, in two segments at right angles. |

**Ova (Eggs)**

| | |
|---|---|
| APPEARANCE. | Small, oval. |
| COLOUR. | White. |
| ARRANGEMENT. | Laid singly. |
| LOCATION. | In the centre of blossom bud. |
| APPEAR IN | April. |
| HATCHES IN | 5 to 7 days. |

**Larva (Grub)**

| | |
|---|---|
| APPEARANCE. | Small, legless (about $\frac{3}{16}$ inch when mature). |
| COLOUR. | White with dark brown head and brown spiracles. |
| LOCATION. | Curved up inside blossom bud. |
| APPEARS IN | April and May. |
| DURATION. | 1 to 3 weeks. |
| REMARKS. | The length of the larval period depends upon weather conditions. |

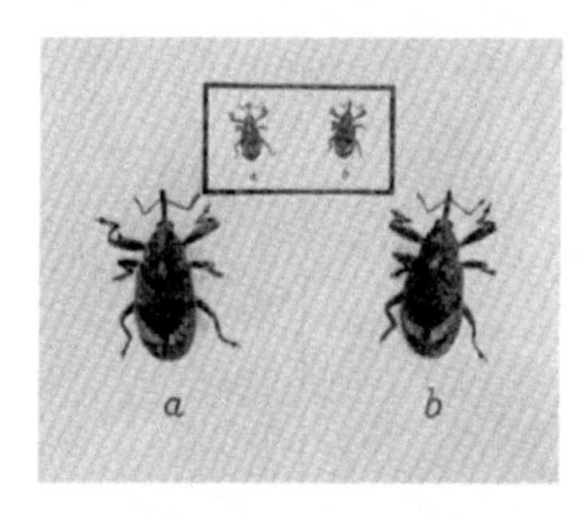


Fig. 118. Two specimens of Apple-Blossom Weevil. Magnified (× 3). Inset, natural size.

**Pupa (Chrysalis)**

| | |
|---|---|
| APPEARANCE. | With long snout folded under head. |
| COLOUR. | Pale brown—eyes black. |
| LOCATION. | In dying or dead blossom buds. |
| DURATION. | 7 to 10 days. |

**Distribution**

Widespread.

**Life-history**

The adult weevils come out from their winter hiding-places on the arrival of the first warm days of spring. The females are winged, but usually prefer to crawl up the trees. The males are often seen flying among the branches in sunny weather in search of the females, which must be fertilised before egg-laying takes place.

The actual process of the egg-laying is very curious and ingenious. The female first takes considerable trouble in selecting the blossom-bud. By means of her long snout, she then bores a deep hole reaching to the centre of the bud, where she produces a small cavity. She then lays an egg in the hole and pushes this down the channel so formed till it

Fig. 119. Diagram of grub and pupa of Apple-Blossom Weevil. Natural size and enlarged. Also of capped blossom.

reaches the cavity in the centre. Finally the entrance is sealed up with saliva, and the insect proceeds to attack another bud in a similar manner. A single female can lay 20 to 60 eggs in as many blossoms, providing these do not expand too rapidly.

The egg hatches out in a few days to a small white legless grub with a brown head, which feeds upon the unopened bud for 1 to 3 weeks, changing finally into a pale, yellowish-brown pupa, with a long snout (proboscis) folded under the head.

The blossom is now practically dead ("capped"), and in a few days the mature weevil emerges, leaving a large round hole in the base of the blossom. It feeds for the rest of the summer upon the leaves of the apple, and, in the autumn,

Fig. 120. View of "capped blossom" caused by the Apple-Blossom Weevil.

seeks out shelter in crevices on the bark, or under rubbish. Here the winter is spent. Many of the weevils fly to other trees to seek their sheltering-places and return to the apples in the spring.

## 2. Economic

### Trees Attacked

Apples, and occasionally pears.

### Frequency of Pest

Common.

## Nature of Attack

The attacked blossom withers (becomes "capped"), the bud never opening, and the petals soon withering, but remaining on the tree till later. The attack on the leaves by the weevils is not serious.

Fig. 121. As fig. 120, but "cap" removed to show pupa within.

## Degree of Damage

This depends almost entirely upon the weather conditions. If, after blossoming, a cold spell intervenes, and the blossoms are prevented from rapidly expanding, a great many may be attacked. The grub cannot live in an opened blossom and many eggs are in this way rendered harmless by the expansion of the flower buds before they are hatched.

## Natural Enemies

Woodpeckers, tits, and certain parasitic flies (ichneumons).

**Preventive Measures**

1. When blossoms are noticed unopened, and brown at the tips, the tree should be shaken. These will then readily fall, and may be collected and burnt, as they will contain either the grub or the pupa of the weevil. This treatment will lessen a future attack.
2. Later on the weevils may be jarred off on to boards, covered with tar or banding composition. A suitable surface is also made by covering a light frame of wood, say 6 feet square, with tarred canvas, and holding amongst the branches by means of a long wooden handle, whilst jarring the tree.

**Remedies**

1. The most effective remedy at present known is to keep the opening buds covered with THICK LIME WASH (see page 428). This may be continued right up to the appearance of the blossom trusses (see fig. 244, page 616). The weevil is thus prevented from laying its egg in the bud.
2. Poison sprays, such as arsenate of lead, have not been found of much use, as it is probable that the weevil eats only a minute portion of the surface of the bud when laying the egg.
3. SOAP[1] sprays with NICOTINE[2] have been found of more service *if the weevils are actually on the trees*, which is difficult to ascertain.
4. The author has found *distinct benefit* resulting from a late LIME-SULPHUR[3] spray, just before the blossom-buds are commencing to open. A little scorching of the leaves and petals resulted, but this had no injurious effect, as the *stamens*[4] were not reached by the spray. The presence of the lime-sulphur probably rendered the blossoms distasteful to the weevil.

[1] See page 452. [2] See page 436.
[3] See page 612. [4] See page 21.

**Calendar of Treatment**

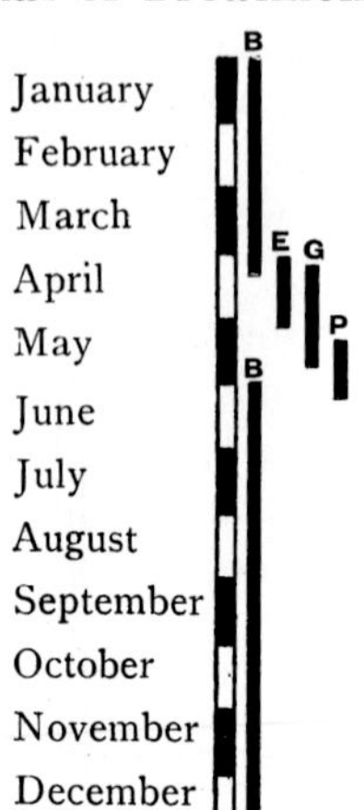


Covering buds with thick lime wash found effective[1]; to a less extent soap[2] and nicotine[3]; and late lime-sulphur spraying[4].
Shake off and burn the withered buds.
Jar weevils off on to tarred boards.

B = beetle stage
E = egg stage
G = grub stage
P = pupa stage

## LEAF WEEVILS ("Oblong," "Green" etc.)

Names *Phyllobius, various species* Class *Rhynchophora* (*Weevils*) Order *Coleoptera* (*Beetles*)

### 1. General

**Description**

**Adult Insect (Weevil)**

| | |
|---|---|
| SIZE. | Small, $\frac{1}{6}$ to $\frac{1}{5}$ inch. |
| COLOUR. | Oblong—black with brown scales.<br>Green—brown, with bright green to yellow scales. |
| APPEARS IN | May. |
| DURATION. | Till July. |

**Ova (Eggs)**

| | |
|---|---|
| LOCATION. | In the ground. |
| APPEAR IN | June and July. |

[1] See page 428.
[2] See page 452.
[3] See page 436.
[4] See page 612.

**Larva (Grub)**

| | |
|---|---|
| APPEARANCE. | Footless, with a few hairs; curved. |
| COLOUR. | White with brown head. |
| LOCATION. | In soil. |
| APPEARS IN | June and July. |
| DURATION. | Throughout winter till following spring. |
| REMARKS. | The maggots live upon roots of plants. |

**Pupa (Chrysalis)**

| | |
|---|---|
| LOCATION. | In soil. |
| APPEARS IN | April. |

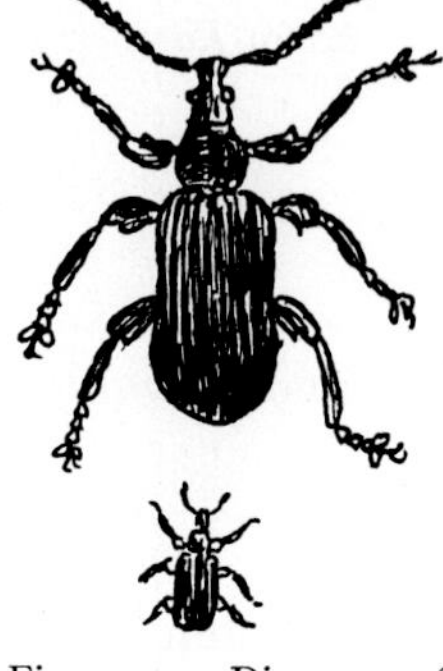

Fig. 122. Diagram of Oblong Leaf Weevil. Natural size and magnified.

**Life-history**

These leaf weevils appear from the soil in late spring and attack the opening flower buds, later feeding upon the leaves. They lay their eggs in the soil and the grub lives there till the spring when it pupates.

## 2. Economic

**Trees Attacked**

Pear, apple, plum, apricot, peach, strawberry.

**Susceptible Varieties**

Grafted plants (Ormerod).

**Frequency of Pest**

Occurs in large numbers locally.

**Nature of Attack**

The blossom-buds and later the leaves are gnawed away by the weevil. Sometimes the roots are also attacked by the grubs.

**Degree of Damage**

Often serious.

**Remedies**

1. Jar the weevils off the trees on a dull day placing tarred boards beneath, or band the trees with COMPOSITION[1] and prevent the weevils regaining the leaves.

[1] See page 404.

2. Spray the leaves and buds with ARSENATE OF LEAD[1], especially if there is also caterpillar attack.

**Calendar of Treatment**

As for *Raspberry Weevil* (see page 205).

## NUT WEEVIL

Name *Balaninus nucum* Class *Rhynchophora* (*Weevils*)
Order *Coleoptera* (*Beetles*)

### 1. General

**Description**

**Adult Insect (Weevil)**

| | |
|---|---|
| SIZE. | About $\frac{1}{3}$ inch long. |
| APPEARANCE. | With *long curved snout.* |
| COLOUR. | Variable—tawny brown covered with yellowish or greyish down. |
| APPEARS IN | June. |
| REMARKS. | Distinguished by its characteristic snout. |

**Ova (Eggs)**

| | |
|---|---|
| ARRANGEMENT. | Singly in interior of nut. |
| HATCH IN | 8 to 10 days. |

**Larva (Grub)**

| | |
|---|---|
| APPEARANCE. | Legless, fat and fleshy, and curved at tail end, about $\frac{1}{3}$ inch long. |
| COLOUR. | Cream coloured. |
| LOCATION. | Inside kernel of nut till mature, afterwards in the ground. |
| APPEARS IN | July and August. |
| DURATION. | Throughout winter (in soil) till following spring. |

[1] See page 397.

**Pupa (Chrysalis)**

| | |
|---|---|
| Colour. | Creamy white. |
| Location. | In ground. |
| Appears in | March and April. |
| Duration. | Till late May or June (variable). |

**Distribution**

Widespread and common.

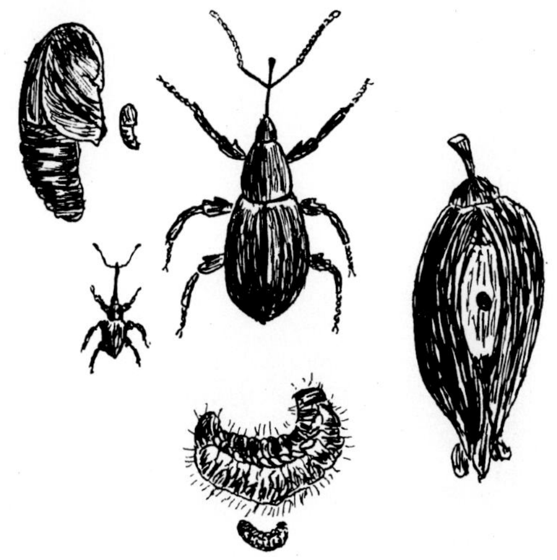

Fig. 123. Diagram of various stages of Nut Weevil, natural size and enlarged, also damaged nut.

**Life-history**

The weevil bores a hole in the soft young nut with her long curved snout, deposits one egg in the hole formed, and then pushes it into the centre of the nut. In a few days the grub appears and lives on the flesh of the kernel till mature. Sometimes the nut drops and at others remains on the tree. In any case the mature caterpillar forces its way out of the nut and winters in the ground in an inactive condition, pupating the following spring.

## 2. Economic

**Trees Attacked**

Nuts (filberts, cobs, wild hazels).

**Frequency of Pest**

Common.

Fig. 124. Grub of Nut Weevil and holes caused by incision of the adult insect. Natural size.

**Nature of Attack**

The kernel of the nut is eaten and spoilt by the grub. (The well-known "maggotty nuts.")

**Degree of Damage**

Often serious.

**Preventive Measures**

1. Remove and burn all nuts which fall before they are ripe (as these contain live grubs).
2. Frequent stirring of the soil under the bushes during the winter exposes the tender grubs to frost and birds.

**Remedies**

1. Choose a DULL day, and beat the trees, after placing tarred boards or sacks underneath, or giving the surface of the ground a good sprinkling of gas-lime.
2. Spraying the bushes in late May with ARSENATE OF LEAD[1] has been found effective.

**Calendar of Treatment**

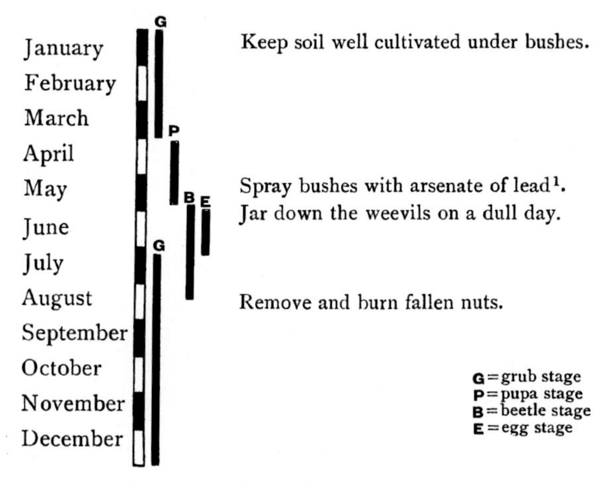


[1] See page 397.

## RASPBERRY (CLAY-COLOURED) WEEVIL

Name *Otiorhyncus picipes* Class *Rhynchophora* (*Weevils*)
Order *Coleoptera* (*Beetles*)

### 1. General

**Description**

**Adult Insect (Weevil)**

| | |
|---|---|
| SIZE. | $\frac{1}{4}$ to $\frac{1}{3}$ inch. No wings. |
| COLOUR. | Brown, covered with patches of brown or grey scales, giving it an *appearance of mottled clay*. |
| APPEARS IN | May. |
| REMARKS | This weevil feeds at night and hides during the day-time. It falls to the ground on being the least disturbed, or with a light. It is extremely difficult to distinguish on the ground as it remains quite motionless after falling. |

**Ova (Eggs)**

| | |
|---|---|
| LOCATION. | Just under the ground. |
| APPEAR IN | August. |
| HATCH IN | Few days. |

**Larva (Grub)**

| | |
|---|---|
| APPEARANCE. | Dull, footless, skin much wrinkled: head brown. |
| COLOUR. | Yellowish white. |
| LOCATION. | In the ground. |
| APPEARS IN | August. |

| | |
|---|---|
| DURATION. | Throughout winter till following April. |
| REMARKS. | The maggot feeds upon the roots of the plant attacked. |

**Pupa (Chrysalis)**

| | |
|---|---|
| APPEARANCE. | Like weevils in shape, but with legs folded beneath. |
| COLOUR. | Yellowish white. |
| LOCATION. | In the ground; near surface. |
| APPEARS IN | About April. |
| DURATION. | 2 to 3 weeks. |

Fig. 125. Diagram of grub and adult of Raspberry Weevil. Natural size and enlarged.

## Distribution

Widespread. Destructive epidemics of it have been reported at various times in almost all the fruit-growing districts of England.

## Life-history

The weevils appear from the ground in May. They are unable to fly, being without wings, and protect themselves by dropping to the ground on the least noise or disturbance, and so escaping detection by their close resemblance to the earth. They feed at night upon the young blossoms and leaves, and also attack the shoots, stripping the rind off and puncturing the tender parts with their snouts.

In August the eggs are laid just under the ground, and the footless grubs which hatch out feed upon the roots of the plant throughout the winter, changing in April to inactive pupæ which give rise to the weevils in due course.

## 2. Economic

**Fruit-Plants Attacked**

Raspberry mainly: also gooseberry, hop, plum, apple, damson, nuts, strawberry.

**Frequency of Pest**

Common.

**Symptoms of Attack**

Young leaves appear with small holes, shoots are damaged.

**Nature of Attack**

The weevils attack the young fruit buds and the blossoms. They also eat the leaves, and puncture or skin the shoots. The grubs attack the roots during the winter months.

**Duration of Attack**

Weevil—April to August.
Grub—throughout winter.

**Degree of Damage**

Often extremely serious, especially when the blossom and buds are attacked.

**Preventive Measures**

1. Remove weeds and rubbish from the plots.
2. Hoe in the usual SOIL INSECTICIDES[1] during the autumn to kill the grubs.

[1] See page 433.

## Natural Enemies

Moles, tits, and some wasp-like flies (*Cerceris*).

## Remedies

1. This weevil is very easily dealt with by making use of its extremely timid habits.
   Provide boards smeared over with BANDING COMPOSITION[1], or sticky tar, and proceed *quietly* to the plantation *after dark*. Take a lantern, but be careful to keep its rays from the plants, or no good will result. Hold the boards on each side of the canes (if raspberries) or under the plants, then jar them or flash the light. The weevils will thus be readily caught.
2. Spray the plants well, as soon as the attack is noticed, with ARSENATE OF LEAD[2]. This is an almost certain cure as the weevils soon get a fatal dose of the poison.

## Calendar of Treatment

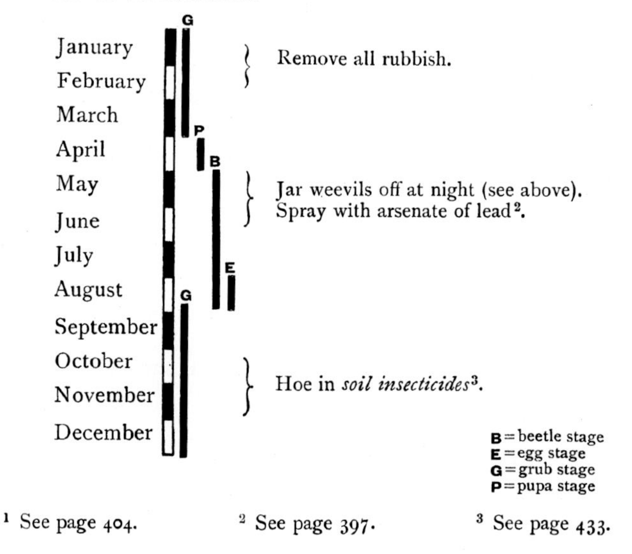


[1] See page 404. [2] See page 397. [3] See page 433.

## RED-LEGGED WEEVIL

Name *Otiorhyncus tenebricosus* Class *Rhynchophora* (*Weevils*)
Order *Coleoptera* (*Beetles*)

### 1. General

**Description**

**Adult Insect (Weevil)**

| | |
|---|---|
| SIZE. | About ½ inch long. |
| COLOUR (BODY). | Shiny black—*legs dull red.* |
| APPEARS IN | May and June. |
| REMARKS. | The weevil feeds at night and hides away during the day. |

**Ova (Eggs)**

| | |
|---|---|
| LOCATION. | In ground. |
| APPEAR IN | June and July. |

**Larva (Grub)**

| | |
|---|---|
| APPEARANCE. | Similar to the preceding species (clay-coloured weevil) but larger. |
| COLOUR. | Yellowish white. |
| LOCATION. | In the ground. |
| APPEARS IN | August. |
| DURATION. | Throughout the winter till following April. |

**Pupa (Chrysalis)**

| | |
|---|---|
| APPEARANCE. | As for clay-coloured weevil, but larger. |
| COLOUR. | Yellowish white. |
| LOCATION. | In the ground, near surface. |
| APPEARS | About April. |
| DURATION. | 2 to 3 weeks. |

**Distribution**

Widespread. Appears suddenly as a pest at various places.

**Life-history**

As for clay-coloured weevil (see page 203).

### 2. Economic

**Trees Attacked**

Most fruit, especially plum and raspberry.

**Nature of Attack**

The weevil attacks the buds and young shoots, and later on feeds upon the leaves.

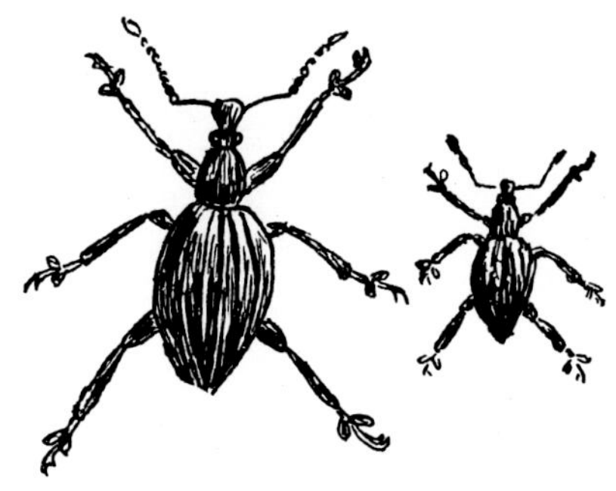

Fig. 126. Diagram of Red-legged Weevil. Natural size and enlarged.

**Degree of Damage**

Often very serious.

**Preventive Measures, Remedies and Calendar of Treatment**

As for clay-coloured weevil.

## TWIG-CUTTING WEEVIL

Name *Rhynchites cæruleus* Class *Rhynchophora* (*Weevils*) Order *Coleoptera* (*Beetles*)

### 1. General

**Description**

**Adult Insect (Weevil)**

| | |
|---|---|
| SIZE. | About $\frac{1}{5}$ inch long. |
| COLOUR. | Shiny blue ; snout and legs black. |
| LOCATION. | On leaves. |

| | |
|---|---|
| APPEARS IN | April and May. |
| REMARKS. | After laying her egg in the shoot, the female severs this just below the puncture, causing the shoot to die and fall to the ground. |

**Ova (Eggs)**

| | |
|---|---|
| APPEARANCE. | Oval. |
| COLOUR. | Yellow. |
| ARRANGEMENT. | Singly placed. |
| LOCATION. | In shoot. |
| HATCH IN | 7 to 10 days. |

Fig. 127. Diagram of Twig-cutting Weevil. Natural size and enlarged.

**Larva (Grub)**

| | |
|---|---|
| APPEARANCE. | Footless, small. |
| COLOUR. | White. |
| LOCATION. | In pith of shoot. |

**Pupa (Chrysalis)**

| | |
|---|---|
| LOCATION. | In ground. |
| DURATION. | Probably till following spring. |

**Distribution**

Noticed chiefly in Kent.

**Life-history**

The weevils appear on the leaves early in the season, but seem to do little damage. After fertilisation, the female punctures a shoot a few inches from the top, lays the egg therein and then CUTS THE SHOOT IN TWO just below the puncture—sometimes the shoot is only cut enough to bend it over without falling. The shoot soon withers and falls to the ground with the contained grub, which feeds upon the pith until mature and then pupates in the soil.

## 2. Economic

**Trees Attacked**

Apples.

**Susceptible Varieties**

Young trees specially attacked (nursery stock and bush fruit).

**Frequency of Pest**

Occasional.

**Nature of Attack**

The young twigs are cut off and withered by the female weevil after laying the egg.

**Degree of Damage**

Occasionally serious.

**Preventive Measures**

Gather and burn the fallen shoots.

**Remedies**

1. Jarring off the weevils when these appear on to tarred boards or sacks (on dull days).
2. Spraying trees with ARSENATE OF LEAD[1]

[1] See page 397.

# CHAPTER 15

## Beetles and Chafers

### FRUIT-BARK BEETLE

Name *Scolytus rugulosus* Family *Scolytidæ*
Order *Coleoptera* (*Beetles*)

#### 1. General

**Description**

**Adult Insect (Beetle)**

| | |
|---|---|
| SIZE. | Small (about $\frac{1}{10}$ inch). |
| COLOUR. | Brownish black to black. |
| APPEARS IN | May to July and October. |
| REMARKS. | Two or three broods occur annually. |

**Ova (Eggs)**

| | |
|---|---|
| LOCATION. | In the wood of tree trunks underneath the bark. |
| HATCHING PERIOD. | A few days. |

**Larva (Grub)**

| | |
|---|---|
| APPEARANCE. | Small, footless. |
| COLOUR. | White. |
| LOCATION. | In "tunnels" of wood. |
| APPEARS IN | November, June and August. |
| DURATION. | A few weeks, except in the case of the last brood, which lasts over the winter. |

**Pupa (Chrysalis)**

| | |
|---|---|
| LOCATION. | Within the tunnels (in wood of tree). |
| DURATION. | Few days. |

**Distribution**

South and West of England.

### Life-history

Two to three broods occur in the year. The grubs which pass the winter in the tree pupate in the spring, and give rise to beetles which appear in May.

The beetles bore into the bark, and then construct vertical "tunnels" or "galleries" between the inner bark and the sap-wood. The eggs are laid here side by side, and the minute grubs feed on the woody material or the inner bark forming further "tunnels." Pupation takes place here, and the beetles escape through the bark, forming many small "shot holes."

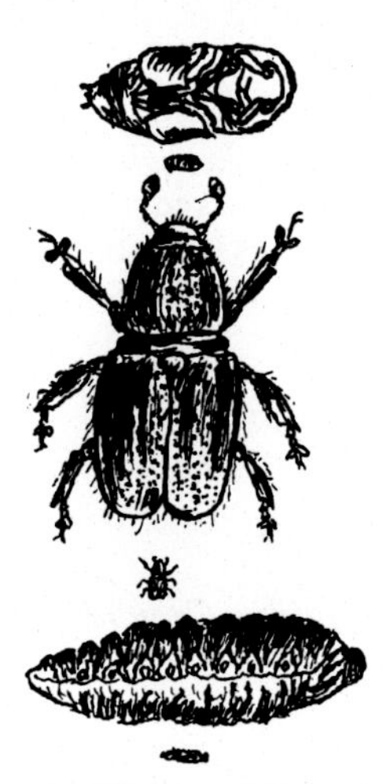

Fig. 128. Diagram of three stages of the Fruit-bark Beetle. Natural size and enlarged.

Fig. 129. Fruit-bark Beetle. Enlarged. Inset, natural size.

## 2. Economic

### Trees Attacked

Apple, plum, pear, peach, apricot.

### Symptoms of Pest

The appearance of numerous SHOT-HOLES, some of which are formed by the entrance of the beetles, and others by the exit of the fresh brood.

**Frequency of Pest**

Occasional.

**Nature of Attack**

The inner bark of the trunk or branches is eaten away by the tunnelling and operations of the beetle and its grubs.

**Degree of Damage**

Often serious, the attacked trees usually dying finally.

**Preventive Measures**

Cut down and destroy the attacked trees before the beetles escape.

**Remedies**

No remedies can be regarded as effective in saving an infested tree, but it should be at once destroyed, so as to prevent further mischief. Coating the trunks and branches with sticky material has been recommended, but involves *so much labour* as to be practically prohibitive.

## GROUND BEETLES

Name *Harpalus etc.* Family *Carabidæ* (*Ground beetles*)
Order *Coleoptera* (*Beetles*)

### 1. General

**Description**

**Adult Insect (Beetle)**

| | |
|---|---|
| SIZE. | From $\frac{1}{4}$ to $\frac{3}{4}$ nch in length. |
| COLOUR (BODY). | Shiny black, legs black or red. Wing cases lined or plain. |
| APPEARS IN | May to July. |
| REMARKS. | About four varieties have been found eating strawberries. |

**Ova (Eggs)**

| | |
|---|---|
| LOCATION. | In soil. |

**Larva (Grub)**

| | |
|---|---|
| APPEARANCE. | Long, with large heads and pronounced jaws. |
| COLOUR. | Yellowish and brown. |
| LOCATION. | In soil. |
| REMARKS. | The grubs are *always beneficial*, living upon other insect life, and upon snails, worms etc. |

Fig. 130. Diagram of a Ground Beetle (*Harpalus ruficornis*). Natural size and enlarged.

**Remarks**

The beetles are not usually pests, but are normally beneficial, feeding upon animal life in the soil.

In the day-time they hide under clods of earth etc. and feed at night, moving with great activity when disturbed.

## 2. Economic

**Fruit-plants Attacked**

Strawberries.

**Frequency of Pest**

Attacks have been reported from many districts in England.

### Nature of Attack

The beetles attack the young strawberries, eating the skin and the seeds and spoil many that are not actually eaten.

### Degree of Damage

Occasionally serious.

### Preventive Measure

Treatment of the soil with SOIL INSECTICIDES[1] during the spring will kill or drive away the grubs.

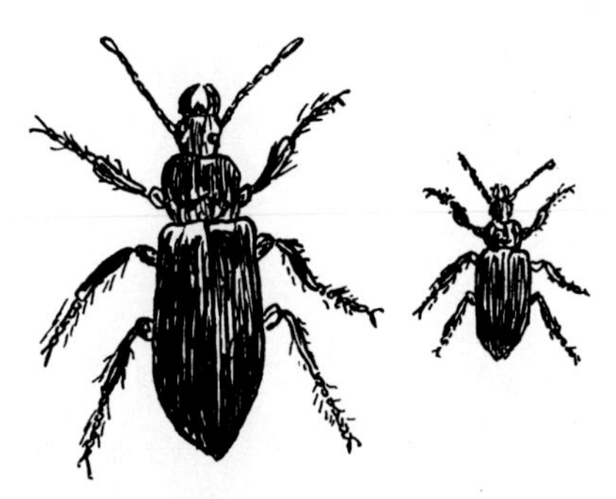

Fig. 131. Diagram of a Ground Beetle found on strawberries (*Calathus*). Natural size and enlarged.

### Natural Enemies

Moles live upon these beetles and will clear a district of them sufficiently to keep down attacks.

### Remedies

The beetles are readily trapped by sinking small vessels (such as flower pots or jam jars) into the ground, placing as bait pieces of meat, and covering over with straw. The trapped beetles are best killed with boiling water.

[1] See page 433.

## **RASPBERRY BEETLE** (Loganberry beetle)

Name *Byturus tomentosus* Class *Phytophaga*
Order *Coleoptera* (*Beetles*)

### 1. General

**Description**

**Adult Insect (Beetle)**

| | |
|---|---|
| SIZE. | About $\frac{1}{6}$ inch long. |
| COLOUR. | Dark brown to yellowish but covered thickly with yellow or grey down. |
| APPEARS IN | May or June. |
| REMARKS. | Old beetles are much darker in colour owing to the down having worn away. |

**Ova (Eggs)**

| | |
|---|---|
| ARRANGEMENT. | Singly. |
| LOCATION. | In the opening blossoms. |
| APPEAR IN | May or June. |

**Larva (Grub)**

| | |
|---|---|
| APPEARANCE. | With three legs in front and two curved spines behind. |
| COLOUR. | Yellowish, with brown plates on the segments of back, and head brown. |
| LOCATION. | (1) in base of flower, (2) in the berry or (3) in sheltering places on the canes or ground. |
| APPEARS IN | June. |
| DURATION. | Till August. |

**Pupa (Chrysalis)**

| | |
|---|---|
| LOCATION. | In crevices, under the rind of canes, or in the ground. |

| | |
|---|---|
| APPEARS IN | August and September. |
| DURATION. | Throughout winter till following spring. |

**Distribution**

Widespread in England and Europe generally.

**Life-history**

The beetles are very active fliers in bright weather, but sluggish on dull days. They feed upon the blossoms, and often bite the young flowers completely off, at other times eating the petals, stamens etc. quite away.

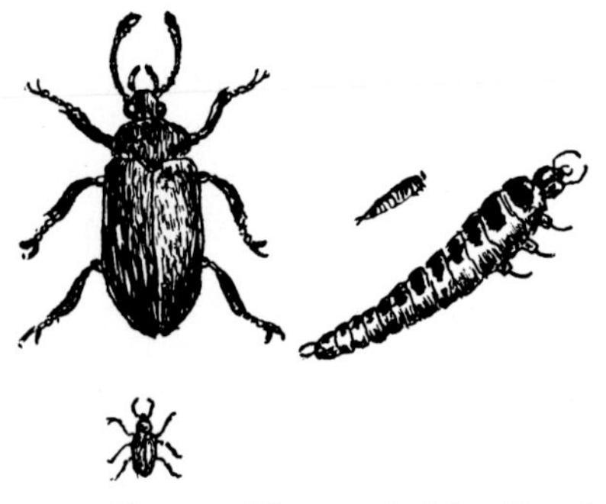

Fig. 132. Diagram of larva and adult of Raspberry Beetle. Natural size and enlarged.

Eggs are then laid in the blossoms. The young grubs feed upon the base of the flower and then enter the young berries. Here they remain till mature when they crawl out, and find shelter on the canes, stakes or in the ground and there pupate. A single grub may attack more than one fruit. The pupa remains all the winter in shelter, and gives rise to the beetle in the following spring.

**Remarks**

These beetles are rather peculiar as passing the winter in the pupal state. In the case of most other beetles, the grub winters in the soil and pupates in early spring.

## 2. Economic

### Fruit-plants Attacked

Raspberry, loganberry, blackberry.

### Frequency of Pest

Very common.

### Nature of Attack

The BEETLE devours the blossoms, or cuts these off.

The GRUB destroys or infests the berries until mature; such berries will not keep and are unfit for the table.

Fig. 133. Raspberry Beetle. Natural size.
Inset, enlarged ( × 3).

### Degree of Damage

Extremely serious; may ruin an entire crop, and prohibit cultivation (especially of loganberries).

### Preventive Measures

1. Cut canes well back in the autumn and burn the cuttings (these will contain some of the pupæ).
2. Work soot and lime into the soil during the winter to kill those in the ground.
3. Remove, as far as possible, all wild brambles from the neighbourhood, as the beetle often breeds in these.
4. Cultivate and turn over ground as much as possible.

## Remedies

1. Spray the young plants with ARSENATE OF LEAD[1] when the blossom buds appear, and also two weeks later.
2. Jar off the beetles on to tarred boards on a dull day when they are inactive.
3. Spray twice at 3 weeks' interval in the following manner:—Spray first with 2 per cent. SOFT SOAP + 2 per cent. PARAFFIN EMULSION[2]. This brings the beetles out of hiding. Follow this up five minutes later with the same emulsion after adding 8 ozs. of pure NICOTINE per 100 gallons of solution. (Professor Lees.)

## Calendar of Treatment

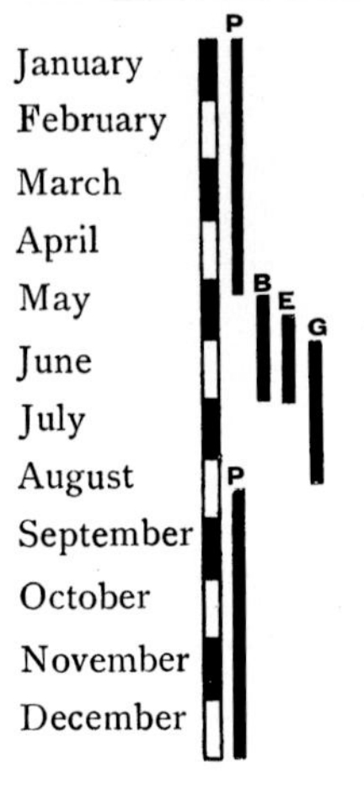


Spray with arsenate of lead[1].
Jar off beetles on dull day.
Spray paraffin emulsion[2] as above.

Prune canes and destroy cuttings.

Hoe in soot and lime.

B = beetle stage
E = egg stage
G = grub stage
P = pupa stage

[1] See page 397. [2] See page 418.

## SHOT-BORER BEETLE

Name *Xyleborus dispar* Family *Scolytidæ*
Order *Coleoptera* (*Beetles*)

### 1. General

**Description**

**Adult Insect (Beetle)**

| | |
|---|---|
| SIZE. | Female about $\frac{1}{8}$ inch long: male much smaller. |
| COLOUR. | Brown to black: wing cases reddish: marked alternately with punctures and hairy interspaces, feelers (antennæ) clubbed and reddish or yellow. |
| APPEARS IN | May and September. |
| REMARKS. | Front part (thorax) of body very large in proportion. |

**Ova (Eggs)**

| | |
|---|---|
| ARRANGEMENT. | In heaps of 7 to 10. |
| LOCATION. | At entrances of tunnels in the wood. |
| APPEAR IN | May or June and August or September. |

**Larva (Grub)**

| | |
|---|---|
| COLOUR. | Pink. |
| LOCATION. | Inside the tunnels. |
| APPEARS IN | June and October. |
| DURATION. | A few weeks. |
| REMARKS. | The grubs appear to feed upon a fungus which lines the interior of the borings. |

**Pupa (Chrysalis)**

| | |
|---|---|
| LOCATION. | In tunnels, often packed closely together. |
| APPEARS IN | June, July and October. |

**Distribution**

Attacks recorded in Cambridge, Surrey and West of England.

**Life-history**

The female beetle, which is much larger than the male and much commoner, bores into the trunk or main branch of the tree, makes a few passages and lays her eggs near the entrance in a heap placed within a small hollowed-out chamber.

The young grubs feed in the tunnels made by the mother, and apparently depend for their food upon a fungus which grows upon the solidified sap which lines the borings (and has been termed *Ambrosia*).

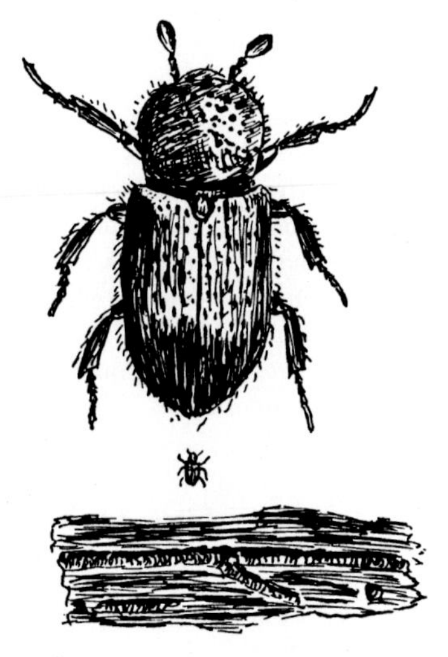

Fig. 134. Diagram of Shot-borer Beetle showing tunnels in wood. Natural size and enlarged.

When mature the grubs turn into pupæ, which are packed like shot in the small tunnels. Beetles soon appear and remain in the tree till the following spring. Whole colonies are thus produced in the trees.

## 2. Economic

**Trees Attacked**

Plum, apple, pear, cherry: also many forest trees.

**Frequency of Pest**

Occurs as a severe attack at intervals locally.

**Nature of Attack**

Long tunnels are made by the beetle into the heart of the wood, and interior rings are also made in the sap-wood, killing young trees. The name is derived from the shot-like openings produced in the bark, and from the close packing of the beetles (like rows of shot) in the holes.

**Duration of Attack**

Practically continuous.

**Degree of Damage**

Young trees are usually killed: older trees seriously weakened owing to the interference with the flow of the sap.

**Preventive Measures**

These consist in

1. Cutting down and destroying all attacked trees in the winter.

   The attacked trees thus act as traps, and in any case they would finally succumb to the attack. This is preferable to blocking up the holes, as has been suggested, which can only be regarded as uncertain in its results.

2. Protecting neighbouring trees with such mixtures as the following smeared on the bark:
   - (*a*) CARBOLIC SOFT SOAP[1].
   - (*b*) SOFT SOAP[1] mixed up into a thick paste with a strong solution of washing soda.
   - (*c*) SOFT SOAP and ARSENATE OF LEAD[2].

     The latter should not be employed if stock are fed in the orchard.

**Remedies**

There are no certain remedies to apply to save infested trees; in any case it is too risky to attempt experiment with uncertain remedies, since, if the beetles became free, the attack is greatly extended.

[1] See page 452. [2] See page 397.

**Calendar of Treatment**

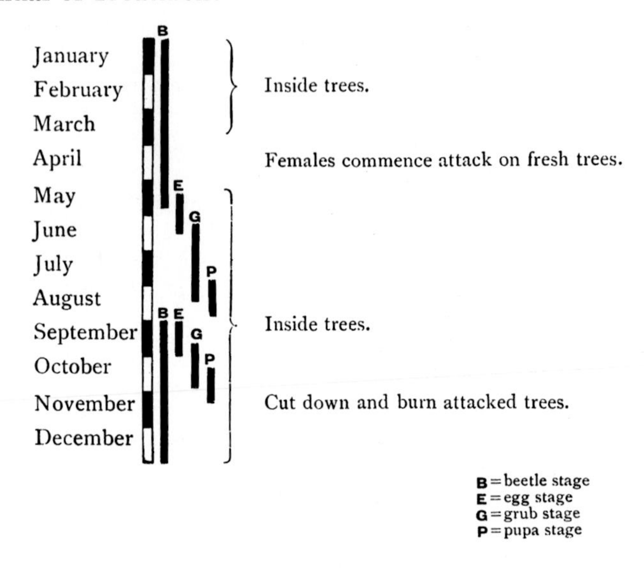


## COCKCHAFER

Name *Melolontha vulgaris* Class *Lamellicornia* (*Chafers*)
Order *Coleoptera* (*Beetles*)

### 1. General

#### Description

##### Adult Insect (Beetle)

| | |
|---|---|
| SIZE. | Upwards of one inch in length. |
| COLOUR. | Black—wing cases reddish brown, legs brown and hairy feelers (antennæ) with 7 "leaves" (male), 6 (female), which open and close like a fan. |
| APPEARS IN | May and June. |

**Ova (Eggs)**

| | |
|---|---|
| APPEARANCE. | Large. |
| COLOUR. | Cream-coloured. |
| LOCATION. | Fairly deep in the soil. |
| APPEAR IN | June and July. |

**Larva (Grub)**

| | |
|---|---|
| APPEARANCE. | About 1½ inches long when mature, similar to Rose chafer, but feet are not pointed. The tail-end is swollen and bladderlike. |

Fig. 135. Cockchafer. Natural size.

| | |
|---|---|
| COLOUR. | Creamy white, except the tail-end, which is dark owing to the contained excrement showing through. |
| LOCATION. | In the soil. |
| APPEARS | About August. |
| DURATION. | 3 *years*. |

**Pupa (Chrysalis)**

| | |
|---|---|
| APPEARANCE. | In earth covering. |
| COLOUR. | Pale brown. |
| LOCATION. | Deep in the ground. |
| APPEARS IN | April. |
| DURATION. | Few weeks. |

### Life-history

These interesting and well-known insects appear, flying at night with a loud humming noise, in May and June. During the day they are in hiding amongst the foliage. Their food is chiefly the leaves of various plants. Eggs are laid in the ground at a considerable depth, and the grubs which hatch out feed voraciously upon the roots of plants during three years before they mature. The chrysalis stage is passed lower in the ground in a covering of earth.

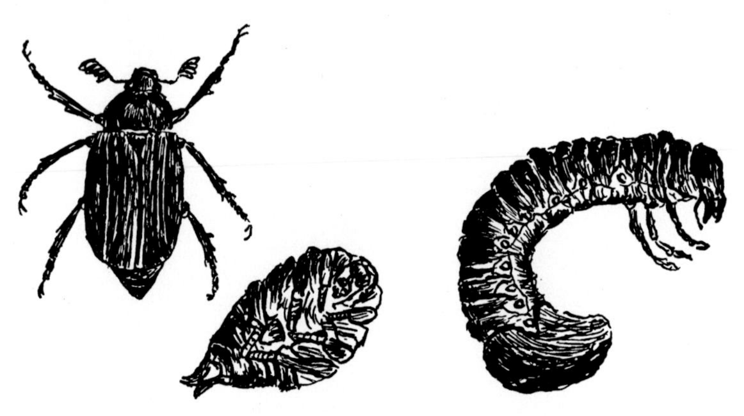

Fig. 136. Diagram of grub, pupa and adult of Cockchafer. Natural size.

## 2. Economic

### Fruit-plants Attacked

Raspberry, currant, strawberry.

### Frequency of Pest

Occurs with periodic frequency about every four years.

### Nature of Attack

The beetles devour the foliage of the raspberry and other plants and the grubs feed upon the roots and frequently kill the plants.

### Degree of Damage

Often serious.

## Remedies

1. Jarring the sluggish beetles during the daytime on to tarred boards, or into bags, is very effective.
2. Mixing SOIL INSECTICIDES[1] with the earth around the roots will kill the grubs.
3. Spraying the plants with ARSENATE OF LEAD[2] will kill the beetles devouring the foliage.

## Calendar of Treatment

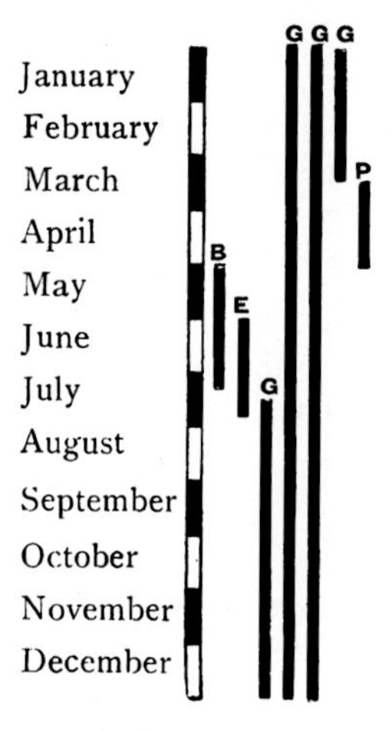


Jar off beetles on to tarred boards or spray with *arsenate of lead*[2].

Mix *soil insecticides*[1] into earth around plants.

B = beetle stage
E = egg stage
G = grub stage
P = pupa stage

[1] See page 433.

[2] See page 397.

## **ROSE CHAFER** (Green chafer)

Name *Cetonia aurata* Class *Lamellicornia* (*Chafers*)
Order *Coleoptera* (*Beetles*)

### 1. General

**Description**

**Adult Insect (Beetle)**

| | |
|---|---|
| SIZE. | About $\frac{3}{4}$ inch long. |
| COLOUR (BODY). | Brilliant metallic green; has a golden sheen in sunlight. Coppery below. |
| APPEARS IN | May or later. |
| REMARKS. | The beetles fly on sunny days. |

**Ova (Eggs)**

| | |
|---|---|
| LOCATION. | Fairly deep in the soil. |
| APPEAR IN | July and August. |
| HATCHING PERIOD. | 12 to 14 days. |

**Larva (Grub)**

| | |
|---|---|
| APPEARANCE. | Thick, fleshy, wrinkled; tail-end much swollen (full-grown = $1\frac{1}{2}$ inches long). |
| COLOUR. | Yellowish white; a brown spot on each side of first segment. |
| LOCATION. | In the soil. |
| APPEARS IN | July and August. |
| DURATION. | *2 to 3 years.* |
| REMARKS. | Resembles the Cockchafer grub, but distinguished from this by the short hairs on the body and legs, and the pointed feet. |

**Pupa (Chrysalis)**

| | |
|---|---|
| APPEARANCE. | In a case of stones, earth etc. (about $\frac{2}{3}$ inch long). |
| COLOUR. | Whitish yellow, eyes dark. |
| LOCATION. | Some distance below the surface of the ground. |
| APPEARS IN | June or earlier. |

**Life-history**

These very fine beetles appear on fine summer days flying amongst the foliage, and feeding upon the leaves and blossom. The eggs are laid in the ground and the grubs pass 2 to 3 years in the soil before they are mature, living upon roots of plants or on decaying vegetable matter. They pupate in a case formed of stones, earth, etc., glued together by a secretion from the insect's body.

Fig. 137. Rose Chafer. Larva. Natural size.

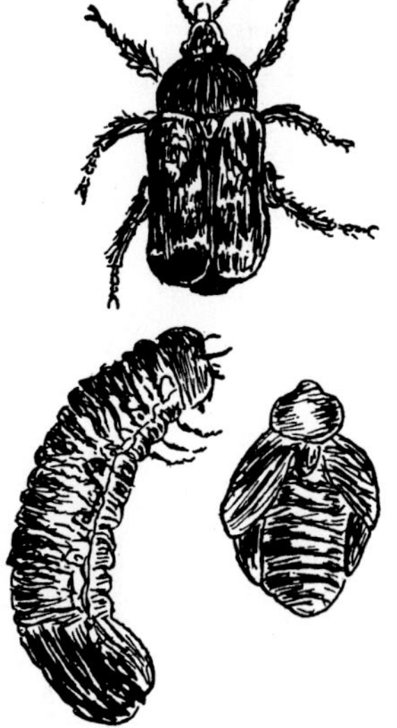

Fig. 138. Diagram of larva, pupa and adult of Rose Chafer. Natural size.

## 2. Economic

**Trees Attacked**

Apple, raspberry, currant, pear, etc., also roses, etc.

**Frequency of Pest**

Fairly common.

**Nature of Attack**

1. The beetle ravenously devours the blossoms and leaves.
2. The grub eats and cuts the roots of plants, often causing these to wilt and die.

## Degree of Damage

Serious, if the beetle appears in any numbers.

## Preventive Measures

1. Sods, placed grass-side downwards on the ground, will attract the grubs, which can then be destroyed.
2. Hand picking the grubs after deeply forking about the roots of the attacked plants is advisable.

## Remedies

1. Shake the beetles off the trees and bushes in dull weather.
2. Spray the plants with ARSENATE OF LEAD[1] when the beetles first appear.

## Calendar of Treatment

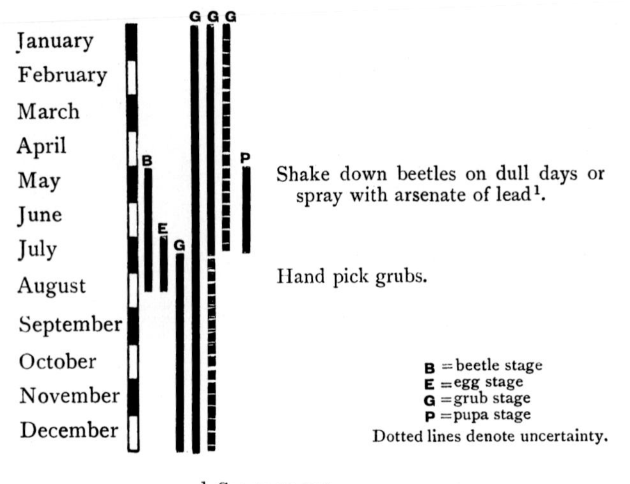


[1] See page 397.

# FLIES

SAWFLIES (*HYMENOPTERA*)

TWO-WINGED FLIES (*DIPTERA*)

# CHAPTER 16

## Flies

(HYMENOPTERA and DIPTERA)

### 1. General

All the pests belonging to these groups are harmful in the maggot (larva) stage only. The adult insects do not injure the fruit.

The injury is of two kinds:

INJURY TO THE LEAVES (mainly).

Gooseberry and Currant sawfly.
Slug worm.
Nut sawfly.
Plum leaf sawfly.
Social Pear sawfly.

INJURY TO THE FRUIT.

Apple sawfly.
Plum fruit sawfly.
Pear midge.

With the exception of the curious *slug worm* the larvæ which feed upon the leaves strongly resemble the caterpillars (Lepidoptera). They can, however, be readily distinguished from these by the number of "sucker feet" (*prolegs*). All caterpillars have *four* of these in the middle of the body (except the *Geometridæ*) while the sawfly larvæ have more than this number, e.g. the Apple sawfly larva has six.

The remedy for these is a stomach poison, as for caterpillar. Arsenate of lead is efficient, but is generally dangerous to use when the fruit is forming, especially in the case of small fruit, as gooseberry, currant, etc. Comparatively non-poisonous insecticides are however available, e.g. Hellebore and Pyrethrum, and are admirably suited for this purpose.

With regard to the maggots of the Apple sawfly and Pear midge, which are hatched and live protected inside the fruitlet, there is no satisfactory remedy. It is at present only possible to mitigate the pest by picking off the attacked fruit and so prevent further damage.

In the case of the Plum fruit sawfly, an early spraying of the blossom with arsenate of lead is of proved value, since the maggot is somewhat exposed in its early stages.

Fig. 139. Apple Sawfly Maggot entering young fruit. Natural size.

The ichneumon flies, which are highly beneficial insects, destroying annually vast numbers of injurious insects, belong to this group. These are elsewhere described[1].

[1] See page 471.

## 2. Scientific

The *Sawfly larvæ* differ very much. They may be naked, or hairy, and the fruit eaters resemble maggots in form, being pale and fleshy. The mature insect resembles the bees and wasps (which are the higher members of this order) in having two pairs of membranous wings.

They are remarkable for the cutting instruments, possessed by the females, which strongly resemble saws in shape. These are employed for making incisions in the tissues of plants for the purpose of inserting the eggs in a safe situation.

The *Pear midge* belongs to a different order (Diptera). It is a two-winged fly belonging to the gall-fly group. In this order there are also a large number of beneficial insects, the maggots of which live inside the bodies of many insect pests, especially caterpillars. These are referred to in a later chapter[1].

*Classification*

### Order **Hymenoptera**

There is a full series of transformations from egg to adult insect, and a definite pupal stage. There are two pairs of membranous wings. Most of the females possess *ovipositors* for placing the eggs in suitable positions. These are also often provided with a saw, a borer, or a sting.

Of the three sub-orders, only the first has members which are of injury to fruit. The other two contain a large number of beneficial insects, viz. those which are parasitic on insect pests and the bees and wasps, which assist fertilisation of the flowers.

Sub-order I. SESSILIVENTRES.

Family *Siricidæ* (wood wasps).

No pests.

Family *Tenthredinidæ* (sawflies).

The female furnished with a pair of saws, for egg-laying. The pests include:

Apple sawfly, *Hoplocampa testudinea.*
Gooseberry and Currant sawfly, *Nematus ribesii.*
Slug worm, *Eriocampa limacina.*
Nut sawfly, *Crœsus septentrionalis.*
Plum fruit sawfly, *Hoplocampa fulvicornis.*
Plum leaf sawfly, *Cladius padi.*
Social Pear sawfly, *Pamphilus flaviventris.*

[1] See Section VI, page 465.

Sub-order 2. PARASITICA.

Most of these are parasites, many of them on destructive caterpillars and pupæ, and others on aphis, etc. (see page 465).

Family *Cynipidæ* (gall flies).

Family *Proctotrypidæ*.

Many of these are parasites on insects.

Family *Chalcididæ*.

The females of this family are winged and the males wingless—a very curious feature.

Family *Ichneumonidæ* (the ichneumon flies).

About all parasitic on insects.

Family *Braconidæ*.

Mostly parasitic.

Sub-order 3. ACULEATA.

Family *Chrysididæ* (ruby wasps).

Family *Apidæ* (bees).

Friendly insects of great value as fertilising agents.

Family *Fossores* (digging-wasps).

Family *Diploptera* (true wasps).

Occasionally pests on ripe fruit, but mainly beneficial insects.

Family *Formicidæ* (ants).

Probably unfriendly to the extent of their care and culture of aphis pests.

## Order **Diptera**

Of this large order of insects, only the PEAR MIDGE, *Diplosis pyrivora*, is a destructive pest on fruit. The *Tachina* flies are parasitic on caterpillars. They do not puncture the skin, but lay the eggs on the surface. The larvæ then bore their way into the caterpillar.

Sub-order 1. NEMOCERA.

The gall midges belong to this group. The only pest is the Pear Midge, *Diplosis pyrivora*.

Sub-order 2. BRACHYCERA.

Horse flies and Gadflies are in this group.

The Robber flies kill and suck insects of all kinds and are beneficial.

Sub-order 3. ATHERICERA.

This includes three large families.

Family *Syrphidæ*.

Many resemble bees and wasps, and are useful as agents in fertilisation of flowers. Some of the larvæ devour aphides.

Fig. 140. Gooseberry Sawfly mounted and magnified, showing especially mouth parts and saws (at end of body).

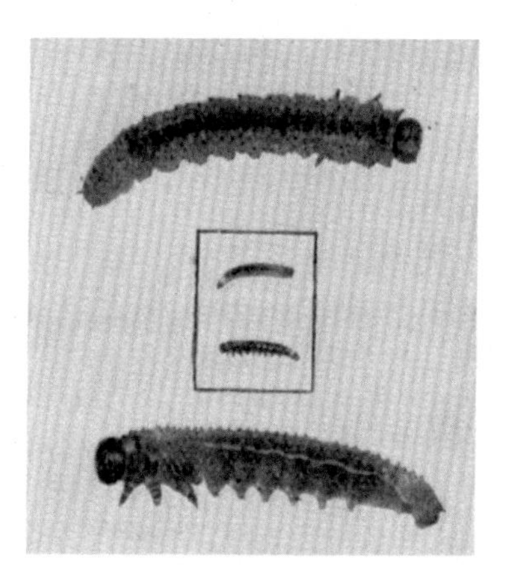

Fig. 141. Gooseberry and Currant Sawfly, enlarged. Inset, natural size. Showing the six prolegs.

Family *Muscidæ.*

An enormous family, all more or less resembling the house-fly. The *Tachina flies* are beneficial parasites of insects. Other groups are destructive to vegetable crops.

Family *Estridæ.*

This includes the *Bot flies* the larvæ of which are parasitic on animals.

Sub-order 4. PUPIPARA.

This includes many parasites as *Sheep ticks,* etc.

Sub-order 5. APHANIPTERA.

Wingless flies, degraded by parasitism, includes the fleas.

# CHAPTER 17

## Sawflies and Midges

## APPLE SAWFLY

Name *Hoplocampa testudinea* Order *Hymenoptera* (*4-winged flies*)

### 1. General

**Description**

**Larva (Maggot)**

| | |
|---|---|
| APPEARANCE. | With *six* middle pairs of sucker feet. |
| COLOUR. | Cream coloured with brown head and final segment. |
| LOCATION. | (1) Inside the young fruits. (2) In the ground in cocoons (throughout winter). |
| APPEARS IN | May onwards. |
| DURATION. | Throughout winter. |
| REMARKS. | Distinguished from the caterpillar of the *codling moth* by its two extra pairs of sucker feet, and its dull appearance. |

**Pupa (Chrysalis)**

| | |
|---|---|
| APPEARANCE. | In cocoon, coated with soil. |
| COLOUR. | Cocoon yellowish. |
| LOCATION. | In ground, a few inches deep. |
| APPEARS IN | March. |
| DURATION. | 2 to 3 weeks. |

**Adult Insect (Sawfly)**

| | |
|---|---|
| SIZE. | $\frac{1}{4}$ to $\frac{1}{3}$ inch. |
| COLOUR (WINGS). | Transparent, with brownish rims. |
| „ (BODY). | Reddish yellow: head black. |
| APPEARS IN | End of April to middle of May. |
| REMARKS. | Female lays eggs in the flowers on sunny days. Two broods may occur in the year. |

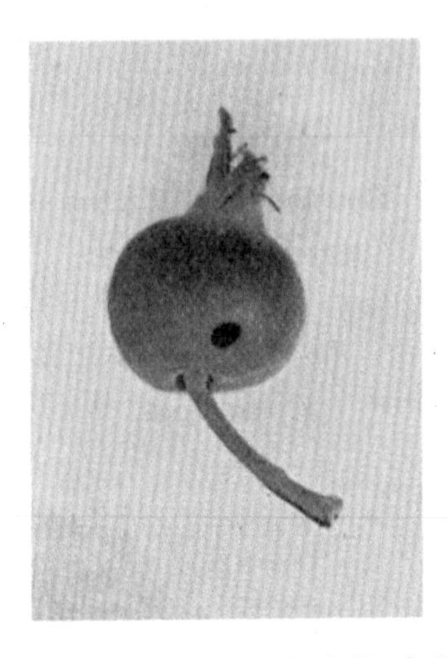

Fig. 142. Fruitlet infested by Apple Sawfly Maggot. Natural size.

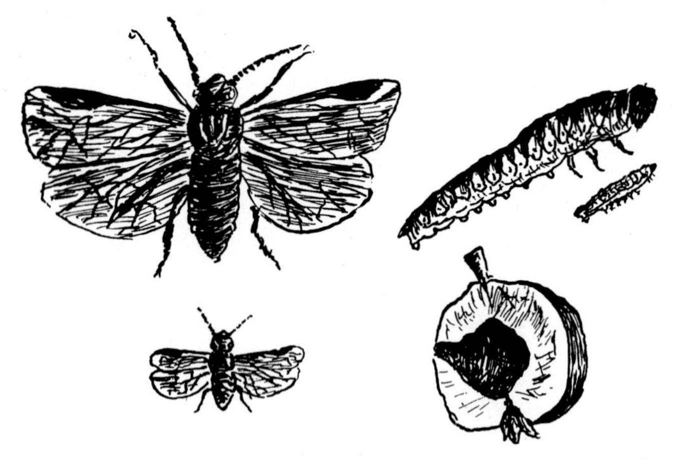

Fig. 143. Diagram of larva (maggot) and adult of Apple Sawfly, showing damage. Natural size and enlarged.

**Ova (Eggs)**

| | |
|---|---|
| ARRANGEMENT. | Singly (as a rule). |
| LOCATION. | At base of flower. |
| APPEAR IN | May. |
| HATCHING PERIOD. | 1 to 2 weeks. |

**Distribution**

Widespread.

**Life-history**

The flies appear from the ground in late April or May, and lay their eggs, generally in the forenoon of sunny days, on the flowers. The young maggots enter the fruitlets and eat out large holes, causing "frass" to exude from the hole on the side of the apple.

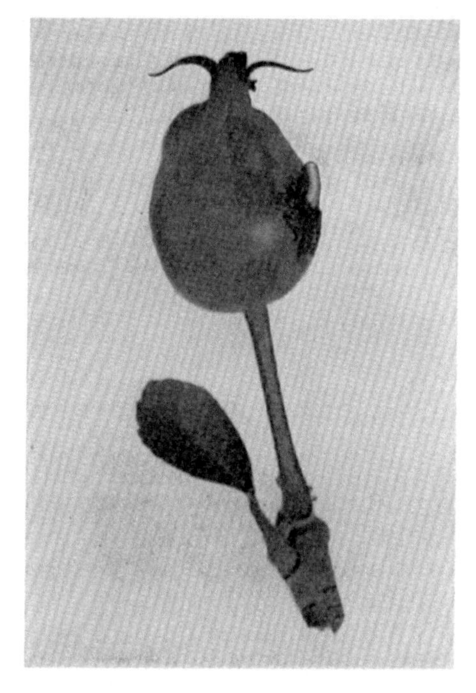

Fig. 144. Maggot of Apple Sawfly entering young apple. Natural size.

Maggots often leave one fruit and enter another, and thus one grub may be responsible for damage to several apples.

The maggots are mature in 4 to 5 weeks, and then fall to the ground (if the fruit has not already fallen) and penetrate a few inches into the soil where the winter is passed, pupation taking place in the spring.

## 2. Economic

### Trees Attacked

Apples.

### Frequency of Pest

Common in certain localities.

### Nature of Attack

The maggots eat out large holes in the fruit, and there is always a hole on the side of the apple, out of which much "frass" appears. A black cavity is formed, in the interior. Apples are occasionally scarred on the surface, when the maggot has failed to enter the apple.

Fig. 145. Fruitlet cut open to show damage caused by maggot of Apple Sawfly. Natural size.

## Degree of Damage

Often serious.

## Preventive Measures

1. The use of SOIL INSECTICIDES[1] in winter, under the trees.
2. Removal of the surface soil for a few inches deep. This is only possible in small plantations and gardens. The soil should be buried deep or baked.

## Remedies

Hand pick the infested fruits as soon as these are seen (they can readily be recognised by the holes on the side of the fruit exuding moisture and "frass") is the only remedy that can be used. This will prevent other fruits being attacked. Spraying with arsenate of lead has not proved successful.

## Calendar of Treatment

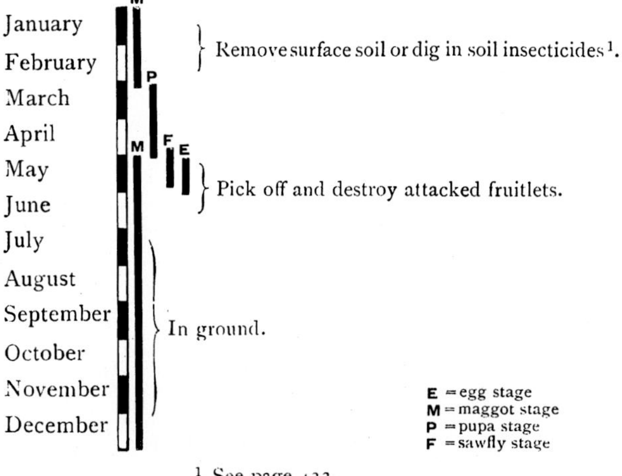


[1] See page 433.

## GOOSEBERRY AND CURRANT SAWFLY

Name *Nematus ribesii* Order *Hymenoptera*
(*4-winged flies*)

### 1. General

### Description

#### Larva (Maggot)

| | |
|---|---|
| SIZE AND APPEARANCE. | Naked, with numerous black pimples. |
| COLOUR. | Green to bluish-green—growing darker with age. |
| LOCATION. | On leaves. |
| APPEARS IN | (First brood) May or earlier: other broods throughout summer. |
| DURATION. | 5 to 6 weeks. |
| REMARKS. | The last brood winters in the soil and pupates in the following spring. |

Fig. 146. Maggot (larva) of Gooseberry and Currant Sawfly. See also fig. 141, page 235.

**Pupa (Chrysalis)**

| | |
|---|---|
| APPEARANCE. | In cocoons covered with particles of earth. |
| COLOUR. | Cocoons: yellow to brown. |
| LOCATION. | In the ground (except some of the summer broods). |
| DURATION. | Few weeks in early spring. |

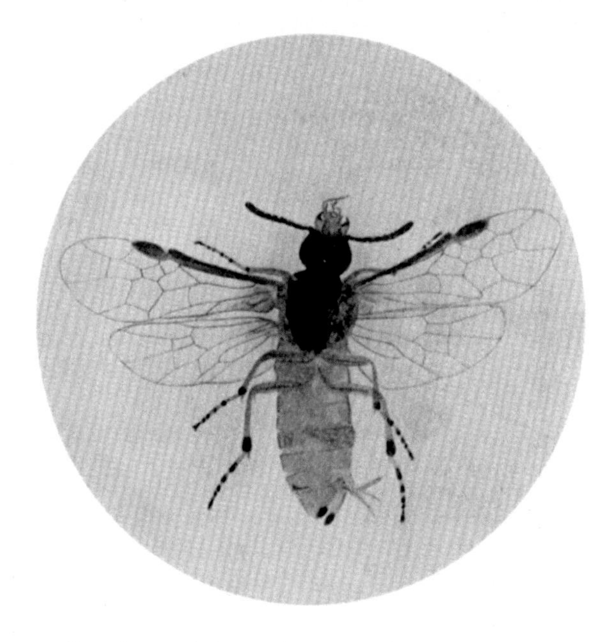

Fig. 147. Gooseberry and Currant Sawfly, showing various parts. Magnified.

**Adult Insect (Sawfly)**

| | |
|---|---|
| SIZE. | About $\frac{1}{3}$ inch long. |
| COLOUR (WINGS). | Two pairs, transparent. |
| ,, (BODY). | Variable { Female—yellowish. Male—black, with yellow band on thorax, and sides yellow. |
| TIME OF APPEARANCE. | April or May, and later (three broods during the summer). |

**Ova (Eggs)**

| | |
|---|---|
| APPEARANCE. | Oval. |
| COLOUR. | Pale greenish-white. |
| ARRANGEMENT. | In rows along ribs of leaf. |
| LOCATION. | Under surface of leaves. |
| HATCHING PERIOD. | 5 to 12 days. |
| REMARKS. | Each egg is placed in a slit made by the "saw" of the fly. |

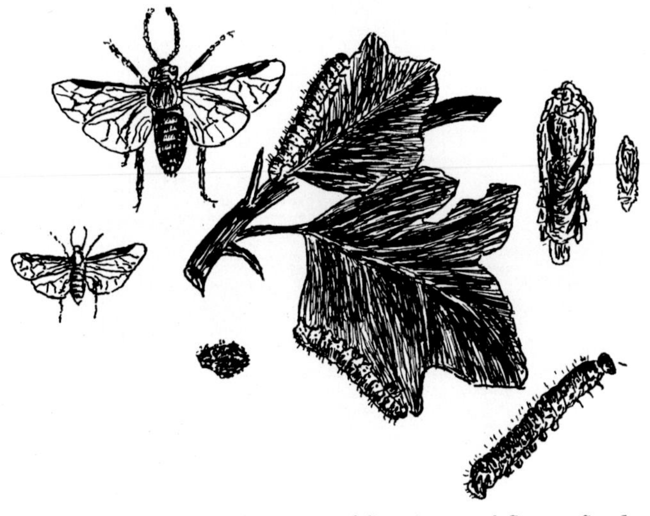

Fig. 148. Diagram of various stages of Gooseberry and Currant Sawfly. Natural size and enlarged.

## Distribution

Occurs all over Britain.

## Life-history

The sawflies appear in April or May and lay their eggs on the under surface of the leaves, in rows on the ribs, each egg being placed in a slit made by the "saw" of the insect for this purpose. The maggots feed on the outer skin of the leaf first, then make holes through and finally spread over the plant, and voraciously devour the entire leaves. When mature, the maggots fall to the ground, form a cocoon just under the earth, pupate, and produce the next brood. This occurs all through the summer, and as many as four broods may be produced. The mature maggots from the last brood winter in their cocoons in the ground and pupate in the following spring.

## 2. Economic

### Trees Attacked

Gooseberry and red currant, occasionally black currants.

### Frequency of Pest

Common.

### Symptoms of Attack

Holes appear in the leaves of the plant, due to small, very active, caterpillar-like maggots, distinguished from caterpillars by their smaller size, pimply naked bodies and by having more "sucker legs" (14 in all).

### Nature of Attack

The leaves are eaten and usually stripped by the maggots, and the gooseberries are also devoured.

### Degree of Damage

Serious.

### Preventive Measures

1. Where it is possible or practicable, e.g. in small gardens, the surface soil may be removed to a depth of 5 to 6 inches, and either baked and replaced or buried deeply and replaced by fresh soil from some other part. The hibernating maggot is thus destroyed, or prevented from reaching the surface.
2. The use of SOIL INSECTICIDES[1] is of benefit, if these are carefully applied, but is not so effective as the above treatment.

### Natural Enemies

Many have been recorded, but few appear to be of much service in coping with attacks.

### Remedies

1. HELLEBORE[2] powder either dusted on dry or applied as described on page 422 is most suitable and may be used within a few days of gathering the fruit. PYRETHRUM[3] is also suitable.

[1] See page 433. [2] See page 421. [3] See page 444.

2. Spraying with ARSENATE OF LEAD[1] as soon as the sawfly grubs appear is a remedy. Owing however to its poisonous nature, it must not be used within *a full month* of the time when the fruit is picked.
3. In small plantations, hand picking of the young maggots (or destroying by crushing on the leaves) is very serviceable if these are taken sufficiently early, i.e. before they spread over the plant. At this stage they make small shot holes in the leaves, and these should be carefully watched for the first signs of trouble.

**Calendar of Treatment**

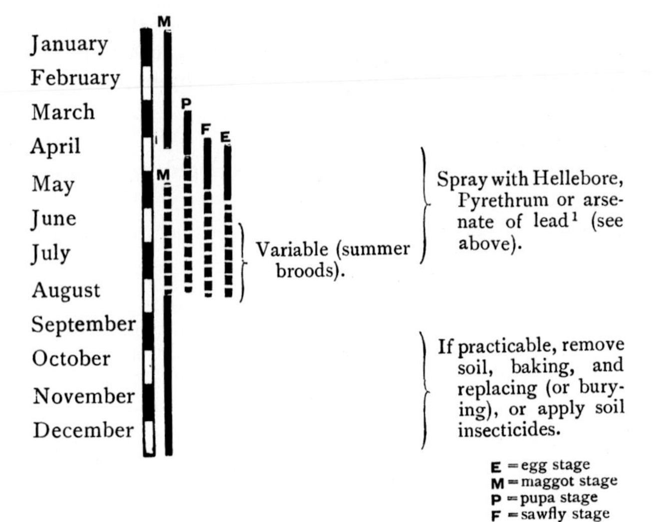


[1] See page 397.

## SLUG WORM ("Snag")

Name *Eriocampa limacina* Order *Hymenoptera* (4-*winged flies*)

### 1. General

**Description**

**Larva (Maggot)**

| | |
|---|---|
| APPEARANCE. | Slimy and shiny (repulsive), later dull and wrinkled. Wide in front, tapering towards tail end. |
| COLOUR. | Green-yellow below. |
| LOCATION. | On upper surface of leaves. |
| APPEARS IN | June. |
| DURATION. | Throughout the winter in cocoons in soil. |

**Pupa (Chrysalis)**

| | |
|---|---|
| APPEARANCE. | In cocoons formed by grubs. |
| LOCATION. | In soil. |
| APPEARS IN | April—May. |

**Adult Insect (Sawfly)**

| | |
|---|---|
| SIZE. | $\frac{1}{4}$ to $\frac{1}{3}$ inch. |
| COLOUR (WINGS). | Transparent. |
| „ (BODY). | Shiny black. |
| APPEARS IN | Early June. |

**Ova (Eggs)**

| | |
|---|---|
| APPEARANCE. | Oval. |
| COLOUR. | White. |
| ARRANGEMENT. | In slits cut by the "saws." |
| LOCATION. | Under surface of leaves. |
| APPEAR IN | June. |
| HATCH IN | About a week. |

**Distribution**

Widespread.

**Life-history**

The sawfly appears in small numbers in early summer. The eggs produce very repulsive-looking maggots, slimy and shiny in appearance. These produce blotchy spots on the leaves, feeding on the upper surface. In the final stage the grub becomes wrinkled and loses its slimy appearance. Winter is passed in the soil in cocoons and pupation takes place in spring.

Fig. 149. Maggot of Slug Worm and cast skin.
Natural size.

## 2. Economic

**Trees Attacked**

Cherry, pear, plum.

**Susceptible Varieties**

Morello, frequent among cherries.

**Frequency of Pest**

Common.

**Nature of Attack**

The leaves are attacked: these are eaten from the *upper* surface and the *under skin* is left intact, producing blotchy spots on the leaves.

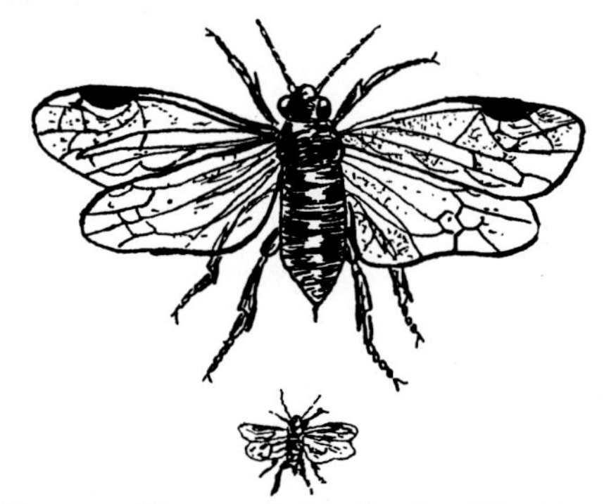

Fig. 150. Diagram of adult Sawfly of Slug Worm. Natural size and enlarged.

**Degree of Damage**

Often serious, since the tree is often induced to produce late leaf-buds and so is greatly weakened.

**Preventive Measures**

SOIL INSECTICIDES hoed into the soil during winter will destroy the hibernating maggots.

**Remedies**

The best remedy is spraying the attacked trees with ARSENATE OF LEAD[1]. (Care should be taken to avoid spraying fruit within one month of picking.) HELLEBORE[2] or PYRETHRUM[3] may also be used and will not poison the fruit.

[1] See page 397. [2] See page 421. [3] See page 444.

**Calendar of Treatment**

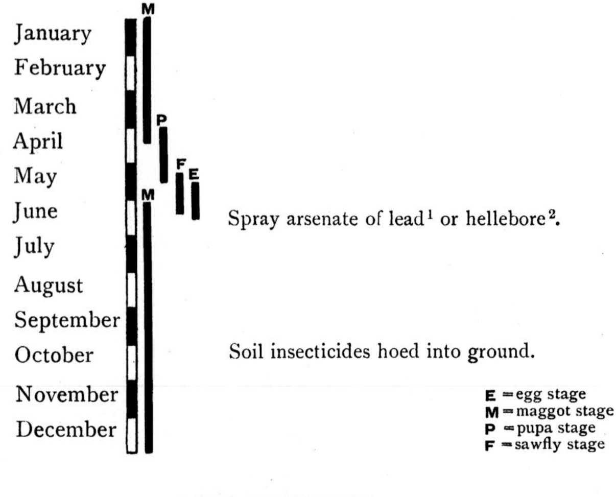


## NUT SAWFLY

Name *Crœsus septentrionalis* Order *Hymenoptera* (4-*winged flies*)

### 1. General

**Description**

**Larva (Maggot)**

COLOUR. Young—very pale: later greenish-blue: second and last segment, yellow, with black spots along sides.

LOCATION. Near edges of leaf in rows.

APPEARS IN July and August (second brood in September).

DURATION. The last brood pass the winter in the ground and pupate in the following spring.

REMARKS. The grubs maintain their hold on the leaf by their true legs (on first three segments of the body) and twist the rest of the body in all positions.

[1] See page 397. [2] See page 421.

**Pupa (Chrysalis)**

| | |
|---|---|
| APPEARANCE. | In spindly cocoon. |
| COLOUR. | Brown. |
| LOCATION. | In the ground. |
| APPEARS IN | April and May (later brood July). |

Fig. 151. Maggots (larvæ) of Nut Sawfly at work. $\frac{2}{3}$ natural size.

**Adult Insect (Sawfly)**

| | |
|---|---|
| SIZE. | About $\frac{1}{2}$ inch long. |
| COLOUR (WINGS). | Transparent. |
| „ (BODY). | Glossy black. |
| APPEARS IN | End of May—June. |
| REMARKS. | The hind pair of legs, unusually long, and the segments flattened. |

**Ova (Eggs)**

| | |
|---|---|
| LOCATION. | Placed in slits on the leaves made by the "saws." |
| APPEAR IN | June. |

### Distribution

Chiefly in South and West of England and widespread in Europe.

## 2. Economic

### Trees Attacked

Nuts (cob, filbert, hazel), occasionally gooseberry.

### Frequency of Pest

Fairly common.

### Nature of Attack

The leaves are ravenously devoured.

### Degree of Damage

Variable: occasionally serious.

### Natural Enemies

Many parasites attack this sawfly and consequently it does not usually appear in large numbers in two successive years.

### Remedy

Spraying with ARSENATE OF LEAD[1].

### Calendar of Treatment

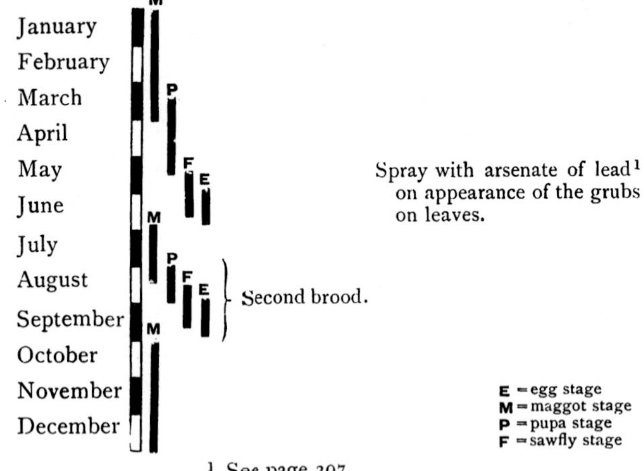


[1] See page 397.

## PLUM FRUIT SAWFLY

Name *Hoplocampa fulvicornis* Order *Hymenoptera* (4-*winged flies*)

### 1. General

#### Description

##### Larva (Maggot)

| | |
|---|---|
| APPEARANCE. | Curved: about $\frac{1}{3}$ inch long. |
| COLOUR. | Cream, with pink tinge, with brown head. |
| LOCATION. | Inside fruitlet. |
| APPEARS IN | May and June. |
| DURATION OF STATE. | Throughout winter, in cocoons, in soil. |
| REMARKS. | A single grub may attack several fruits. The grub has six pairs of "sucker" legs and a tail pair and is thus distinguished from a true caterpillar. |

Fig. 152. Plum cut open to show larva of Plum Fruit Sawfly: also the maggot enlarged (× 3).

##### Pupa (Chrysalis)

| | |
|---|---|
| APPEARANCE. | In the cocoons spun by the grubs. |
| LOCATION. | In soil. |
| APPEARS IN | February. |

**Adult Insect (Sawfly)**

| | |
|---|---|
| SIZE ACROSS WINGS. | About $\frac{1}{3}$ inch. |
| COLOUR (WINGS). | Transparent. |
| „ (BODY). | Black with yellow to brown legs. |
| APPEARS IN | April and May. |
| REMARKS. | Only one brood has been noticed in the year. |

**Ova (Eggs)**

| | |
|---|---|
| APPEARANCE. | Small and transparent. |
| COLOUR. | Greenish. |
| ARRANGEMENT. | Singly. |
| LOCATION. | In the blossom buds. |
| APPEAR IN | April and May. |
| HATCH IN | 7 to 14 days. |

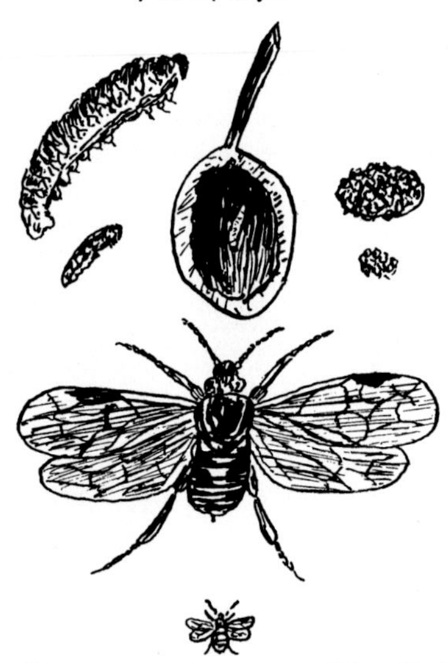

Fig. 153. Diagram of various stages of Plum Fruit Sawfly. Natural size and enlarged.

**Life-history**

The egg is laid in the fruit bud, inside a deep saw-cut. The maggot hatches out and feeds in the developing fruitlet. It is paler than the *plum fruit moth caterpillar* (see page 155), and has two pairs more of sucker feet. Plums are often quite hollowed out by the maggot, which may attack several fruits in turn.

When mature, the maggot enters the ground, spins a brown cocoon and winters in this, pupating in the following February.

Only one brood apparently occurs.

## 2. Economic

### Trees Attacked

Plums, chiefly greengages, but others are attacked.

### Frequency of Pest

Not uncommon.

### Symptoms of Attack

Fruitlets remaining undeveloped, falling, or (later) appearing with small holes made by the grub on leaving the fruit.

### Nature of Attack

The fruit is injured and often almost destroyed by the maggot. Attacked plums will not keep.

### Degree of Damage

Serious.

### Preventive Measures

Destroy all fallen fruit and gather all attacked plums and burn them where possible.

### Remedy

Early spraying of the blossoms with ARSENATE OF LEAD[1] should poison the young maggots.

### Calendar of Treatment

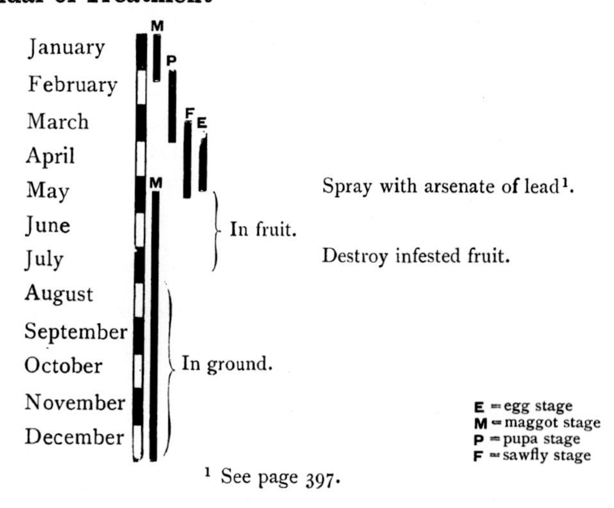


[1] See page 397.

## PLUM LEAF SAWFLY

Name *Cladius padi* Order *Hymenoptera* (*4-winged flies*)

### 1. General

**Description**

**Larva (Maggot)**

| | |
|---|---|
| COLOUR. | Greenish with white sides: head dark brown. |
| LOCATION. | On leaves. |
| APPEARS IN | May, July (sometimes also September). |
| DURATION. | About three weeks (except last brood which winter in the ground in silken cocoons). |

**Pupa (Chrysalis)**

| | |
|---|---|
| APPEARANCE. | In the cocoons of the grubs. |
| COLOUR. | Pale grey. |
| LOCATION. | In soil. |
| DURATION. | 9 to 12 days. |
| APPEARS IN | April, June (August). |

**Adult Insect (Sawfly)**

| | |
|---|---|
| COLOUR (WINGS). | Transparent. |
| „ (BODY). | Black, with two grey spots: legs white. |
| APPEARS IN | May to June and later (2 to 3 broods). |
| REMARKS. | In the female, the feelers (antennæ) are specially long. |

**Ova (Eggs)**

| | |
|---|---|
| ARRANGEMENT. | Irregular. |
| LOCATION. | Under surface of leaves. |
| APPEAR IN | May and June. |
| HATCH IN | 7 to 8 days. |

**Distribution**

Widely distributed.

## 2. Economic

### Trees Attacked

Plums, also roses, sloes, etc.

### Frequency of Pest

Common, but not often in dangerous numbers.

Fig. 154. Diagram of various stages of Plum Leaf Sawfly. Natural size and enlarged.

### Nature of Attack

The leaves are holed, and often shrivel up.

### Remedies

1. Spray with arsenate of lead[1].
2. Shake off the maggots (which readily fall) and prevent re-ascent by banding the trees (or collect and destroy).

[1] See page 397.

## SOCIAL PEAR SAWFLY

Name *Pamphilus flaviventris* Order *Hymenoptera* (*4-winged flies*)

### 1. General

**Description**

**Larva (Maggot)**

| | |
|---|---|
| APPEARANCE. | Without sucker feet, about 1 inch (mature). |
| COLOUR. | Yellowish: orange-yellow later with two brown stripes. |
| LOCATION. | Amongst leaves in *tents* of silk. |
| APPEARS IN | May and June. |
| DURATION. | Throughout winter in cocoons in soil. |
| REMARKS. | Distinguished from the *Lackey tents* (page 124) by greater width and dirty appearance, also the grub is very different from Lackey caterpillar. |

**Pupa (Chrysalis)**

| | |
|---|---|
| APPEARANCE. | In grub cocoons. |
| LOCATION. | In soil. |
| APPEARS IN | April. |
| REMARKS. | Are said to occasionally remain in the ground throughout summer till next season. |

**Adult Insect (Sawfly)**

| | |
|---|---|
| SIZE ACROSS WINGS | $\frac{1}{3}$ to $\frac{1}{2}$ inch. |
| COLOUR (WINGS). | Transparent. |
| " (BODY). | Black: legs yellow. |
| APPEARS IN | May and June. |

**Ova (Eggs)**

| | |
|---|---|
| APPEARANCE. | Oval and sticky. |
| COLOUR. | Yellowish. |
| ARRANGEMENT. | In rows: about 50 in all. |
| LOCATION. | On under surface of leaves. |
| APPEAR IN | May and June. |
| HATCH IN | About a week. |

Fig. 155. Social Pear Sawfly larvæ, showing also damage. Natural size.

**Distribution**

Widespread.

**Life-history**

The sawflies appear in May and June, and lay eggs under leaves on sunny days. The maggots spin webs and gradually increase the size of these until they may reach considerable proportions. They resemble the *Lackey* tents, but appear darker, and the maggots are easily distinguished from the striking Lackey caterpillars (see page 124). When frightened they expel a clear, brown fluid.

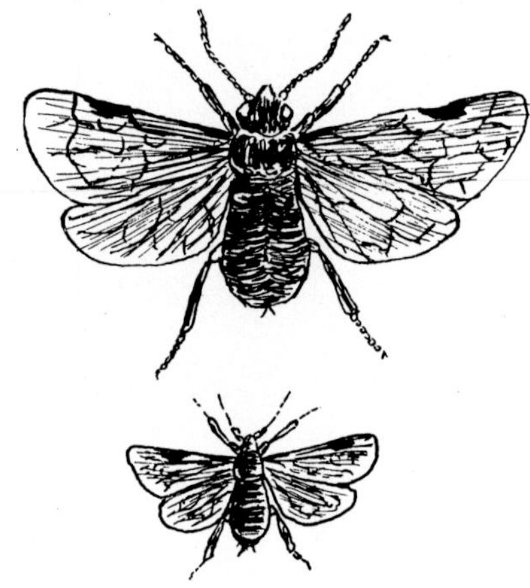

Fig. 156. Diagram of adult of Social Pear Sawfly. Natural size and enlarged.

They move with difficulty on a smooth surface, having no sucker feet. When mature, they fall to the ground and spin cocoons in which the winter is passed in the soil, pupating the following spring.

## 2. Economic

**Trees Attacked**

Pears.

**Frequency of Pest**

Fairly common.

**Nature of Attack**

The maggots devour the leaves inside the "webs" or "tents.'

Fig. 157. Eggs of Social Pear Sawfly hatching out on leaf. Enlarged (× 3). Inset natural size.

## Degree of Damage

Serious : trees may be totally stripped of leaves.

## Preventive Measures

SOIL INSECTICIDES[1] hoed into the ground around trees will destroy the hibernating caterpillars and prevent attack the following year.

## Remedies

As for the *Lackey caterpillar* (page 126).

1. The destruction of the tents is by far the best means of dealing with the pest.
   Since the maggots readily fall to the ground when disturbed, care must be taken that none escape in this way.
   Pails of hot lime are recommended for placing the tents in after cutting out from the trees.
2. If ARSENATE OF LEAD[2] is used, it must be sprayed on the trees before the larvæ have had time to construct tents.

## Calendar of Treatment

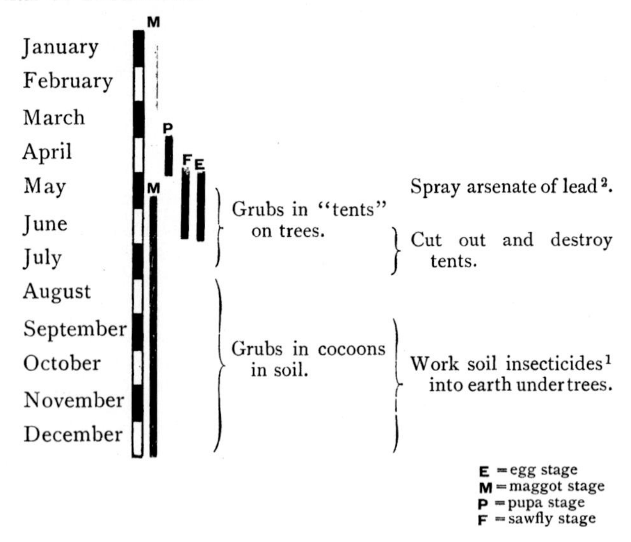


[1] See page 433. [2] See page 397.

## PEAR MIDGE

Name *Diplosis pyrivora* Class *Cecidomyrdæ* (*Gall flies*)
Order *Diptera* (*2-winged flies*)

### 1. General

**Description**

**Larva (Maggot)**

| | |
|---|---|
| APPEARANCE. | Small (not over $\frac{1}{7}$ inch), legless. |
| COLOUR. | Yellowish white: brown head. |
| LOCATION. | In young fruits. |
| PROGRESSION. | By "jumping." |
| APPEARS IN | April and May. |
| DURATION. | Usually till late autumn. |
| REMARKS. | The maggots leave the pears when mature in June and enter the ground, forming small cocoons. |

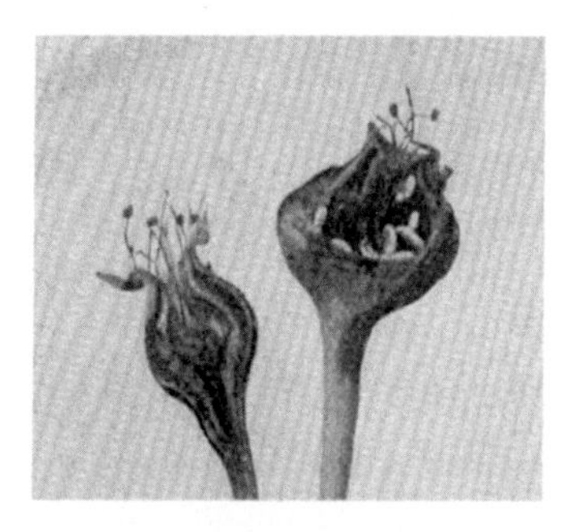

Fig. 158. Healthy and infested fruitlets cut open to show maggots and damage. Natural size.

**Pupa (Chrysalis)**

| | |
|---|---|
| APPEARANCE. | Inside the grub-cocoons. |
| LOCATION. | In the ground. |
| APPEARS IN | Late autumn or winter. |
| DURATION. | Till following spring. |

**Adult Insect (Fly)**

| | |
|---|---|
| SIZE. | $\frac{1}{10}$ to $\frac{1}{8}$ inch. |
| COLOUR (WINGS). | Two: transparent. |
| „ (BODY). | Dark grey to black: legs and feelers long and brownish. |
| APPEARS IN | Middle April to middle May. |
| REMARKS. | The female has very long egg-depositor. |

Fig. 159. Diagram of various stages of Pear Midge, showing also injury to fruitlet. Natural size and enlarged.

**Ova (Eggs)**

| | |
|---|---|
| APPEARANCE. | Long and pointed. |
| COLOUR. | White. |
| ARRANGEMENT. | 10 to 30 in a group. |
| LOCATION. | In blossoms (usually on the *anthers*) |
| APPEAR IN | April and May. |
| HATCHING PERIOD. | 4 to 6 days. |

**Distribution**

Widespread in Britain, Europe and the States.

**Life-history**

The flies appear in spring, and after fertilisation the female lays her eggs with her long egg-depositor in the flowers. These hatch into minute maggots which infest the young fruitlets, eating out the interior. By June they are mature and leave the decayed fruit, exhibiting the curious jumping movements characteristic of the midges. They penetrate the soil to a depth of 1 to 2 inches and, spinning a slender white cocoon, remain till autumn or winter when pupation takes place.

## 2. Economic

**Trees Attacked**

Pears.

**Susceptible Varieties**

*Comice* not often attacked.

**Frequency of Pest**

Common.

**Symptoms of Attack**

The attacked fruitlets swell rapidly and appear to be growing at a faster rate than normal: later they become mis-shapen and assume many curious forms.

**Nature of Attack**

The fruitlets are eaten out in the interior by the grubs, and much black "frass" is produced. The pears often fall to the ground, but occasionally remain on the trees. They crack and rapidly become rotten.

**Degree of Damage**

Very serious. In many places the cultivation of pears has been abandoned from this cause.

**Preventive Measures**

1. In gardens, and where the facilities are forthcoming, removal of the soil to a depth of 3 to 4 inches, and burying this or baking it in the autumn or winter kills the maggots and pupæ, and prevents attack.
2. KAINIT[1] laid upon the soil at the rate of about 5 cwt. per acre,

[1] See page 426.

when the maggots were entering, has been found very beneficial. This should be applied by the first week in June to be effectual.

3. Run poultry in the orchards from June till August and again in spring.
4. Spraying the unopened blossom with SOFT SOAP[1] and NICOTINE[2] wash has been found to prevent egg-laying.
5. Wherever possible, all infested fruitlets should be picked and burnt as soon as they are detected. The maggots are thus destroyed.

Fig. 160. One infested and several healthy pear fruitlets. Natural size.

Where the midge has established a hold, it is a wise, if somewhat drastic, plan to kill the blossoms and so lose the fruit for a season. By this means the midge is starved out. This is best done by spraying heavily with PARIS GREEN[3] just as the blossoms burst. This plan is however of little avail if growers in surrounding plots do not take the same measures, and is an instance of the necessity for combined action, or for compulsory spraying, discussed in a previous chapter[4].

[1] See page 452. [2] See page 436. [3] See page 442. [4] See page 7.

**Calendar of Treatment**

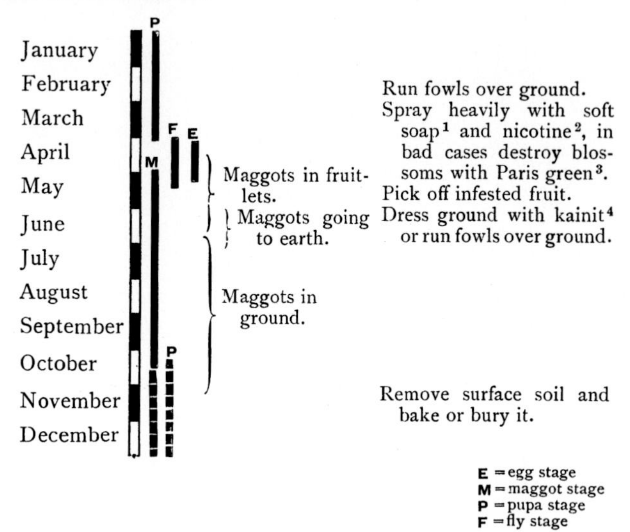


[1] See page 452. [2] See page 436.
[3] See page 442. [4] See page 426.

# APHIDES

AND

# PLANT BUGS

(*HEMIPTERA*)

# CHAPTER 18

## Aphides and Plant Bugs

(HEMIPTERA)

### 1. General

All fruit trees and plants have one or more species of these small, but very injurious, pests. They all agree in living upon the sap of the trees which is obtained by means of a jointed beak. This is inserted into the tissues of the leaf, etc.

The aphides ("green flies") vary a good deal in colour but are otherwise, in many cases, very similar.

They practically all secrete a sticky fluid, termed *Honey-dew*, on which often grows a black fungus which greatly disfigures the leaves. They often cluster on the leaves and stems so closely as to form a continuous covering.

Five forms occur throughout though not always on the same host plant, thus:

(*a*) STEM-MOTHER: large wingless aphis, produced from the egg, or wintering in sheltered spots.

(*b*) WINGLESS FEMALE (unsexed[1]) or *louse*, the progeny of (*a*).

(*c*) WINGED FEMALE (unsexed) often migratory to other plants.

(*d*) WINGLESS EGG-LAYING FEMALE (sexual) laying the eggs after fertilisation by the

(*e*) MALE (either with or without wings).

In addition to these, so-called pupæ (more correctly *nymphs*) are produced with wing-buds, before the flies appear.

The unsexed winged forms of the insect are produced in the summer, begetting living young, and these continuously produce living broods of "lice" all the summer. These, in the autumn, give rise to a second

[1] The word **unsexed** is used to denote the form which, although capable of bearing young, has no sexual organs for reception of the male. (The scientific term is *viviparous* and for the sexual egg-laying female the word *ovigerous* is employed.)

brood of unsexed winged females. Males and females are now produced and eggs laid which remain over the winter and hatch in the spring, giving rise to the *stem mothers* or *mother queens*.

In some cases, only the winged forms and the "lice" occur on fruit: in others all forms occur, while quite frequently, the winged females fly away to some other host plant in the summer and similar winged females return in the autumn. These give rise to sexual forms and eggs are laid upon the fruit plants.

Aphides are capable, under suitable conditions, of extremely rapid increase. Conversely, in unsuitable conditions, they may all rapidly disappear.

Fig. 161. Hop aphides. Enlarged and natural size (inset).

Aphides usually attack the leaves mainly. A few also infest the flower buds, and one species, the *Woolly Aphis*, attacks the trunk, branches and roots.

Many aphides are covered with a fine "meal" or so-called *wax*. This resists sprays of all kinds, and is difficult to penetrate. In order to kill aphides it is necessary to employ a *contact poison* such as NICOTINE[1], combined with a wetting substance, of which SOFT SOAP[2]

[1] See page 436. 5 ozs. per 100 gallons is a suitable quantity to use.
[2] See page 452.

Fig. 162. Two varieties of aphis on an apple tree. Woolly Aphis (below), Blue Apple Aphis (above) in curled leaves.

is the most successful (see chapter 25) and to use a high pressure spray, rather coarse—see fig. 250, page 634.

Many claims have been made for various proprietary sprays to destroy the eggs of the aphis and the apple sucker during the winter. Careful investigation has shown however that this is not feasible, the limit of resistance being practically equal to that of the tree itself.

The APPLE SUCKER (Psylla) infests the flower buds and is a very troublesome pest. Blooms attacked invariably die. The treatment is as for aphis attacks.

In addition to a *nicotine and soap* spray, benefit has been found from spraying in the spring with LIME[1] which hinders the hatching of the eggs. This treatment may be given RIGHT UP TO THE APPEARANCE OF THE BLOSSOM TRUSSES without injury (see figs. 244, 245, pages 617, 618).

The CAPSID BUG is an extremely destructive pest on apples in the Cambridge district and in the western counties. It is a curious instance of change of food plants by an insect. Until recent years, the apples were not attacked, the bug feeding upon willows.

Soft soap and nicotine wash (the latter in higher strength than for aphis) is the best remedy.

The Scale Insect belongs to this order of insects, but as it offers an entirely different problem from the grower's point of view, it is separately dealt with (chapter 21, page 355).

None of these insects have a definite *pupa* stage. After a certain number of moults, wing-buds appear, but the insect in this stage is quite active. It is termed a *nymph*. After a few more moults, the winged fly appears.

## 2. Scientific

Aphides are all of a similar form. They possess a sucking jointed beak, prominent eyes and antennæ (feelers), three pairs of jointed legs, and curious "horns" termed *cornicles*, one on each side of the body at the hinder end.

They may be light to dark green, brown, purple or black in colour, naked and transparent or opaque and "mealy." The Woolly Aphis secretes a tangled mass of white threads which cover its body and hang down in masses on the branches.

[1] See page 428.

*Classification*

## Order **Hemiptera**

Sub-order HETEROPTERA (with dissimilar wings).

The fore-wings are tough and usually opaque and fold over the under membranous pair.

Capsid-bug, *Plesiocoris rugicollis.*

Sub-order HOMOPTERA (with similar wings).

Two pairs of membranous wings. Usually erected over the back when at rest.

Family *Aphidæ.*

This includes all the aphides attacking fruit as described in this chapter.

Family *Psyllidæ.*

The chief pest on fruit is the Apple Sucker, *Psylla mali.*

Family *Circidæ.*

Leaf-hoppers, species of *Typhocybidæ*, *Chlorita*, etc.

Family *Coccidæ.*

These are the "scale insects." The chief pests are:

Mussel Scale, *Lepidosaphes ulmi.*

Brown Scale, *Lecanium persicæ.*

# CHAPTER 19

## Aphides

## BLUE APPLE APHIS

Name *Aphis malifoliæ*[1] Family *Aphidæ* Order *Hemiptera*

### 1. General

### Description

#### Adult Insect

All five forms are known as follows:

(*a*) STEM-MOTHER (from the egg) produces living young.
Colour—usually blue, or grey-blue, but very variable, may be brownish, reddish, or almost black: mealy.

(*b*) WINGLESS FEMALE (unsexed), as for (*a*).

(*c*) WINGED FEMALE (unsexed) produced by (*b*) in summer (brood 1). These fly to some other host plant at present not located, and return (brood 2) in autumn to the apple.
Colour—thorax and head black, abdomen dull reddish, feelers and horns black.

(*d*) WINGLESS EGG-LAYING FEMALE (sexual), from (*c* 2), lays eggs on the apple after fertilisation by (*e*).
Appearance—small (½ size of (*a*)).
Colour—yellow to greenish-yellow.

(*e*) MALE (winged).
Appearance—½ size of (*c*). Thorax dark, abdomen dull red with dark patches.

#### Ova (Eggs)

| | |
|---|---|
| APPEARANCE. | Oval. |
| COLOUR. | Shiny black. |
| ARRANGEMENT. | Singly or small groups. |
| LOCATION. | Shoots, angles of buds, or trunk. |
| APPEAR IN | Late autumn. |

[1] Fitch.

**Larva (Louse)**

| | |
|---|---|
| COLOUR. | Very variable: green, pink, reddish, becoming bluish later, and mealy. |
| LOCATION. | On or in buds or leaves. Leaves tightly curled. |

**Distribution**

Widely distributed in Britain and abroad.

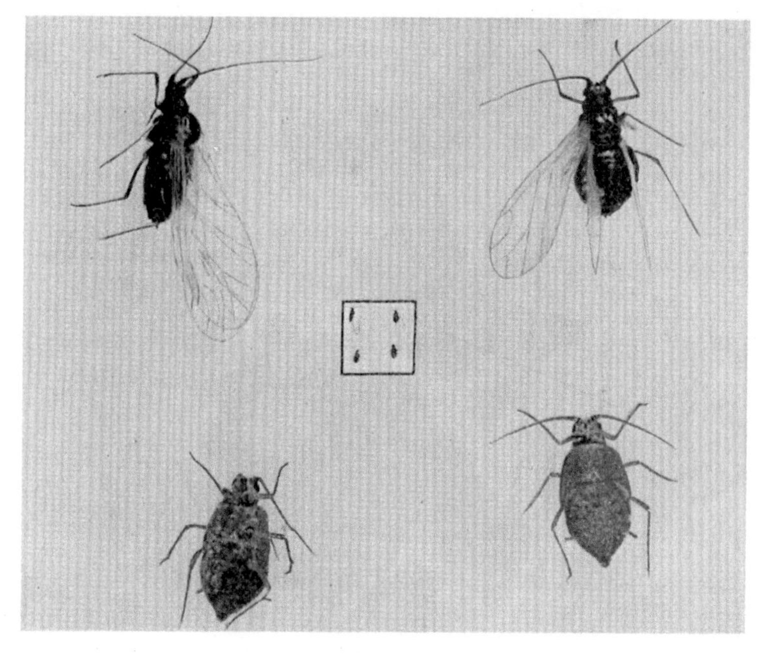

Fig. 163. Blue Apple Aphis. Magnified. Inset natural size.

**Life-history**

This "blue bug" breeds by means of eggs laid upon the trees by the sexual females in the autumn. These hatch out in April and the young lice gather upon the buds and enter them as they open. Later they feed upon the leaves, and cause these to curl very tightly. Flies are produced in the summer which leave for some

Fig. 164. Blue Apple Aphis, showing curled leaves and cast skins. ($\frac{3}{4}$ natural size.)

other plant (at present undetected) and similar flies return to the apples in autumn and produce sexual females and males when the eggs are laid.

## 2. Economic

### Trees Attacked

Apples, and sometimes pears.

### Frequency of Pest

Very common.

### Nature of Attack

The leaves and shoots are attacked, and the former become tightly curled. Much honey-dew is produced and this appears to poison the leaves which often turn brown and fall off. The shoots and often the young fruit become distorted and stunted in growth.

### Duration of Attack

Until mid-July when the aphides commence to disappear, being gone entirely usually by the end of the month.

### Degree of Damage

Usually very serious.

### Preventive Measures

1. Spraying in *autumn* with SOFT SOAP[1] and NICOTINE[2], or other efficient insecticide, kills the egg-laying females and prevents future attack. Carefully prepared PARAFFIN EMULSION may be used at this period.
2. LIME SPRAYING[3] in March has been found beneficial in preventing the successful hatching out of eggs. It has not been found injurious to spray, even as late as when the blossom trusses are appearing (see fig. 244, page 616).

### Natural Enemies

Ladybirds and other beneficial insects appear in summer, but too late to prevent damage.

### Remedies

Spray *early* in the spring[4] with SOFT SOAP[1] and NICOTINE[2] (no other insecticide is as efficient for this purpose) when as little

[1] See page 452. [2] See page 436. [3] See page 428.
[4] See fig. 193, page 341.

curling of the leaves has occurred as possible. 4 to 5 ozs. of nicotine per 100 gallons of soapy wash is required (see page 438).

### Remarks

In spraying, a somewhat *coarse spray* with a *high pump-pressure* (over 100 lbs. per square inch) should be used. The adjustment of nozzle shown on fig. 250, page 634 is suitable.

### Calendar of Treatment

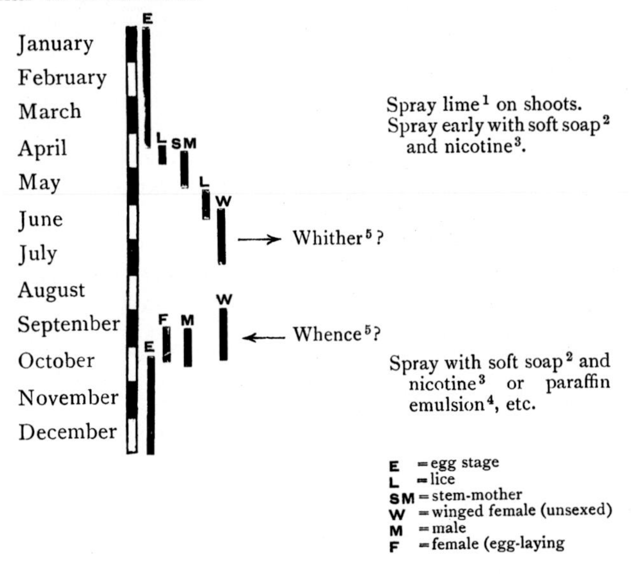


[1] See page 428. [2] See page 452. [3] See page 436. [4] See page 417.
[5] Professor Theobald has recently shown that one, at any rate, of the host plants to which the winged females fly is the *plantin* (rib-wort).

## THE APPLE-OAT APHIS

Name *Siphocoryne avenæ*[1] Family *Aphidæ* Order *Hemiptera*

### 1. General

**Description**

The egg stage is passed on the apple and pear, and the flies migrate to various cereals, especially oats and barley, in the summer, returning in the autumn. The usual series of adult forms appear:

(*a*) STEM-MOTHER (from egg), produces living young on the apple and pear.

Appearance—green to yellowish-green. Feelers green with dark tips. Horns greenish-brown to brown.

Appears—about end of April.

(*b*) WINGLESS FEMALE (unsexed), the adult product of (*a*), produces living young: appearance as for (*a*).

(*c*) WINGED FEMALE (unsexed), occurs in May and June (1) and in autumn (2), flies to, and returns from, various cereals (see above).

Appearance—head and thorax black, abdomen green, with black spots on sides, feelers black, horns brown to greenish-brown, short.

(2) smaller than (1).

(*d*) WINGLESS EGG-LAYING FEMALE (sexual), produced from offspring of (*c* 1) late in autumn: lays its eggs after fertilisation by (*e*).

Appearance—small, yellowish-green.

(*e*) MALE (winged), appears at same time as (*d*): similar in appearance to (*c* 2) but smaller.

**Ova (Eggs)**

| | |
|---|---|
| APPEARANCE. | Oval, shiny black. |
| LOCATION. | At junction of buds and on spurs. |
| APPEAR IN | Late autumn. |
| HATCH IN | Spring. |
| DURATION. | Throughout winter. |

**Distribution**

General in Europe.

[1] *Aphis prunifoliæ* (Fitch).

**Life-history**

This aphis does not differ in its history from the green apple aphis (page 283) except that it leaves the apples and pears in summer and flies to various cereals, to return in autumn.

Fig. 165. Apple-oat Aphis (from the apple). Magnified and natural size (inset).

## 2. Economic

**Trees Attacked**

Apples, pears.

**Frequency of Pest**

Not so common as the other two species (Green and Blue).

**Nature of Attack**

The blossoms, as well as the leaves and shoots, are attacked and these, in bad cases, turn brown and die.

**Remedies**

As for green apple aphis.

Fig. 166. Aphis (from corn), probably same as fig. 165 (?). Magnified and natural size (inset).

## GREEN APPLE APHIS

Name *Aphis pomi*[1] Family *Aphidæ* Order *Hemiptera*

### 1. General

**Description**

**Adult Insect**

Five forms occur as follows:

(*a*) STEM-MOTHER. This is the mature form from the eggs after hatching in spring. She produces living young continuously during the spring and summer.

Colour—greenish or greyish, mottled yellow: black "horns," feelers black at tip.

Shape—round.

Appears in—spring.

[1] De Geer.

(*b*) Wingless female (unsexed), the adult product of (*a*). Produces living young continuously.
Colour—green with two dark spots on head: "horns" long and dark.
Shape—oval.
Appears—throughout spring and summer months.

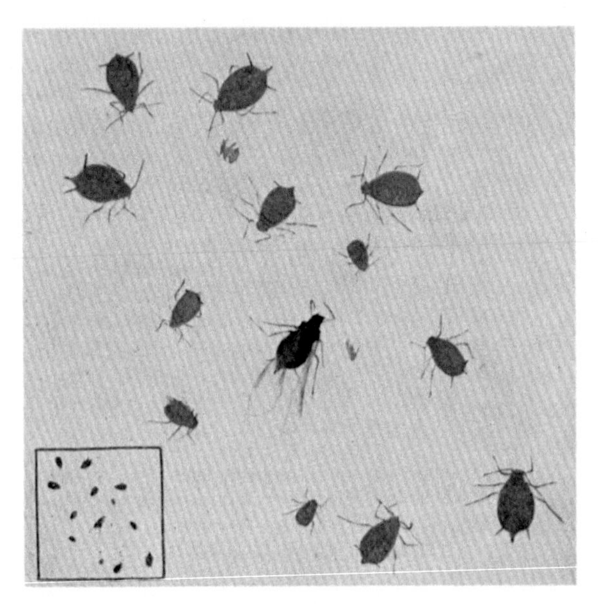

Fig. 167. Green Apple Aphis magnified. Inset natural size.

(*c*) Winged female (unsexed), produced by (*b*) from pupæ (nymphs) in certain numbers in July and August. Produces living young.
Colour—brownish-black thorax, green abdomen, wings transparent.
Shape—slender.
Appears in—July and August.

(*d*) Wingless egg-laying female (sexual), produced by offspring of (*c*) late in autumn: lays its eggs after fertilisation by (*e*).
Colour—green-yellow.
Shape—smaller than foregoing.
Location—firmly attached to underside of leaves.

(*e*) WINGLESS MALE, appears at same time as (*d*).
Colour—yellow-brown.
Shape—extremely small.

### Ova (Eggs)

| | |
|---|---|
| APPEARANCE. | Oval, small. |
| COLOUR. | Shiny black. |
| ARRANGEMENT. | In large masses. |
| LOCATION. | On shoots, or base of buds. |
| APPEAR IN | Usually November. |
| HATCHING PERIOD. | Late April. |
| DURATION. | About 5 to 6 months. |
| REMARKS. | Eggs which are not fertile gradually shrivel up. |

### Larva or "Louse"

| | (1) From eggs. | (2) From "stem-mothers." |
|---|---|---|
| COLOUR. | Bright yellowish-green. | Green with dark marks on head. |
| LOCATION. | On shoots or leaves. | On shoots or leaves. |
| APPEARS IN | Late April. | Throughout summer. |
| NUMBER OF MOULTS. | Three. | Three. |
| DURATION. | Varies according to weather. | Variable. |
| REMARKS. | Grows slowly in cold, rapidly in warm weather: in both cases the adult females are produced and some of (2) become winged females about July and August. | |

### Nymph (Pupa)

| | |
|---|---|
| APPEARANCE. | Covered with mealy powder: with wing-buds. |
| COLOUR. | Greenish-yellow. |
| LOCATION. | On leaves. |
| APPEARS IN | July and August. |
| NUMBER OF MOULTS. | One. |
| DURATION. | Few days (variable). |
| REMARKS. | These produce the winged females. |

**Distribution**

Widespread.

**Life-history**

The eggs hatch out in late April and produce the lice which grow into the *stem-mothers* (*a*). These produce living young producing adult *unsexed females* (*b*), some of the progeny of which turn to nymphs (pupæ) and produce the *winged unsexed females* (*c*). From these (which do not leave the trees) living young are produced, some of which give rise later to *wingless sexual females* (*d*) and *males* (*e*), both very small, especially the males. Eggs are laid in early winter which hatch out the following spring.

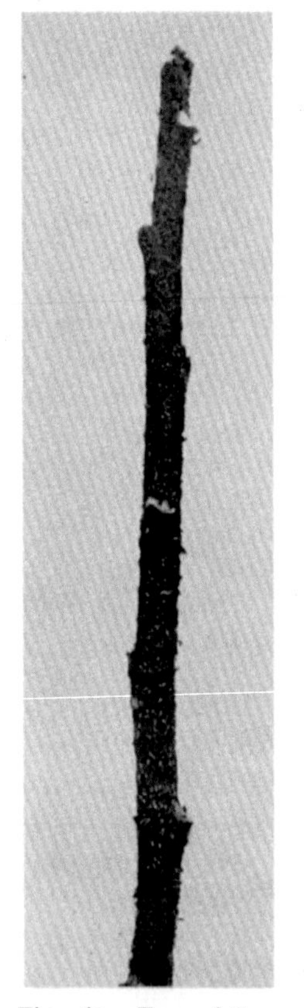

Fig. 168. Eggs of Green Apple Aphis on stem. Natural size.

### 2. Economic

**Trees Attacked**

Apples.

**Frequency of Pest**

Very common.

**Nature of Attack**

The young leaves, and later on the older leaves and shoots are attacked, but mostly the top parts of the branches. The leaves curl up very soon and enclose the aphis in a protected pocket. "Meal" and "honey-dew" is produced.

**Duration of Attack**

Throughout spring, summer, and autumn.

**Degree of Damage**

Sometimes serious, but much less so than the "blue aphis." The leaves are sometimes destroyed, and the shoots distorted and prevented from growing.

Fig. 169. Green Apple Aphis on apple shoots, slightly reduced. (Early summer.)

Fig. 169 *a*. Sexual females and males of Green Apple Aphis in autumn. Eggs are being deposited on the branch.

### Preventive Measures

1. Destroy prunings, as these will probably have many eggs on them.
2. Spraying in early spring with thick LIME WASH[1], so as to cover the shoots and prevent the eggs from successfully hatching, has had a considerable measure of success (see also page 274).

### Remarks

Many statements have appeared in the past as to the possibility of destroying aphis eggs by caustic and other winter washes. It has now been fairly definitely established that *no winter wash can be relied upon to kill the eggs*[2].

### Remedies

Spray with a SOAP WASH[3] containing an efficient insecticide (chapter 26) (nicotine[4] is very suitable) *in the autumn*. Reliable compound washes may also be used.

Spraying in spring is better than not spraying at all, but it is difficult to reach the aphides, on account of the tightly curled leaves. This would not matter if the trees were sprayed just as the eggs hatched, before the leaf curling started, but unfortunately, the hatching period is usually a long one, and many sprayings would have to be given to be sure of killing all the aphides. If spray washing is used, a nicotine[4] spray is most suitable.

### Remarks on Remedies

A usual percentage of nicotine to employ for this aphis is 4 to 5 ozs. per 100 gallons of finished wash. Nozzle adjustment for spraying, as fig. 250, page 634. A high pump-pressure is necessary (100 lbs. per square inch upwards).

[1] See page 428.

[2] A proportion of eggs always shrivel up because they are unfertile, and this has mislead many observers.

[3] See page 452. [4] See page 436.

**Calendar of Treatment**

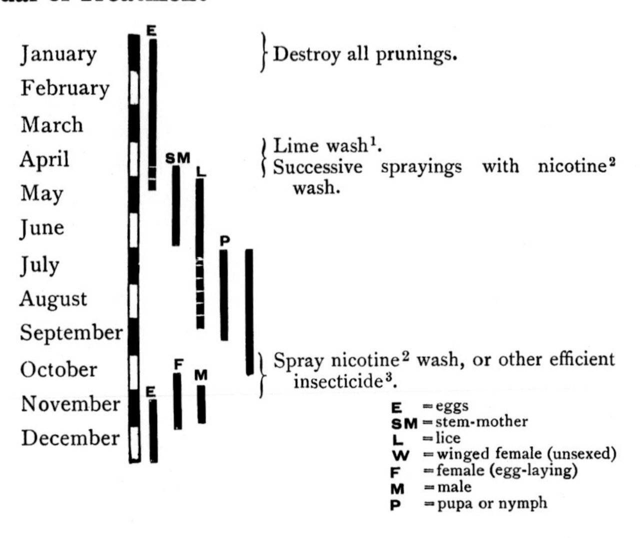


## BLACK CHERRY APHIS

Name *Myzus cerasi*[4] Family *Aphidæ* Order *Hemiptera*

### 1. General

**Description**

**Adult "Fly"** (winged unsexed female)

| | |
|---|---|
| SIZE. | About $\frac{1}{12}$ inch long. |
| COLOUR (WINGS). | Transparent. |
| ,, (BODY). | Thorax black, abdomen green, horns black. |
| APPEARS IN | July and autumn. |

[1] See page 428. [2] See page 436.
[3] See page 390. [4] Fabricius.

| | |
|---|---|
| REMARKS. | All the usual forms of adult, (*a*) to (*e*), occur, the winged unsexed females disappearing in summer and returning in autumn like the plum leaf-curling aphis (see page 315). The flies leave the cherries for some other host-plant and return in the autumn, when the male and egg-laying females appear. |

**Ova (Eggs)**

| | |
|---|---|
| COLOUR. | Shiny black. |
| LOCATION. | On shoots (or suckers near the ground). |
| APPEAR IN | September to November. |
| HATCHING PERIOD. | Spring. |
| DURATION. | Throughout winter. |

**Larva or "Louse"**

| | |
|---|---|
| COLOUR. | Shiny black (pale when young). |
| LOCATION. | Leaves and shoots. |
| APPEARS IN | Spring. |
| DURATION. | Till late July. |

**Pupa or Nymph**

| | |
|---|---|
| COLOUR. | Olive green. |
| LOCATION. | Leaves. |
| APPEARS IN | July. |

**Distribution**

Widespread.

**Life-history**

The "cherry black fly" hatches from eggs in spring upon the cherry, and the lice breed throughout the spring and early summer. Flies appear in July and vanish to some other plant returning in autumn and producing the wingless sexual females and winged males.

## 2. Economic

**Trees Attacked**

Cherries; also red and black currants.

**Frequency of Pest**

Common.

**Nature of Attack**

The aphis infests the shoots and leaves, and sometimes destroys these and kills the shoots. The leaves usually curl up, but not invariably. Much sticky honey-dew is produced which may ruin the fruit, as it forms a medium for the growth of black fungus which soon covers it.

**Degree of Damage**

Variable.

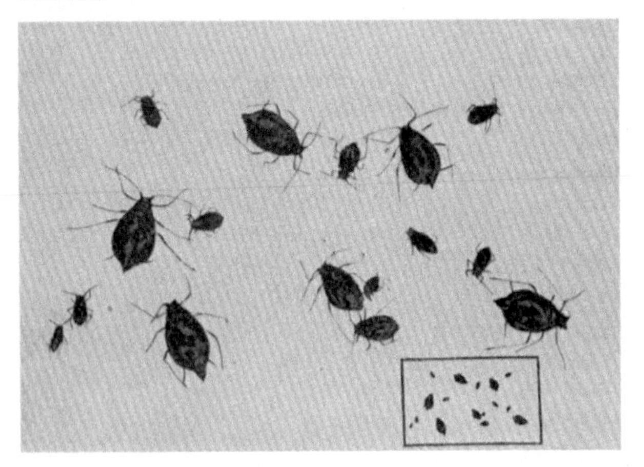

Fig. 170. Black Cherry Aphis, magnified. Inset natural size.

**Preventive Measures**

Autumnal spraying kills the sexual females and prevents egg-laying. NICOTINE[1] and SOFT SOAP[2] or PARAFFIN EMULSION[3] may be used. Thick LIME[4] wash applied in the spring prevents the successful hatching of the eggs (see page 274).

**Remedies**

Spraying with a good SOFT SOAP[2] and INSECTICIDE[5] wash (e.g. NICOTINE[1]) is a pretty certain cure. 4—5 ozs. of nicotine to 100 gallons is required and plenty of soft soap.

**Remarks**

A high pressure on the spraying pump (upwards of 100 lbs. per square inch) is necessary. Adjustment of nozzle as fig. 250 page 634.

[1] See page 436. [2] See page 452. [3] See page 417.
[4] See page 428. [5] See page 390.

**Calendar of Treatment**

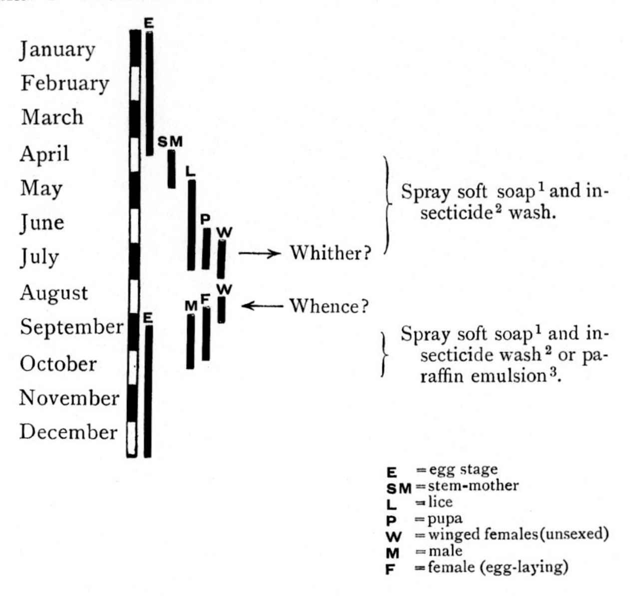


## LEAF-BUNCHING CURRANT APHIS

Name *Aphis grossulariæ*[4] Family *Aphidæ* Order *Hemiptera*

### 1. General

**Description**

**Adult Insect** (winged unsexed female)

REMARKS. This is apparently one of the migratory aphides, the only forms on the currant being the (unsexed) winged females, wingless females, and pupæ. Obviously the flies come from and return to some other host-plant, but this is at present unknown.

[1] See page 452. [2] See page 390. [3] See page 417. [4] Kaltenbach.

| | |
|---|---|
| APPEARS IN | Middle to end May. |
| DURATION. | Till middle July. |

**Ova (Eggs)** Unknown.

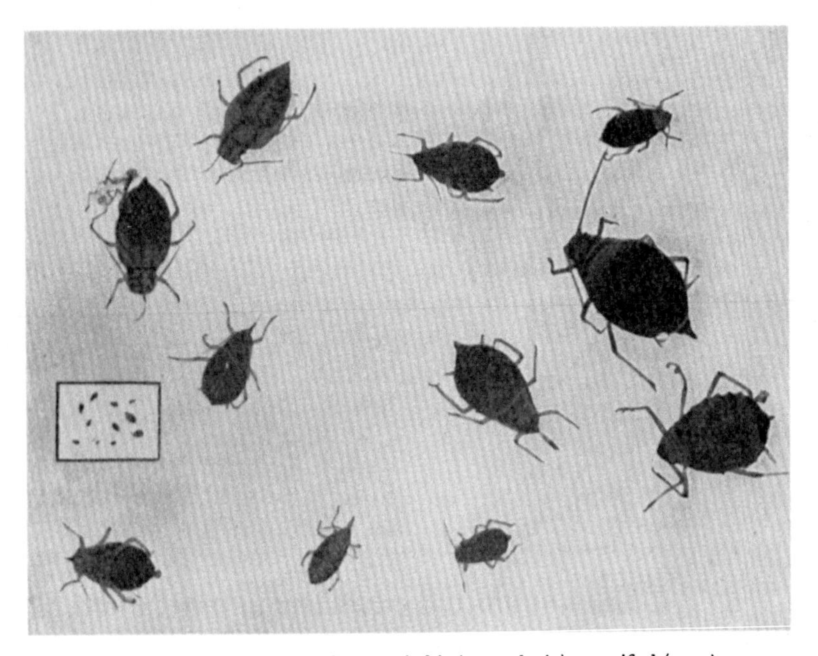

Fig. 171. Leaf-bunching Currant Aphis (*grossulariæ*) magnified (× 10). Inset natural size.

**Larva or "Louse"**

| | |
|---|---|
| SIZE. | Fairly large. |
| COLOUR. | Very dark green. |
| LOCATION. | Inside tuft of young leaves at top of plant. |
| APPEARS IN | May. |
| DURATION. | Till end of June. |
| REMARKS. | "Horns" pale and outline of body "pimply" (from above) due to a row of small projections at the sides of the abdomen (see fig. 171). |

**Distribution**

Common

**Life-history**

This aphis appears on the currants and gooseberries about the middle or end of May. and continues usually till the middle of July. Winged forms then appear and leave the plants, flying probably to the *Guelder Rose*[1]. It is easily distinguished by its very dark colour, and by the characteristic bunching together of the top leaves of the plant.

## 2. Economic

**Trees Attacked**

Red and black currants and gooseberries.

**Frequency of Pest**

Common.

**Nature of Attack**

The aphis causes a dense bunching together, or clustering, of the top leaves, and practically prevents the growth of the plant. The aphis is thus completely hidden and protected.

**Duration of Attack**

May to July.

**Degree of Damage**

Serious.

**Remedies**

1. It has been suggested to dip the bunched heads of the plants into pails of paraffin emulsion. It is however doubtful if by this method the air could be removed sufficiently to actually wet the aphis.
2. A method which the author has found very successful, is to tie thin bladders of skin or rubber round the heads, passing into these hydrocyanic acid gas or nicotine fumes, and then securing by tying round the stem and leaving for several hours.

[1] Theobald, *Journ. Econ. Biol.* VII. 1912, page 100.

Fig. 172. Showing the leaf-bunching effect produced by Aphis. ($\frac{3}{4}$ natural size.)

**Calendar of Treatment**

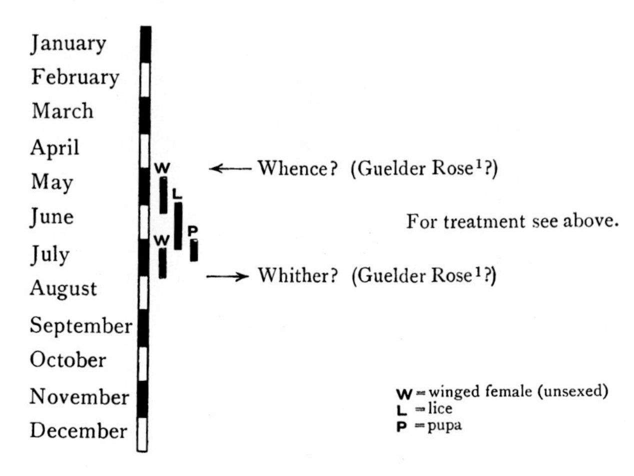


## LEAF-BLISTER CURRANT APHIS

Name *Myzus ribis*[2] Family *Aphidæ* Order *Hemiptera*

### 1. General

**Description**

**Adult Insect** (winged unsexed female)

| | |
|---|---|
| APPEARANCE. | Greenish-yellow or yellow: head darker: eyes red. |
| REMARKS. | Only the winged sexless females (*c*) and lice appear on the currants. The plant on which the egg-laying occurs is not at present known. |
| APPEARS IN | June to August. |
| **Ova (Eggs)** | Unknown. |

[1] Professor Theobald has practically demonstrated that this is the normal host-plant.
[2] Linnæus.

Fig. 173. Leaf-blister Currant Aphis on currant leaf. Natural size and magnified.

**Larva or "Louse"**

| | |
|---|---|
| APPEARANCE. | Almost transparent. |
| COLOUR. | Yellow or yellowish-green. |
| LOCATION. | In the red blisters of the leaves. |
| APPEARS IN | About middle of May. |
| DURATION. | Till end July. |

**Distribution**

Widespread.

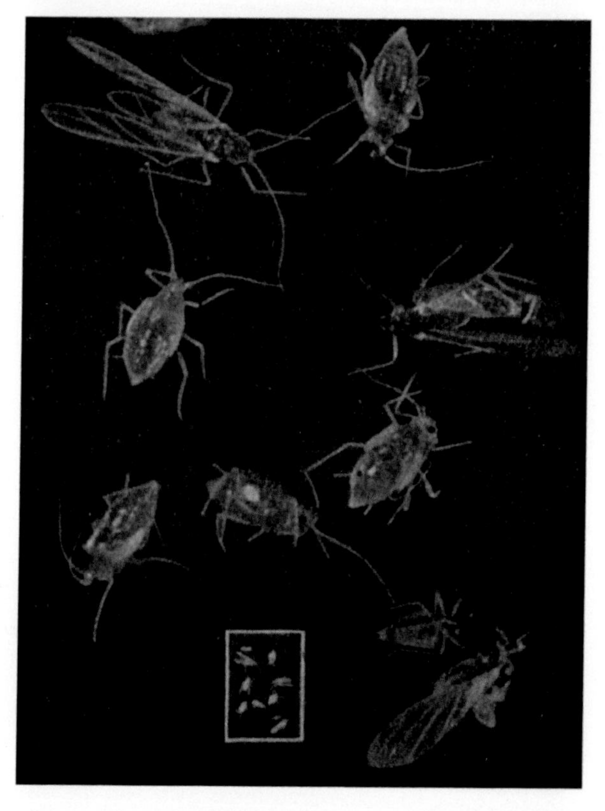

Fig. 174. Currant Aphides magnified (× 10) and natural size (inset) from interior of blisters on leaf.

**Life-history**

The lice appear, sheltering under the red blisters, in May and remain till August. The plant on which the sexual insects are produced, and the eggs laid, is not yet known.

**Remarks**

Some "lady-birds" and hover-fly larvæ feed upon the aphis and keep it in check, but too late to prevent weakening of the plant.

Fig. 175. Blisters produced on currant leaves in which the aphis shelters.

## 2. Economic

**Trees Attacked**

Currants, gooseberries.

**Frequency of Pest**

Very common.

**Nature of Attack**

The yellowish lice swarm on the inside of the red blisters on the leaves, and seriously weaken the plant.

**Degree of Damage**

Variable.

**Natural Enemies**

Some "lady-birds" and hover-fly larvæ feed upon the aphis and help to control it.

**Remedies**

No preventive remedies are possible, since it is not known where the aphides lay their eggs, and in what plant the sexual forms appear.

1. Spray with SOFT SOAP[1] and NICOTINE[2] wash. This is by far the most successful wash to use. Plenty of soap should be used and if a ready mixed wash is employed, growers should insist on a high soap content (see page 454).
2. PARAFFIN EMULSION[3] is good, but less effective.

**Calendar of Treatment**

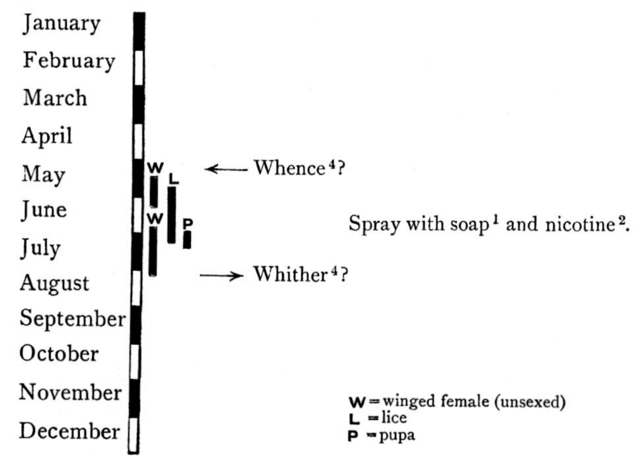


[1] See page 452. [2] See page 436. [3] See page 417.
[4] Probably Horehound and other plants of the natural order *Labiatæ* (Theobald).

## CURRANT-LETTUCE APHIS

Name *Rhopalosiphum lactucæ*[1] Family *Aphidæ*
Order *Hemiptera*

### 1. General

**Description**

**Adult Insect**

| | |
|---|---|
| APPEARANCE. | Shiny green. Thorax, head and feelers black; rest green or greenish-yellow. |
| REMARKS. | The usual series of adults are produced, viz. stem-mothers ("Queens") (*a*), wingless females (*b*), winged females (*c*), sexual wingless females (*d*), males (*e*). The winged females fly to lettuce, sow-thistle, and other plants, breed there during the summer and return in the autumn to the currants. |
| APPEARS IN | (*a*) April, (*b*) May, (*c*) May—June, (*d*) and (*e*) September—October. |

**Ova (Eggs)**

| | |
|---|---|
| APPEARANCE. | Oval. |
| COLOUR. | Brown. |
| ARRANGEMENT. | In groups. |
| LOCATION. | On or under bark of twigs. |
| APPEAR IN | Autumn. |
| HATCH IN | April. |
| DURATION. | Throughout the winter. |

**Larva or "Louse"**

| | |
|---|---|
| SIZE. | Variable according to age. |
| COLOUR. | Green or yellowish-green: eyes red. |
| LOCATION. | Under the red blisters of the leaves. |
| APPEARS. | During the spring and early summer. |
| REMARKS. | The blisters are not apparently caused primarily by the aphides. |

[1] Kaltenbach.

**Nymph (Pupa)**

| | |
|---|---|
| APPEARANCE. | With wing-buds. |
| COLOUR. | Green or yellowish-green. |
| LOCATION. | On leaves. |
| APPEARS. | Irregularly in summer. |
| DURATION. | Few days. |
| REMARKS. | Gives rise to the winged sexless female (*c*). |

Fig. 176. Currant-lettuce Aphis, magnified. Inset natural size.

**Distribution**

Widespread.

**Remarks**

The winged flies (*c*) produced in summer from the pupæ visit other plants, chiefly lettuces, sow-thistles, etc.

## 2. Economic

**Trees Attacked**

Currants (red, black and white), gooseberries.

**Frequency of Pest**

Common.

**Nature of Attack**

The aphis feeds on the under surface of the leaves inside the red blisters; like the following variety, it is not actually responsible for these, but makes use of the protection so afforded.

**Degree of Damage**

Variable.

**Preventive Measures**

Hard pruning and destruction of the cuttings will destroy many eggs.

**Remedies**

Spray early, directly the aphis is seen, with SOFT SOAP[1] wash and NICOTINE[2] (or other insecticide[3]), dilute PARAFFIN EMULSION[4], or an efficient compound wash.

**Remarks on Remedies**

A type of spray nozzle must be employed to reach the under surfaces of the leaves.

**Calendar of Treatment**

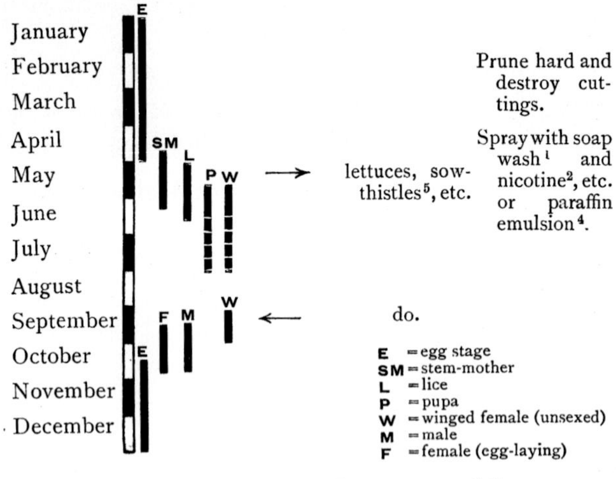


[1] See page 452. [2] See page 436. [3] See page 390.
[4] See page 417. [5] Theobald.

## **HOP APHIS** (Hop-damson aphis), Hop "blight," Hop "fly," etc.

Name *Phorodon humuli*[1] Family *Aphidæ* Order *Hemiptera*

## 1. General

### Description

#### Adult Insect

(*a*) STEM-MOTHER (wingless), derived from eggs on *damson trees* and from a few which pass the winter on or near *hops*. Bears living young.
Appearance—oval, yellowish-green, eyes red, legs green.
Remarks—special formation at *base of feelers* is distinguishing feature of this aphis (see fig. 177).

(*b*) WINGLESS FEMALE (unsexed), derived from (*a*) and from (*c* 1). Bears living young on damsons in spring.
Appearance—as for (*a*).
Duration—about 3 generations on *damsons* in spring, and many generations on *hops* in summer.

(*c*) WINGED FEMALE (unsexed), derived from pupæ from (*b*) in (1) spring on *damsons* flying to hops, and (2) summer on *hops* flying to damsons.
Appearance—green with black heads and spots, eyes reddish. The second form (2) on hops is larger than (1).
Appears in—(1) spring on damsons, (2) summer on hops.

(*d*) WINGLESS SEXUAL FEMALE, derived from (*c* 2) on the *damsons only*.
Appearance—small, white at first, green later.
Appears in—autumn on damsons.

(*e*) WINGED MALE. In this species the male is *winged* contrary to most aphides. It is derived from (*c* 2) and usually occurs on the damsons only.
Appearance—light green with patches of greenish-brown, *very long feelers*.
Appears in—autumn on damsons and rarely from hops.

[1] Schrenk.

**Ova (Eggs)**

| | |
|---|---|
| APPEARANCE. | Shiny—spindle shaped. |
| COLOUR. | Black. |
| ARRANGEMENT. | In heaps sticking together. |
| LOCATION. | At *junction* of small buds and boughs. |
| APPEAR IN | Late autumn. |
| HATCHING PERIOD. | Spring. |
| DURATION. | Throughout winter. |

Fig. 177. Hop Aphis. Various forms, magnified (× 10). Inset natural size.

### Larva or "Louse"

| | |
|---|---|
| COLOUR. | White to yellowish-green according to age. |
| LOCATION. | On damsons and hops. |
| APPEARS | During spring (damsons) or summer (hops). |

### Nymph (Pupa)

| | |
|---|---|
| APPEARANCE. | With wing-buds developed. |
| COLOUR. | Greenish-yellow with brown wing-buds. |
| LOCATION. | (1) on damsons, (2) on hops. |
| APPEARS IN | Spring on damsons: late summer on hops. |
| DURATION. | Variable. |

## Distribution

Occurs wherever hops are grown.

## Life-history

The life-history has been fully worked out, and is very complicated, there being two host-plants (damson and hop) and at least eight different forms, viz. two wingless females (unsexed), two winged females (unsexed), two pupæ, one male winged and one sexual female wingless.

In brief, the aphis produced from eggs on the damson flies to the hops, and after propagating there during the summer, further winged forms appear and return to the damsons.

## Remarks

The exact history can be best studied from the diagram on page 312. It takes place as follows:

1. Eggs are laid on the twigs of the damson in autumn.
2. These hatch in April and produce "stem-mothers" (*a*).
3. The stem-mothers produce three generations of lice growing into sexless females (wingless) (*b*).
4. In May and later, winged females (*c* 1) are produced which fly to the hops and there bear living young.
5. Several generations of sexless females (wingless) (*b*) are produced on the hops.
6. Towards the end of the summer, winged females appear on the hops (*c* 2) and fly back to the damsons.

7. From the winged females, wingless sexual females (*d*) and *winged* males (*e*) are produced, and on fertilisation, eggs are laid on the damsons. In addition, some stem-mothers (*a*) pass the winter under crevices and rubbish near the hops.

This aphis is distinguished from all other pests by the curious tooth-like projection at the base of the feelers.

Fig. 178. Badly infested hop leaf. Natural size and magnified.

The flight of the winged females in early summer from the damsons to the hops, and the return in the autumn, appears to be mainly directed by the prevalent winds so that only a small fraction are usually able to reach their objective. "Swarms" of these winged females often occur in hop districts (even in the towns) after a bad "blight."

The effect of weather conditions on the hop aphis is very marked. Thus, a bad attack may clear off in a few days—almost hours, if the weather is unsuitable (cold, heavy rain, thunderstorms) and, conversely, the aphides increase with marvellous rapidity under favourable conditions (warm bright days).

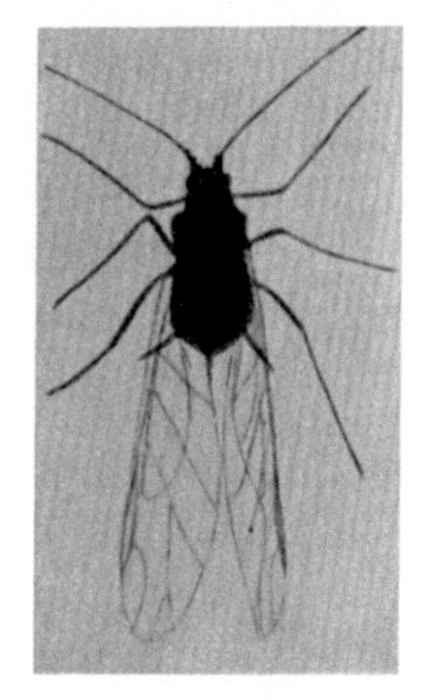

Fig. 179. Hop-damson Aphis from damson. Highly magnified.

## 2. Economic

**Trees Attacked**

Hops, damsons.

**Frequency of Pest**

Very common, occurring in varying degrees of severity each year in the hop districts (Kent, Worcester, Hereford, Sussex, etc.).

**Nature of Attack**

The aphis lives on the young leaves of the damson, and later upon the hops, often persisting until the formation of the flower (burr) and then doing great havoc to its growth.

**Duration of Attack**

Damsons—spring.

Hops—throughout summer.

**Degree of Damage**

The damsons are much weakened by the attack especially if young.

The damage to hops is very variable: in bad attacks the

plants are entirely withered and blackened. Since it increases with such rapidity there is always a grave risk to the crop unless spraying is at once carried out. Further, as the cost of cultivation of hops is relatively high, the failure of even a proportion of the crop may result in serious financial loss. It is therefore good policy to spray as soon as any sign of aphis appears and to keep spraying till the hops are quite free.

**Natural Enemies**

The most important are the lady-birds[1] and their larvæ or "niggers."

**Preventive Measures**

1. Spray the damsons with thick LIME WASH[2] just previous to the swelling of the buds and so prevent successful hatching out of the eggs. Late spraying with lime—even up to the appearance of the flower trusses—has not been found injurious (see fig. 245, page 618).
2. Spray the damsons in the early spring with a SOFT SOAP[3] and INSECTICIDE[4] wash, and so prevent migration to the hops.

This is another instance of the desirability of compulsory spraying or for co-operation. If damsons were everywhere sprayed in the neighbourhood of hop gardens, subsequent "blight" on the hops would be much mitigated. It is useless for one or two growers to spray if they are surrounded by neighbours who do not trouble.

**Remedies**

Fortunately, the hop aphis does not curl up the leaves, and so is easily accessible to sprays.

SOFT SOAP[3] is the basis of all hop-sprays, and pure potash soap is preferable to the soda-soft soap substitute[5] sold during the war. Growers should specify freedom from *caustic*[6] and *carbonate of soda*[7]. In addition it is best, though not absolutely necessary, to add a proportion of an EFFICIENT INSECTICIDE[4], such as nicotine[8]. Quassia[9] is less powerful but of proved value. (For further information on spraying see Section IX.)

[1] See page 469. [2] See page 428. [3] See page 452. [4] See page 390.
[5] See page 454. [6] See page 453. [7] See page 453. [8] See page 436.
[9] See page 446.

Fig. 180. Appearance of a hop garden badly infested with aphis.

## Remarks on Remedies

It is essential to use *plenty of soap* so that the aphis shall be *thoroughly wet.* Especially is this the case when soap is used alone. The hardness of the water should be known[1], and this allowed for in making up the wash (see remarks on pages 454, 455, 458, 460).

It is important to use only the *best makes* of soft soap on hops, since they are notably delicate and likely to be scorched or injured with inferior materials.

Early in the season, when the hops are small, a rather fine spray should be used. Later, when the leaves are well grown, a fairly coarse, powerful spray, with a good pressure behind it, is essential. Nozzle adjustment as fig. 250, page 634.

## Caution

Towards the end of the season, when the "burr" or "pin" is forming, a too powerful spray should be avoided, as the "pin" is very delicate and may be damaged. At this time also it is well to reduce the percentage of insecticide used in the wash, and to avoid an excess of soft soap, even if this is of the highest quality.

## Calendar of Treatment

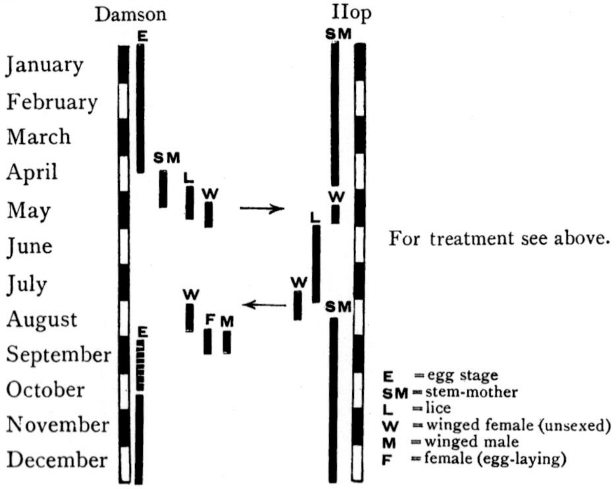

Damson | Hop

January
February
March
April
May
June
July
August
September
October
November
December

Damson: E, SM, L, W, W, F M, E

Hop: SM, W, L, W, SM

For treatment see above.

E = egg stage
SM = stem-mother
L = lice
W = winged female (unsexed)
M = winged male
F = female (egg-laying)

[1] One well-known firm analyses samples of water free for growers.

## PLUM LEAF-CURLING APHIS[1]

Name *Aphis pruni*[2] Family *Aphidæ* Order *Hemiptera.*

### 1. General

### Description

#### Adult Insect

The usual series of adults appear in the normal order:

(*a*) STEM-MOTHER. Dull purple.—On young leaves (from eggs) in early spring.

(*b*) WINGLESS FEMALE (unsexed). Green to olive brown—on leaves, which rapidly curl up. Produced during spring and summer.

(*c*) WINGED FEMALE (unsexed). Vivid green with black head, thorax and feelers.—It flies away in June and leaves the plums, returning in autumn, and giving rise to:

(*d*) SEXUAL FEMALE (wingless). Small, pale green.—Autumn.

(*e*) WINGED MALE. Small, dull yellow to black.—Autumn.

Remarks.—A large amount of sticky "honey-dew" is secreted on the leaves by the aphis, and the leaves rapidly curl up when the young lice appear.

#### Ova (Eggs)

| | |
|---|---|
| APPEARANCE. | Small, shiny. |
| COLOUR. | Black. |
| ARRANGEMENT. | Singly or in groups. |
| LOCATION. | On twigs and at base of buds. |
| APPEAR IN | October or November. |
| HATCHING PERIOD. | Early spring. |
| DURATION. | Throughout the winter. |

#### Larva or "Louse"

| | |
|---|---|
| COLOUR. | Green (brownish later) feelers, "horns" and legs brown. |
| LOCATION. | Inside tightly curled leaves. |
| APPEARS IN | Early spring. |
| DURATION. | Throughout spring and early summer. |
| REMARKS. | Much sticky "honey-dew" is secreted on the leaves, and some "meal." |

[1] See also Appendix XII, p. 710. [2] Fabricius.

Fig. 181. Plum Leaf-curling Aphis. ($\frac{3}{4}$ natural size.)

**Nymph (Pupa)**

| | |
|---|---|
| APPEARANCE. | With wing-buds. |
| COLOUR. | Greenish-yellow. |
| LOCATION. | In the curled leaves. |
| APPEARS IN | June or July. |
| DURATION. | Few days. |
| REMARKS. | Change into the winged sexless females (*c*) which fly away and return in autumn (see above). |

**Distribution**

Widespread.

**Life-history**

Similar to the other aphides. The eggs hatch out very early in spring, and the lice produced, at first green, grow into the large purple "mother queens." These produce lice at a surprising rate, and the leaves commence rapidly to curl up. The winged flies appearing in June or July fly away to some unknown plant and return in autumn, giving rise to sexual forms which produce the eggs.

## 2. Economic

**Trees Attacked**

Plums, damsons, and occasionally apples, peaches and apricots.

**Frequency of Pest**

Common.

**Nature of Attack**

The edges of the leaves gradually curl up and roll tightly round, due to the attack of the aphis. In bad attacks the leaves soon die, and the young fruits fall.

The aphis attacks damsons severely, and is distinguished from the Hop-damson aphis by the absence of the tooth-like projection at the base of the feelers (see fig. 179).

**Duration of Attack**

Spring and early summer.

**Degree of Damage**

Often very severe.

### Preventive Measures

A LIME COVER WASH[1] applied in March before buds swell has been found effective in preventing successful hatching of the eggs. This may be applied quite late on without injury.

### Remedies

Spraying is pretty certain to be successful if done thoroughly at the right time *before the curling of the leaves.* This should be not later than the fall of the blossom, and preferably before it opens. The stem-mothers are thus killed.

SOFT SOAP and an EFFICIENT INSECTICIDE are necessary ingredients of the wash, and these may be made up or a compound wash of known efficiency employed.

### Remarks on Remedies

Plenty of soap should be used, and a high pressure of the spraying pump maintained. Adjustment of nozzle as fig. 250, page 634.

### Calendar of Treatment

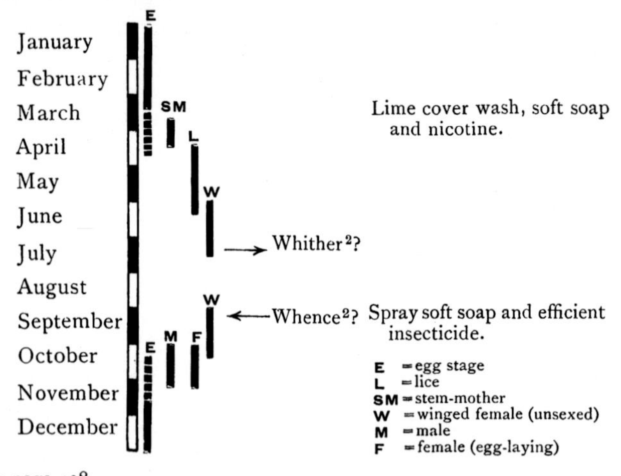


[1] See page 428.
[2] Probably various weeds, such as cow-weed, etc. (Theobald).

## MEALY PLUM APHIS

Name *Hyalopterus pruni*[1] Family *Aphidæ* Order *Hemiptera*

### 1. General

### Description

#### Adult Insect

Only two of the five standard types are at present known :

(*a*) STEM-MOTHER (from egg). Unknown.

(*b*) WINGLESS FEMALE (unsexed), occurs on plum leaves during August, and bears living young.
Appearance—light green, covered thickly with meal.
Duration—from June to September.

(*c*) WINGED FEMALE (unsexed). These fly to the plums from some unknown plant (possibly grasses or reeds). They produce living young on the plum leaves.
In autumn, the same form occurs, and flies off again to the other breeding plant.
Appearance—bright yellowish green, with dark green markings on thorax and abdomen. Slightly mealy.
Occurs—in June (1) and September (2).

(*d*) WINGLESS EGG-LAYING FEMALE. Unknown.

(*e*) MALE. Unknown.

**Ova (Eggs)** Unknown.

**Larva or Louse**

| | |
|---|---|
| COLOUR. | Light green covered with meal. |
| LOCATION. | Crowded together on under surface of leaf. |
| APPEARS IN | June. |
| DURATION. | Till September. |
| REMARKS. | The aphides do not usually curl the leaves but plenty of "honey-dew" is produced. |

### Distribution

Widespread.

[1] Fabricius.

Fig. 182. Mealy Plum Aphides. Magnified (× 10). Inset natural size.

### Life-history

The aphis appears as winged sexless females about June or July. Lice are produced alive on the plum leaves, which are not curled by the attack, and great numbers appear on the under surface of the leaves, sometimes completely covering them. Much grey meal appears on the surface of the bodies and a large amount of honey-dew is secreted.

Towards August, flies are again produced, and the aphis leaves the plum and flies to some unknown breeding plant (possibly grasses).

## 2. Economic

### Trees Attacked

Plums of all kinds, also damsons, peaches, nectarines, etc.

### Susceptible Varieties

Worst on Victorias and Czars.

### Frequency of Pest

Common.

### Nature of Attack

It infests the lower surface of the leaves in great numbers, and weakens the plant.

### Degree of Damage

Not so serious as the leaf-curling aphis, as it seldom attacks the plums before midsummer.

### Remedies

1. SOFT SOAP[1] and NICOTINE[2], or an efficient insecticide.
2. WEAK PARAFFIN[3] emulsion and LIVER OF SULPHUR[4].

### Remarks on Remedies

The soft soap and nicotine wash is preferable, but, owing to the mealy covering being difficult to penetrate, about 50 per cent. excess (½ as much again) of soft soap over the usual amount required for aphis is necessary (see pages 438, 454).

[1] See page 452. [2] See page 436.
[3] See page 417. [4] See pages 431, 624.

Fig. 183. Plum leaves badly infested with the Mealy Plum Aphis.

About 5 ozs. of nicotine for 100 galls. of wash is a suitable quantity. If the paraffin emulsion is used, care must be taken that it is freshly prepared and that a thorough emulsion is produced (see page 418).

A medium fine spray should be used, the adjustable nozzle being screwed so as to give a middle position (see page 632 and figs. 250 to 253). A high pump-pressure is essential.

## Calendar of Treatment

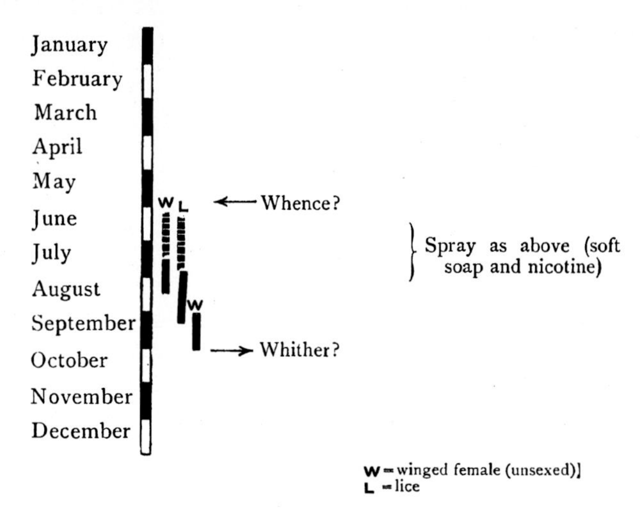


## RASPBERRY APHIS.

Name *Amphonophora rubi*[1] Class *Aphididæ* Order *Hemiptera*

### 1. General

## Description

### Adult Insect

All the usual forms occur :

(*a*) STEM-MOTHER (from egg), produces living young on the leaves. Appearance—shiny green, hairy, eyes red, legs and horns long.

(*b*) WINGLESS FEMALE (unsexed). As above (*a*).

[1] Kaltenbach.

Fig. 184. Raspberry Aphides. Various stages of growth. Magnified ( × 10). Inset, natural size.

(*c*) WINGED FEMALE (unsexed). 1. Summer brood fly away. 2. Autumn brood return and produce sexual males and females.
Appearance—green all over, legs and feelers long.

(*d*) WINGLESS EGG-LAYING FEMALE (sexual), product of (*c*) 2 above.
Appearance—oval, green, long feelers and legs: occurs in late autumn.

(*e*) MALE (winged), product of (*c*) 2 above.
Appearance—green with black head, large wings and legs. Occurs with (*d*) in autumn.

**Ova (Eggs)**

| | |
|---|---|
| COLOUR. | Shiny black. |
| LOCATION. | On and under rind of canes and on brambles. |
| APPEAR IN | Late autumn. |
| HATCHING PERIOD. | Spring. |
| DURATION. | Throughout the winter. |

**Larva or "Louse"**

Description as for "stem-mother" (*a*) above.

**Distribution**

Widely distributed.

**Life-history**

The aphis goes through the usual changes, and disappears from the raspberries in late summer, reappearing in autumn. The other food plant is at present unknown.

## 2. Economic

**Trees Attacked**

Raspberries, blackberries, brambles.

**Frequency of Pest**

Fairly common.

**Nature of Attack**

The aphides attack the leaves in the usual manner, but they do not cause them to curl to any extent.

**Degree of Damage**

Not usually serious. If in large numbers serious weakening of the plant may occur and the crop suffer.

**Preventive Measure**

Destroy all the winter prunings.

**Remedies**

Spray the raspberries with SOFT SOAP[1] and NICOTINE[2] or other efficient insecticide[3] whenever the aphis occurs in harmful numbers.

**Calendar of Treatment**

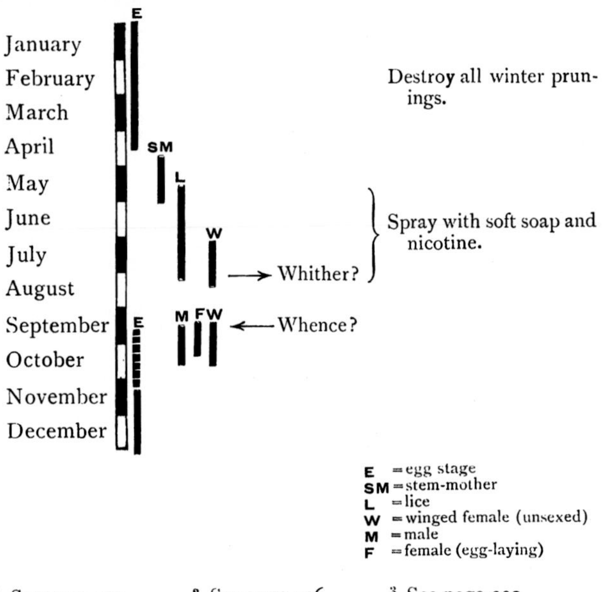


[1] See page 452. [2] See page 436. [3] See page 390.

## STRAWBERRY APHIS

Name *Myzus fragariella*[1] Class *Aphididæ*
Order *Hemiptera*

### 1. General

### Description

#### Adult Insect

All the usual forms occur:

(*a*) STEM-MOTHER (from egg), living young produced on the strawberry leaves.
Appearance—bright green or yellowish, head green, eyes black, horns and feelers long and thin.

(*b*) WINGLESS FEMALE (unsexed), as for (*a*) produces living young throughout summer.

(*c*) WINGED FEMALE (unsexed), does *not* appear to leave the strawberries in the late summer.
Appearance—head and feelers green, eyes brown, horns black.

(*d*) EGG-LAYING FEMALE (sexual), occurs in late autumn and lays eggs on the leaves after fertilisation by the

(*e*) MALE (sexual).

#### Ova (Eggs)

| | |
|---|---|
| COLOUR. | Black shiny. |
| LOCATION. | On leaves near veins. |
| APPEAR IN | Late autumn. |
| HATCHING PERIOD. | Spring. |
| DURATION. | Throughout winter. |

### Distribution

Mainly West of England, so far.

### Life-history

Follows the usual course, the aphis apparently existing throughout all its stages on the strawberry. Eggs hatch out in spring and the lice appear till end of summer. Flies occur and the sexual forms in late autumn when egg-laying takes place on the leaves.

[1] Theobald.

## 2. Economic

### Trees Attacked

Strawberries (and wild strawberries).

### Frequency of Pest

Not common.

### Nature of Attack

The aphis attacks the under-surface of the leaves in the usual manner.

### Degree of Damage

Not serious unless in large numbers.

### Preventive Measures

Cut back all old leaves in winter and burn them. Many eggs will thus be destroyed.

### Remedies

Spray in early spring (beginning of April) with SOFT SOAP[1] and NICOTINE[2] or other efficient insecticide[3]. This will kill the stem-mothers. If left later, spraying becomes proportionately more difficult on account of the thick leafage.

### Calendar of Treatment

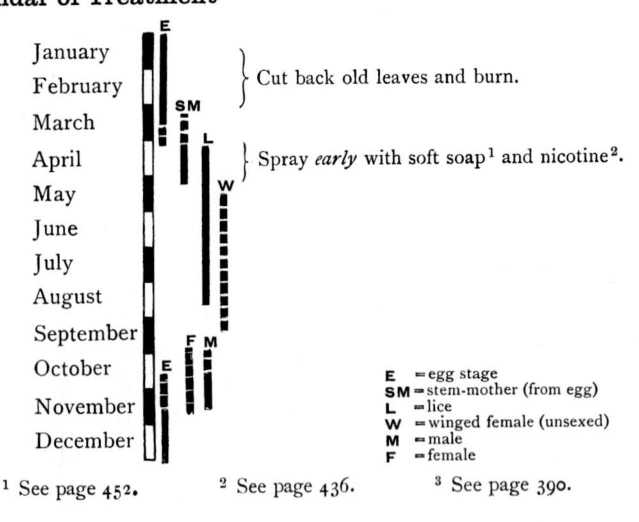


[1] See page 452. [2] See page 436. [3] See page 390.

## WOOLLY APHIS ("American Blight")

Name *Eriosoma lanigera*[1] Class *Aphididæ*
Order *Hemiptera*

### 1. General

## Description

### Adult Insect

Forms occur as follows:

(*a*) MOTHER-QUEEN. This is either the hibernating form (hiding in winter in crevices or below ground on the roots) or the adult from the eggs. They produce living young throughout spring and summer.
Colour—purple brown, with much "wool."
Shape—oval.
Appears in—spring and throughout the year.

(*b*) WINGLESS FEMALE (unsexed), the adult product of (*a*). Produces living young continuously.
Appearance—as (*a*): much "wool" secreted.
Location—on branches or roots.

(*c*) WINGED FEMALE (unsexed). *Uncommon.* Produced from pupæ very irregularly during July to September.
Appearance—chocolate brown in colour, distinguished by the arrangement of the veins on the wings.
Location—above ground only.

(*d*) WINGLESS EGG-LAYING FEMALE. *Very rare.* Produced at long intervals in autumn.
Appearance—very minute: only $\frac{3}{1000}$ inch in size. Yellowish red colour and without mouth. Deposits one egg and dies, shielding it with her body.
Location—above ground only.

(*e*) WINGLESS MALE. *Very rare.* As for (*d*).
Appearance—as for (*d*).
Location—above ground only.

[1] Hausmann.

**Ova (Eggs)**

| | |
|---|---|
| APPEARANCE. | Minute: oval. |
| COLOUR. | Shiny black. |
| ARRANGEMENT. | Singly under dead body of female (*d*). |
| LOCATION. | In crevices of bark near ground. |
| APPEAR IN | Autumn. |
| HATCHING PERIOD. | Spring. |
| DURATION. | Throughout winter. |
| REMARKS. | Comparatively rare. |

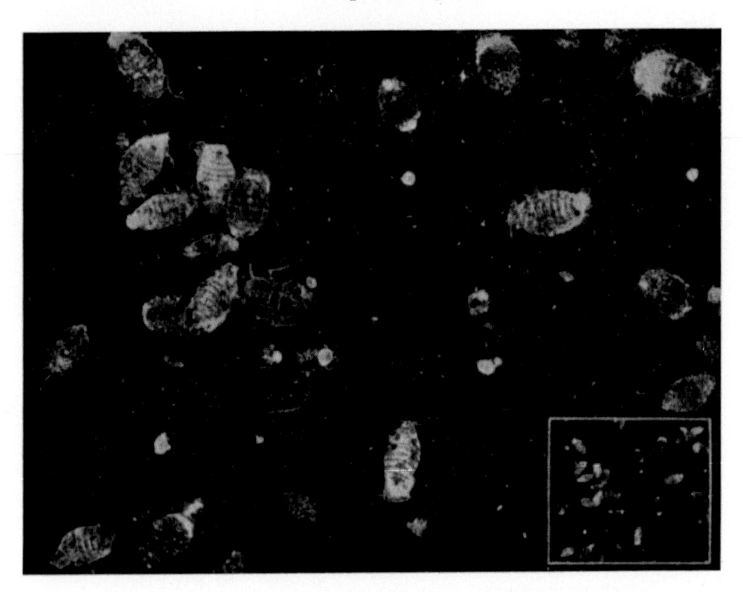

Fig. 185. Woolly Aphides—detached lice. Magnified (× 5) and natural size (inset).

**Larva or "Louse"**

| | |
|---|---|
| APPEARANCE. | Naked at first, later covered over with "wool." |
| COLOUR. | Yellowish: then purple. |
| LOCATION. | (1) On branches, shoots, trunk, etc.; (2) below ground on roots. |

| | |
|---|---|
| DURATION. | (1) Spring and summer above ground; (2) all the year round below ground. |

**Nymph**

| | |
|---|---|
| LOCATION. | Above ground on branches, etc. |
| APPEARS IN | Summer. |
| REMARKS. | Not common. |

**Distribution**

Practically world-wide.

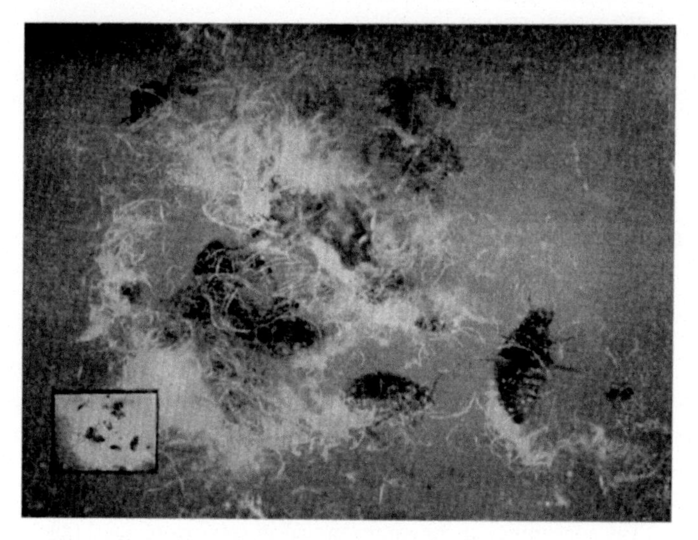

Fig. 186. Woolly Aphides, showing "wool." Magnified ($\times 6$) and natural size (inset).

**Life-history**

This aphis is peculiar in living both ABOVE AND BELOW ground, and to pass the winter the above-ground individuals descend to the roots and attack these, in company with those which have remained there throughout the summer. In fact, there is probably a constant migration occurring between the roots and branches. In addition, some of the adults spend the winter in crevices of the bark, protected by their WOOLLY COVERING. This is the other peculiar feature of this remarkable insect. The very young are naked, but soon secrete strands of the wool and often to such an extent that it hangs down below the branches.

The mother-queen (*a*) gives rise to living young, which reproduce throughout the year. Only rarely are the pupæ and winged insects (*c*) produced: and only very rarely are the sexual forms seen (in autumn), although these may escape observation on account of their minute size. It would appear that the development of the root form has almost obliterated egg-laying as a means of carrying the species on over the winter.

The insects infest all parts of the trees, but only rarely the leaves and fruit. Its activities appear to be absolutely unaffected by even extreme cold.

The normal host-tree is the *elm*.

## 2. Economic

### Trees Attacked

Apples and occasionally pears.

### Immune Varieties

Apples on the *Northern Sky* and *Majetin* stocks are said to be free from attack.

### Frequency of Pest

Increasingly common in all parts, but particularly in the West of England.

### Symptoms of Attack

The appearance of the white "wool" in crevices of the branches and trunk very quickly reveals its presence.

### Nature of Attack

1. The trunk and boughs are attacked (rarely the leaves and blossoms or fruit), the aphis penetrating to the inner bark and living upon the sap.
2. The roots are attacked in a similar manner. This is the worst form.

### Duration of Attack

1. On the branches, etc., mainly in the spring and summer.
2. On the roots, all the year round.

Fig. 187. Woolly Aphis attack on an apple tree, and cankerous swelling. (See also fig. 162).

### Degree of Damage

1. The tree is exhausted by the LOSS OF SAP both above and below ground. Young trees are often killed.
2. LARGE SWELLINGS and deformities are produced on both branches and roots which after a time dry up and crack.
3. The inner surfaces are exposed to FUNGUS ATTACK, particularly *canker*.

### Natural Enemies

The most useful are birds, especially the TITS.

Lady-birds devour them and several other insects attack them, but the economic results are trifling.

### Preventive Measures

Plant immune varieties. There are about a dozen blight proof New Zealand varieties obtainable.

Fumigate all nursery stock with HYDROCYANIC ACID GAS[1] before planting. A thorough treatment should kill all eggs as well as insects and insure freedom from the disease.

### Remedies

The above- and below-ground lice must be destroyed at the SAME TIME, otherwise one form will infest the other portion of the plant.

1. Spray *in winter* with strong CAUSTIC PARAFFIN EMULSION[2] or CAUSTIC ALKALI WASH[3] alone. Strong LIME-SULPHUR[4] washing should also be of benefit if thoroughly carried out.
2. *Simultaneously* inject CARBON BISULPHIDE[5] into the soil to kill below-ground variety.
3. Wash in the summer with strong SOAP[6] and NICOTINE[7] (5—6 oz. per 100 gall. wash), using in addition a small quantity of paraffin properly emulsified[8] (not exceeding 2 galls. per 100 of the wash).
4. Banding the trees about the middle of May was effective in catching many migrating aphides[9].

   In all cases a *coarse* and *powerful* spray is required. This necessitates a high pressure on the pump, and an adjustment of the nozzle as fig. 250, page 634.

[1] See page 423. [2] See page 420. [3] See page 413.
[4] See page 612. [5] See page 410. [6] See page 452.
[7] See page 436. [8] See page 418. [9] See page 632.

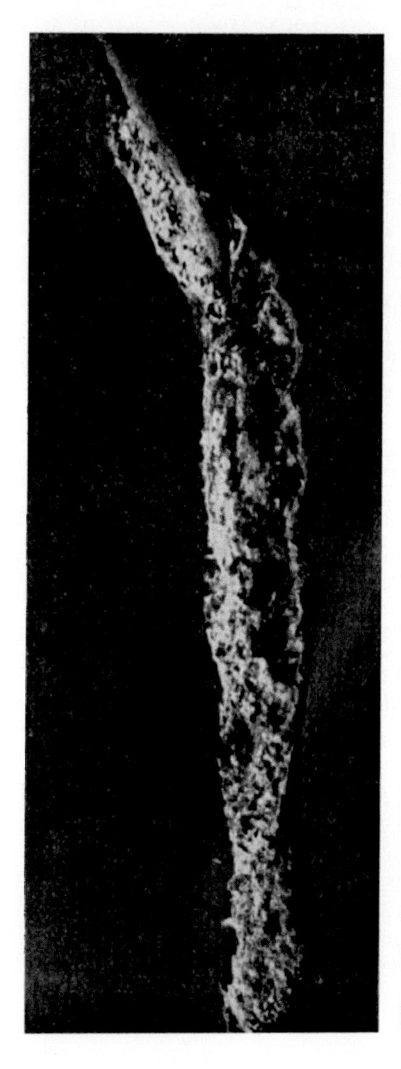

Fig. 188. Root form of Woolly Aphis showing damage to main root, and aphides *in situ*. (About $\frac{1}{3}$ size.)

## Remarks on Remedies

1. The carbon bisulphide injection must be made before April, and preferably in dry weather. (For other details see page 410.)
2. Generally speaking it is useless to employ insecticides for ummer spraying other than nicotine and soft soap unless the "wool" is first removed, as insufficient penetration will be effected. It should, however, be mentioned that Professor Lees has found that a properly prepared emulsion[1] containing 2 per cent. of soft soap and 2 per cent. paraffin oil can be made to penetrate the wool and kill the aphides.

## Calendar of Treatment

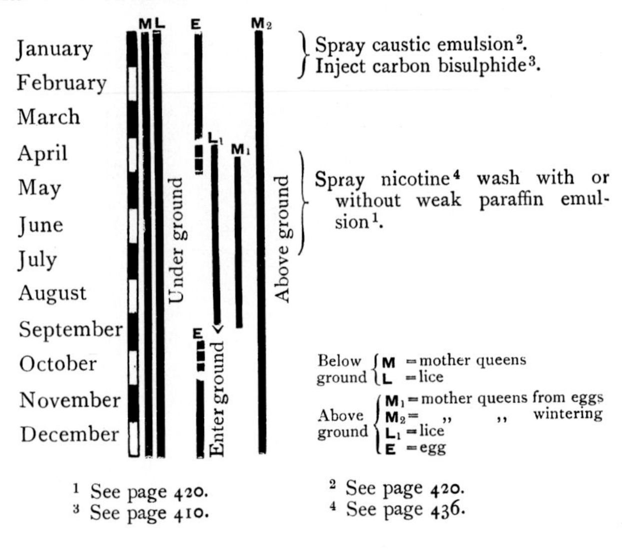


[1] See page 420.
[3] See page 410.
[2] See page 420.
[4] See page 436.

BUGS AND HOPPERS

# CHAPTER 20

## Bugs and Hoppers

### APPLE SUCKER (Psylla)

Name *Psylla mali* Class *Psyllidæ* Order *Hemiptera*

#### 1. General

**Description**

**Adult Insect**

| | |
|---|---|
| SIZE. | From about $\frac{1}{10}$ to $\frac{1}{8}$ inch. |
| COLOUR (WINGS). | Transparent : veins green. |
| ,, (BODY). | Variable : greenish yellow changing to a brownish shade. |
| APPEARS IN | Middle May. |
| REMARKS. | Insects hop off the leaves when disturbed and also use their wings. |

**Ova (Eggs)**

| | |
|---|---|
| APPEARANCE. | Small, oval, with minute curled "tail." |
| COLOUR. | White changing to yellowish red. |
| ARRANGEMENT. | In line or irregular. |
| LOCATION. | On small twigs near blossom buds, or on buds. |
| APPEAR IN | September to November. |
| HATCHING PERIOD. | Middle to end of April. (Variable with variety of tree and season.) |
| DURATION. | Throughout winter. |

**Larva or "Louse"**

| | |
|---|---|
| SIZE. | Very small, flat. |
| COLOUR. | Greenish or yellow, with bright red eyes. |
| LOCATION. | On or inside blossom buds. |
| APPEARS IN | Middle to end of April. |
| NUMBER OF MOULTS. | Two. |

| | |
|---|---|
| DURATION. | About 10 days. |
| REMARKS. | A globule appears from the body attached by a thread after moulting, waxy threads partly cover body. |

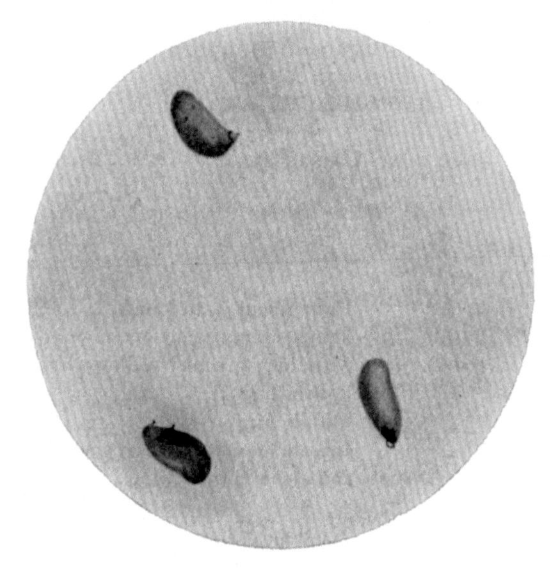

Fig. 189. Eggs of Apple Sucker highly magnified (× 20).

**Nymph**

| | |
|---|---|
| APPEARANCE. | With wing buds and feelers. Abdomen large and broad. Many waxy threads. |
| COLOUR. | Green. |
| LOCATION. | On leaves. |
| APPEARS IN | Early May. |
| NUMBER OF MOULTS. | Three. |
| DURATION. | $2\frac{1}{2}$ to $4\frac{1}{2}$ weeks. |

**Distribution**

General in England.

**Life-history**

The eggs of the Apple Sucker are laid on the twigs or on the buds, and these hatch out in the spring. The date of hatching *varies with the variety of tree*, this apparently being naturally regulated so that the buds are ready for the attack of the newly hatched insect. The young "nits" enter the buds, their flat bodies rendering this easy. Two moults then occur and the "nymph" or "pupal" stage then commences and wing buds appear. This is the form most familiar to growers and may be termed the "louse" stage.

Fig. 190. Winged Apple Suckers. Magnified ($\times 6$) and natural size (inset).

These lice[1] are found on the leaves and the stage lasts a few weeks. At its completion, the insect fixes itself firmly to the leaf with its "beak," the skin bursts, and the winged Psylla emerges, leaving the cast skin adhering to the leaf. These skins are very characteristic of sucker attack. The winged insects (male and female) are seen from May till November. After fertilisation, the female lays her eggs on the spurs and buds.

[1] See illustration on page 711.

## 2. Economic

### Trees Attacked

All varieties of apple.

### Susceptible Varieties

Ecklinville appears to be the worst sufferer: also Lane's, Blenheim, Wellington and Grosvenor are cited as often badly attacked. Worcesters are not commonly affected. In the West of England all apples are attacked annually.

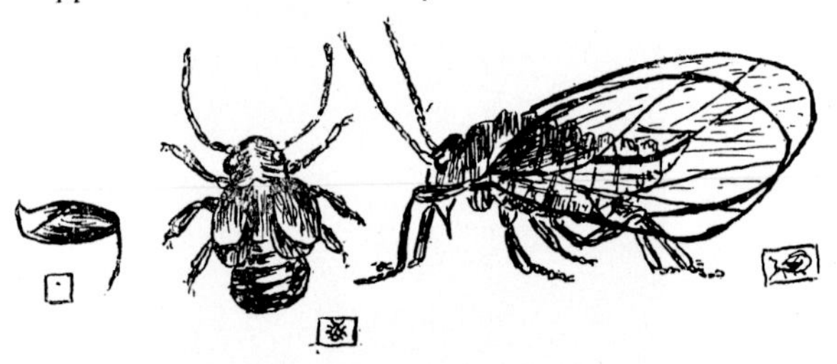

Fig. 191. Diagram of egg, nymph and adult of Apple Sucker (Psylla). Enlarged and natural size.

### Frequency of Pest

Unusually bad attacks occur locally at intervals, but the sucker is present in affected districts each year, and is very persistent. Probably sucker is responsible for much injury attributed to other causes, being difficult to see in its young and most injurious stage.

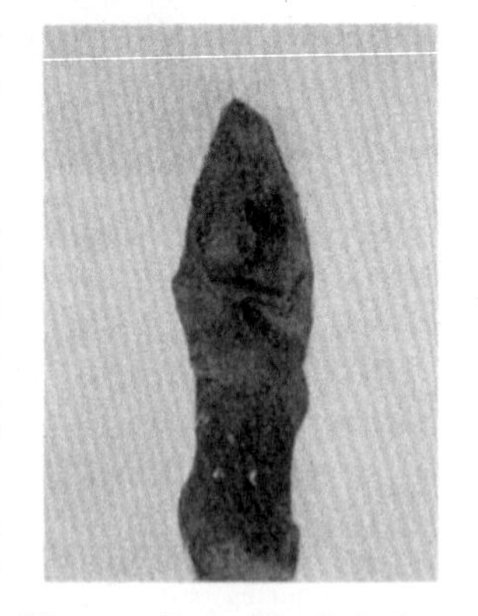

Fig. 192. Eggs of Apple Sucker near bud. Natural size.

### Symptoms of Attack

The flowers may open well if only a few suckers are present, but before the petals are ready to fall, the whole blossom shrivels up, producing a characteristic brown "truss," which often remains on the tree for long periods.

To distinguish from (*a*) *frost*, (*b*) *brown rot*, or (*c*) *weevil*, examine the withered petals, and in the case of sucker attack, one or more young insects will be seen. Also the cast skins will later on be noticed on the leaves.

### Nature of Attack

Chiefly the blossom buds. Often these fail to open if many young suckers are present. If they open, a withered "truss" is formed as above. The leaf buds are also attacked and on expansion present a distorted, blighted appearance. The subsequent attack on the leaves is not so serious.

### Duration of Attack

Worst in early spring but continues to the end of the "louse" period (4 to 6 weeks).

### Degree of Damage

Variable : in serious cases trees are frequently robbed of all fruit for many successive seasons.

### Preventive Measures

1. Spraying with **thick lime**[1] to prevent the newly hatched "nit" from reaching the buds has met with considerable success. The lime may be applied without injury right up to the appearance of the blossom trusses (see page 274).
2. Benefit has also been found from a LIME and SALT wash applied in the late winter.

Fig. 193. Young Apple Suckers just hatched on opening buds. Natural size.

### Natural Enemies

The tits feed on them and occasionally clear off the pest entirely. A fungus and a mite have also been mentioned as parasitic on Psylla.

### Remedies

1. Spray in spring immediately the eggs hatch out (see fig. 193)

[1] See page 428.

with SOFT SOAP[1] and NICOTINE[2] or other efficient insecticide before the "nits" enter the buds.

2. Spray in autumn with NICOTINE[2] and SOFT SOAP[1] or PARAFFIN EMULSION[3] (nicotine is preferred) to kill the egg-laying suckers.

**Calendar of Treatment**

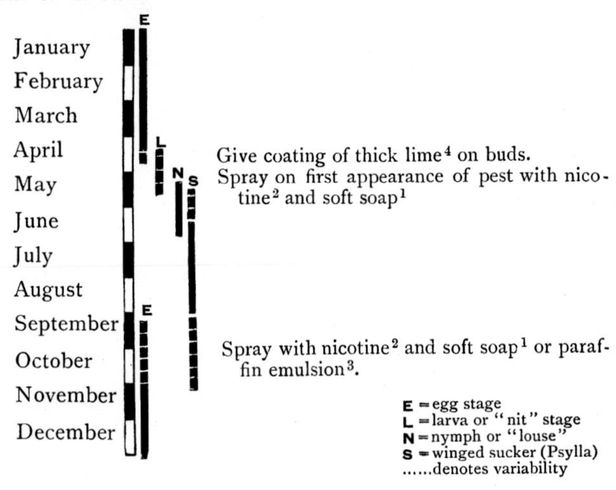


## THE CAPSID BUG

Name *Plesiocoris rugicollis* Class *Capsidæ* Order *Hemiptera*

### 1. General

**Description**

**Adult Insect**

| | |
|---|---|
| SIZE. | About $\frac{1}{4}$ inch: oval in shape: body smooth. |
| COLOUR (WINGS). | Green, with orange-yellow edges. |
| „ (BODY). | Green. |
| REMARKS. | Has a small yellow collar between the head and body. |
| APPEARS IN | Middle till end of May. |
| DURATION. | About 7 weeks (till middle July). |

[1] See page 452. [2] See page 436. [3] See page 419. [4] See page 428.

**Ova (Eggs)**

| | |
|---|---|
| APPEARANCE. | Curved : about $\frac{1}{18}$ inch long. |
| COLOUR. | Cream. |
| ARRANGEMENT. | Singly. |
| LOCATION. | Completely embedded in shoots. |
| APPEAR IN | June and July. |
| HATCHING PERIOD. | Following spring. |
| DURATION. | Throughout winter. |

**Larva or "Louse"** (first stage)

| | |
|---|---|
| SIZE. | About $\frac{1}{18}$ inch. |
| COLOUR. | Yellowish green, with pink tips to feelers. |
| LOCATION. | Between developing leaves and flower buds. |
| APPEARS IN | Latter end of April. |
| NUMBER OF MOULTS. | Two. |
| DURATION. | About 12 days. |
| REMARKS. | Distinguished from *aphis* by its greater *activity.* |

**Nymph or "Louse"** (bug after second moult till wings appear)

| | |
|---|---|
| APPEARANCE. | *Wing pads* appear and develop progressively. |
| COLOUR. | Green. |
| LOCATION. | On leaves or fruit. |
| APPEARS | About middle of May. |
| NUMBER OF MOULTS. | Five (on the fifth moult, mature insect appears). |
| DURATION. | About 18 days. |
| REMARKS. | Most of the damage to the fruit is done during this stage. |

**Distribution**

Chiefly in Cambridgeshire district of Britain.

Has also appeared in Worcester and Hereford for some years and locally in Kent.

**Life-history**

The eggs hatch out from the middle till the end of April, and the young bugs enter the developing buds. They are *very active*, and at once commence to puncture the leaves and suck the sap.

After about 6 days, the first skin is cast, and moults occur at intervals of about 6 days till the FIFTH MOULT when the MATURE

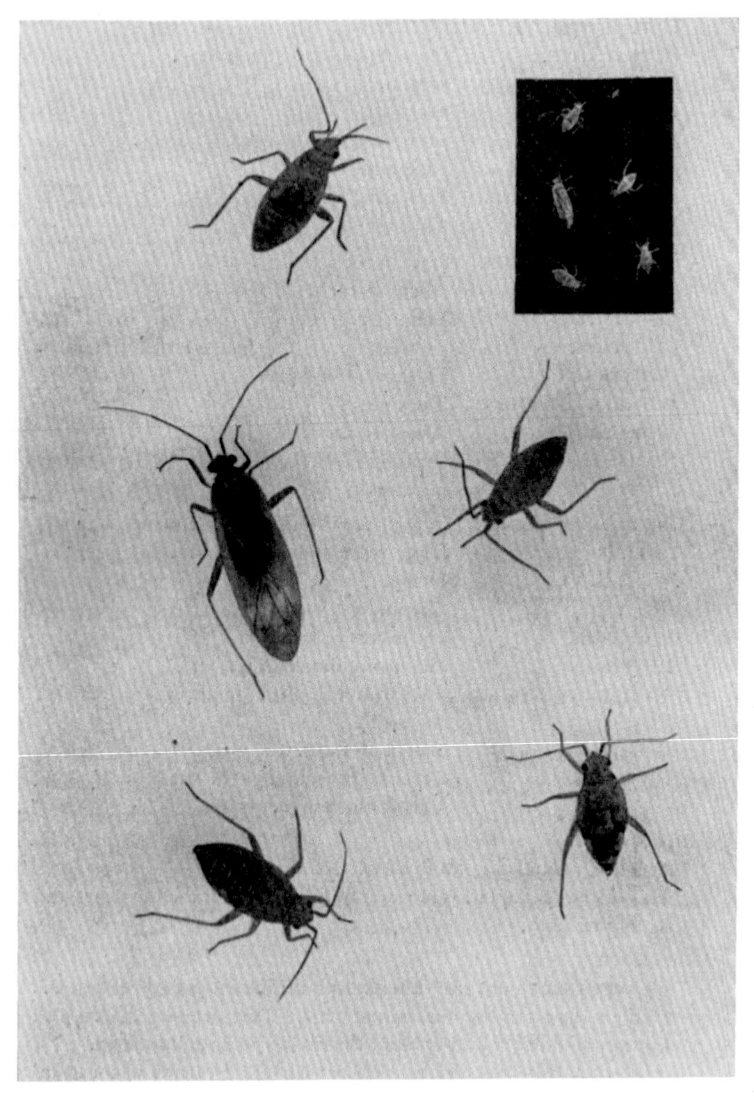

Fig. 194. Capsid Bugs in various stages. Magnified and natural size (inset).

WINGED INSECT appears. The first two moults are similar but the insect increases in size and in depth of colour each time. On the second moult the nymph stage appears with *wing pads* which become more prominent in the third and fourth moults. The fruit is attacked after the third moult, and most of the damage is done by the nymph.

Fig. 195. Damage to young apple fruitlets by Capsid Bugs. Bottom row, unattacked fruitlets of same age.

The mature bug feeds on the young leaves and the tips of the stems, causing, in the latter case, exudation of a brown fluid.
The insects pair, and the female lays her eggs deep in the bark of the young shoots during late June and early July, the eggs hatching the following spring.

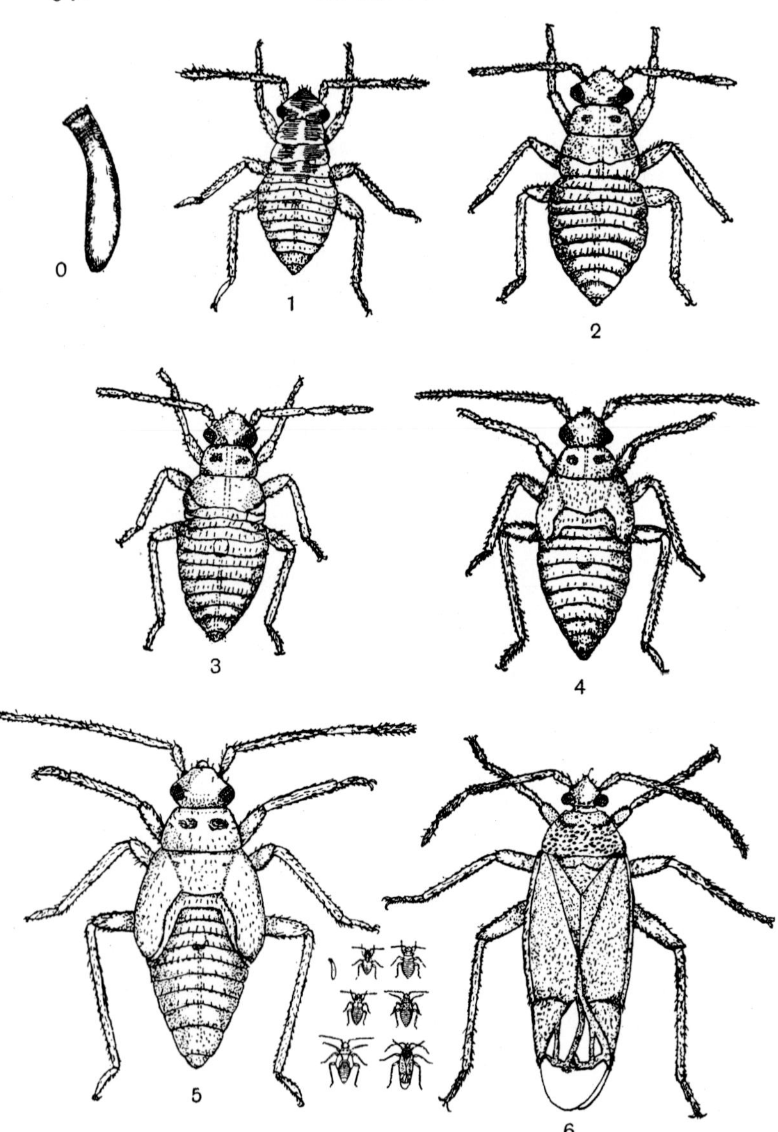


Fig. 196. Stages in the development of the Capsid Bug (after Petherbridge). o, egg. 1—5, various moults ("instars"). 6, winged adult. Greatly enlarged. Inset, natural size.

**Remarks**

Another green capsid, *Orthotylus marginalis*, often occurs on apples, but has not been observed to cause damage[1].

There are also the red capsid, *Atractotomus mali*, and a brown bug, *Psallus ambiguus*, both of which are found on apples from time to time, but also have not been observed to do injury.

## 2. Economic

**Trees Attacked**

Normally the food plant is the WILLOW, but of late years APPLES and CURRANTS have been badly attacked in places, notably in the Wisbech district of Cambridgeshire and in Worcester and Hereford.

**Susceptible Varieties**

Worcester Pearmain, Lord Grovesnor, Lane's Prince Albert, Lady Hollendale and Lord Derby are amongst the worst attacked varieties of apple. In the case of Grenadier and quick-growing varieties like Bramley's Seedling[2] and Early Victoria the fruit has not been injured to the same extent.

**Frequency of Pest**

Apples are continuously attacked in the affected districts.

**Nature of Attack**

The leaves show red or brown spots and holes where they have been punctured. The shoots "bleed" a brown fluid and are stunted or killed. The number of lateral shoots becomes abnormally large. The fruit shows progressive damage. There may be pimples or pits due to hindered development where punctures have been made, or rough discoloured areas. In bad cases, the apple is very distorted, and cracks appear sometimes penetrating to the core.

**Duration of Attack**

Until about end of July.

**Degree of Damage**

Variable, but generally very serious.

[1] Petherbridge, *Journal of Board of Agriculture*, 1918, 1401—1410.

[2] Bramley's are reported severely attacked in Worcester by Mr J. H. W Best.

Fig. 197. Mature apples showing degrees of injury caused by Capsid Bug.

## Preventive Measure

Benefit has been found from a thick coating of LIME WASH[1] applied just before the bursting of the buds, in order to prevent the eggs on the shoots from hatching.

## Remedies

NICOTINE[2] and SOFT SOAP[3] spray, best applied between middle April to end of first week in May, if this is possible. (The time varies with the locality and the variety of apple attacked.)

This is the only sure remedy so far found, but a certain specific proprietary wash has had marked success in the Cambridgeshire district. Amongst many other substances tried, PYRIDINE[4] was the most hopeful, and appeared to give good results one year, but unfortunately this standard of efficiency was not maintained.

## Remarks on Remedies

To ensure a good kill the minimum amount of nicotine[2] to employ for this pest is 7 to 8 ozs. per 100 gallons of water, and the quantity of soft soap[3] should be 8 to 10 lbs. per 100 gallons with soft water, or more in proportion to its hardness (see pages 454–460).

## Calendar of Treatment

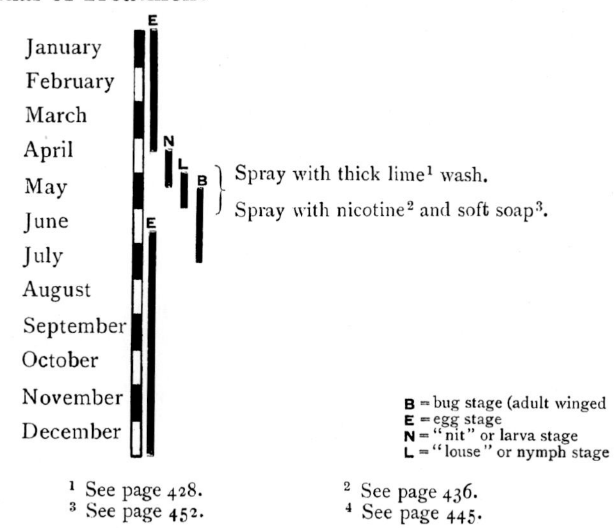


[1] See page 428.
[2] See page 436.
[3] See page 452.
[4] See page 445.

## LEAF HOPPERS

Name *Species of Typhlocybidæ, Chlorita*
Order *Hemiptera*

### 1. General

**Description**

**Adult Insect**

| | |
|---|---|
| SIZE. | About $\frac{1}{8}$ inch. |
| COLOUR (WINGS). | Transparent or milky white with dusky marks and red spots. |
| „ (BODY). | Bright yellow, green or yellowish green. |
| PROGRESSION. | By leaping (using the wings). |
| APPEARS IN | About September. |
| DURATION. | Throughout winter. |

**Ova (Eggs)**

| | |
|---|---|
| APPEARANCE. | Smooth or ribbed. |
| COLOUR. | White. |
| LOCATION. | On under surfaces of leaves (in some species, under the outer skin). |
| APPEAR IN | June (and later ?). |
| HATCH | About August. |

**Larva or "Nymph"**

| | |
|---|---|
| COLOUR. | White at first, greenish or yellow later. |
| PROGRESSION. | Active, running, and in some varieties jumping. |
| APPEARS IN | August or thereabouts. |
| NUMBER OF MOULTS. | Five. |
| DURATION. | About 1 month. |

**Distribution**

Widespread.

## 2. Economic

### Trees Attacked

Most fruit trees and also forest trees.

### Frequency of Pest

Fairly common : sometimes in enormous numbers.

### Nature of Attack

The incisions of the insect cause either a silvery or marked appearance of the leaf (strongly resembling the effect of the fungus "silver-leaf") or jagged holes may appear (particularly on nuts) by the leaf splitting as it grows.

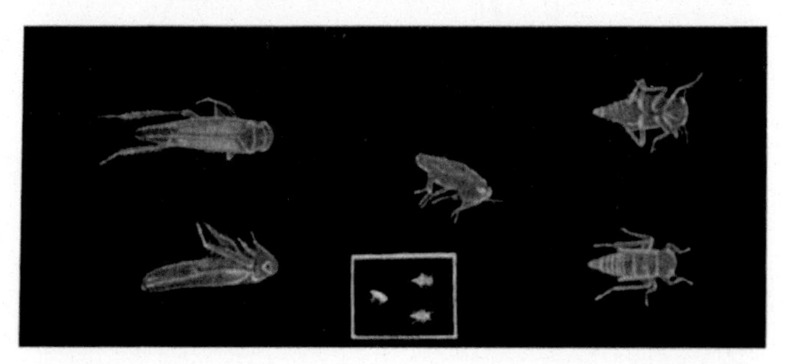

Fig. 198. Leaf Hoppers from nut. Nymphs and adults. Magnified. Inset, natural size.

### Degree of Damage

Variable.

### Remedies

There are three methods employed successfully :

1. Spraying the lice (larvæ) when these appear with SOFT SOAP[1] and NICOTINE[2] or PARAFFIN EMULSION[3].
2. Spraying the adults with WEAK SOFT SOAP first. This causes them to lie upon the ground as if dead. The ground is now sprayed with a STRONG PARAFFIN EMULSION[4] which kills them. (They cannot be killed on the leaves without injuring the latter.)
3. Holding tarred boards for the hoppers to jump on (as for weevils, page 205).

[1] See page 452. [2] See page 436. [3] See page 417.
[4] Say 20 per cent. paraffin (see page 417).

**Calendar of Treatment** (two broods may occur).

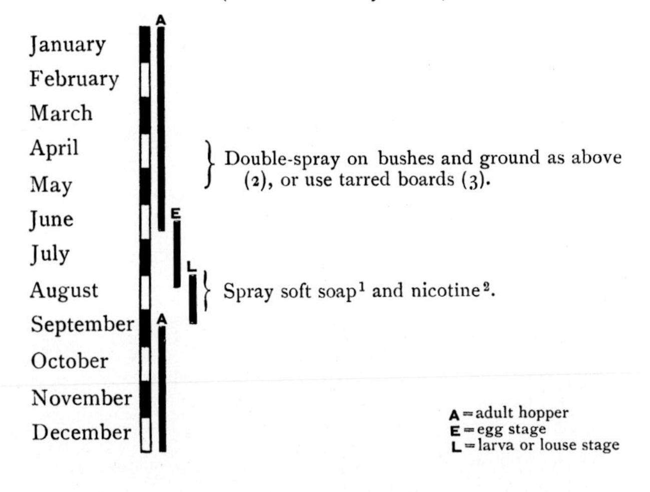


[1] See page 452. [2] See page 436.

# SCALE INSECTS

# CHAPTER 21

## **Scale Insects** (Coccidæ)

These form a very curious group of insects. Like the aphides they suck the plant juices by means of jointed beaks. Unlike any other member of the order, however, they are stationary once the young forms have selected their position. They are covered with a hard, adhesive *scale*, and under this protection the insect lives, performs its functions and then dies and shrivels up, leaving the eggs which it has laid under the protection of the scale. The male insects are winged, but they are rather rare. Fertile eggs apparently are produced for some generations without the assistance of the male insect.

The scale insect is killed by treatment with strong caustic emulsion in the winter. Lime-sulphur is also of proved benefit. Summer treatment is only of use if the young insects are hatched and have not come to rest.

## **MUSSEL SCALE** (Oyster-shell bark louse)

Name *Lepidosaphes ulmi* Class *Coccidæ* Order *Hemiptera*

### 1. General

**Description**

**Adult Insect**

| | Female | Male |
|---|---|---|
| SIZE. | $\frac{1}{6}$ to $\frac{1}{8}$ inch. | About $\frac{1}{10}$ inch across wings. |
| APPEARANCE. | Without legs, wings, or feelers under "scale." | With a single pair of wings, long legs and feelers, and long sexual fertilising tube. |
| COLOUR (BODY). | Whitish. | |
| „ (SCALE). | Dark brown or grey. | |

| | Female | Male |
|---|---|---|
| LOCATION. | Adhering to bark of trunks and branches. | On trees. |
| APPEARS IN | Late summer. | Late summer. |
| DURATION. | Produces eggs under scale and dies in winter. | Dies after fertilising the female. |
| PROGRESSION. | Stationary. | Crawls and flies. |
| REMARKS. | The female can apparently produce fertile eggs without fertilisation in many cases. | |

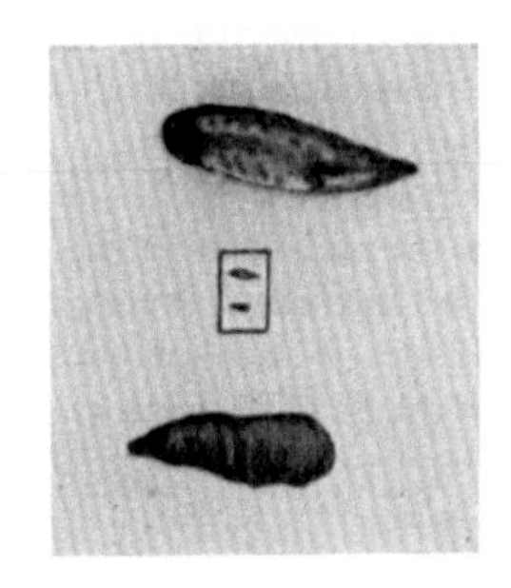

Fig. 199. Mussel Scale, upper and under surfaces, showing eggs. Magnified. Inset, natural size.

**Ova (Eggs)**

| | |
|---|---|
| APPEARANCE. | Minute, oval specks. |
| COLOUR. | Grey. |
| ARRANGEMENT. | In a mass under the scale, alongside shrivelled body of female. |
| LOCATION. | On trunk or twigs. |
| APPEAR IN | Winter. |
| HATCHING PERIOD. | June. |
| DURATION. | Throughout late winter and spring. |

**Larva or "Louse"**

| | |
|---|---|
| SIZE. | Very small. |
| APPEARANCE. | With six legs and two feelers. |
| COLOUR. | Light grey: appear as white specks on tree. |

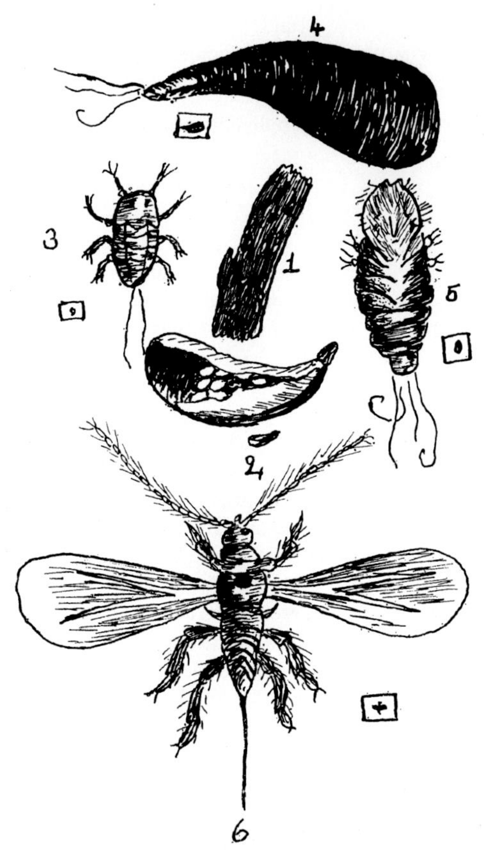


Fig. 200. Diagram of various stages of Mussel Scale insect. 1, scale on twig. 2, reverse side of scale, showing eggs, enlarged and natural size. 3, final nymph stage, enlarged and natural size. 4, female scale, enlarged and natural size. 5, insect removed from shell, enlarged and natural size. 6, male insect, enlarged and natural size.

| | |
|---|---|
| LOCATION. | On all parts of tree. |
| PROGRESSION. | Actively crawling at first, then stationary. |
| APPEARS IN | June. |
| NUMBER OF MOULTS. | One: changes into adult male or female. |
| DURATION. | Few days or weeks. |

**Pupa** No definite pupal stage exists.

## Distribution

World-wide.

## Life-history

The life-history of this insect is very curious and interesting.

The scales are very commonly seen on most fruit trees, especially in old orchards. If one of these scales be examined in spring or late winter, it will be found firmly attached to the trunk or branch, and when carefully detached by the point of a knife, the under surface magnified, a mass of grey dust—the eggs—will be seen together with the shrivelled body of the female insect.

If other scales on the tree be kept in close observation, the eggs will be seen to hatch out (in about June) into very minute, active mites, provided with feelers and six legs. These crawl all over the trees, becoming covered with a greyish white powder.

Finally they come to a standstill, and, plunging their long beaks into the tissue of the tree, cast their skin, and commence the formation of the "SCALE," which gradually envelopes their bodies. Usually they change into the female form, the scale of which is larger and narrower than that of the male. In this case, legs, feelers, etc. are shed, and the eggs formed in the body of the female are forced out under the protection of the scale, and the female dies and gradually shrivels up, the scale remaining as before externally.

In the case of the larva producing the male insect, it settles down as before, but does not produce a scale of the same character as the female. A kind of pupa stage ensues, and the male issues forth as a complete winged insect (see fig. 200). The long tube at the end of the body is now used in inserting under the scales to fertilise the females, after which the insect dies.

It does not appear to be necessary to fertilise the female upon all occasions in order to produce fertile eggs.

Fig. 201. Mussel Scale on apple tree ($\frac{3}{4}$ natural size), also same magnified (inset).

## 2. Economic

### Trees Attacked

Almost all fruit, e.g. apple, pear, plum, damson, cherry, currant, peach, etc.

### Frequency of Pest

Very common.

### Nature of Attack

Feeding upon the sap of the plant, this insect exhausts the tree and is most injurious to its health.

The normal "breathing processes" of the tree are also interfered with by the close incrustation of the scales.

The trunk and branches are mainly attacked, but scales are also found upon the leaves and fruit.

### Degree of Damage

Young trees are frequently so thickly attacked as to be killed.

### Preventive Measures

The most important of these is the *treatment of nursery stock.* It is in this way that the disease has been spread, in the past, to all parts of the world. Stock should be *guaranteed free*; or fumigated with HYDROCYANIC ACID GAS[1].

### Natural Enemies

Certain Chalcid flies destroy the scales, and the tits devour them greedily, but these cannot be relied upon to deal with an attack.

### Remedies

Spray in the *winter* with either of the following:

1. PARAFFIN EMULSION[2] with or without CAUSTIC SODA[3].
2. CAUSTIC SODA[3] alone ($2\frac{1}{4}$ per cent.).
3. LIME-SULPHUR[4] (full winter strength).

In summer, spray with weak PARAFFIN EMULSION[2], with SOFT SOAP[5] and NICOTINE[6], or with 1 in 30 (S.G. 1·01) LIME-SULPHUR[4] late in May or early June to kill the newly hatched larvæ.

[1] See page 423. [2] See page 417. [3] See page 420.
[4] See page 612. [5] See page 452. [6] See page 436.

**Calendar of Treatment**

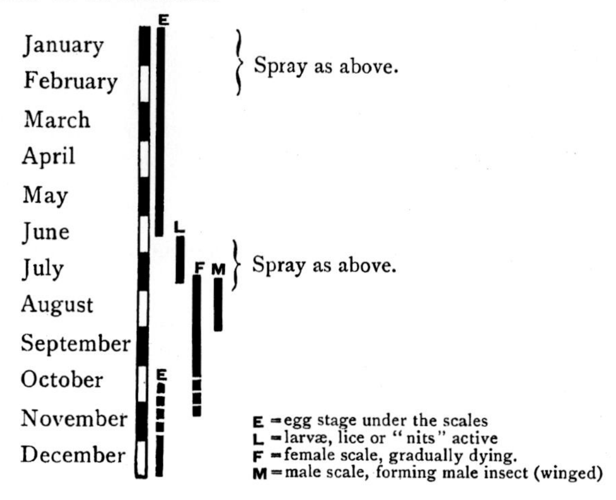


## BROWN SCALE

Name *Lecanium persicæ* Class *Coccidæ* Order *Hemiptera*

### 1. General

**Description**

**Adult Insect (Female)**

| | |
|---|---|
| SIZE. | About $\frac{1}{8}$ inch. |
| APPEARANCE. | Rounded: old scales much grooved. |
| COLOUR. | Brownish yellow. |
| APPEARS | About April or May. |
| REMARKS. | Two or three broods may also occur in one year. The MALE has not been found in this country, and probably appears only once in several generations. |

**Ova (Eggs)**

| | |
|---|---|
| APPEARANCE. | Minute, dust-like, amongst white threads. |
| LOCATION. | Under the scales on young shoots. |

| | |
|---|---|
| APPEAR | About May, occasionally in the autumn. |
| HATCH IN | June. |
| DURATION. | About 1 month. |

**Larva or "Louse"**

| | |
|---|---|
| COLOUR. | Yellowish or reddish, with fine threads. |
| LOCATION. | On all parts of the plant. |
| PROGRESSION. | Active. |
| APPEARS IN | June (or later). |
| DURATION. | Usually throughout winter (in dormant condition). |

**Life-history**

Similar to that of the Mussel Scale (page 358), but the winter is generally passed in the immature larval condition, the egg-laying taking place in late spring.

## 2. Economic

**Trees Attacked**

Currant, gooseberry, also raspberry, plum.

**Frequency of Pest**

Fairly common.

**Nature of Attack**

The insects suck the sap from the young shoots, preventing growth and sometimes killing the shoot.

**Preventive Measure**

Fresh stock should be fumigated with HYDROCYANIC ACID GAS[1].

**Natural Enemies**

Numerous, especially lady-bird, beetles and tits: but of little practical use.

**Remedies**

Spray with strong LIME-SULPHUR[2] in the winter.
This is a specific for this particular insect and cannot fail to kill if properly applied.

[1] See page 423. [2] See page 612.

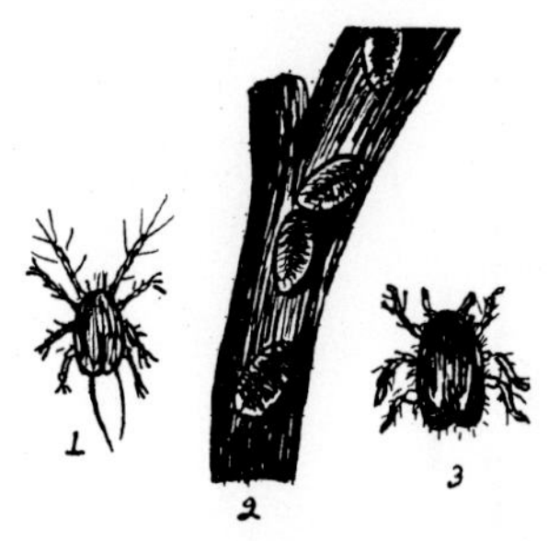


Fig. 202. Diagram of Brown Scale. 1, larva. 2, scale on branch. 3, female. Enlarged (× 2).

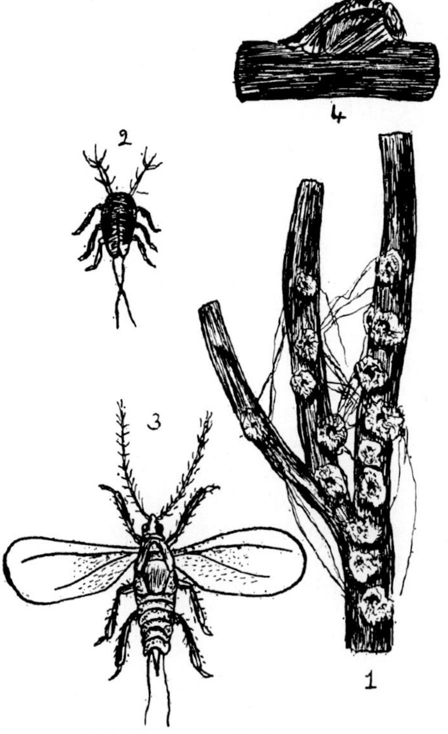


Fig. 203. Diagram of Woolly Currant Scale. 1, scale on branch. 2, larva, magnified. 3, male (adult), magnified. 4, female scale, magnified.

**Calendar of Treatment**

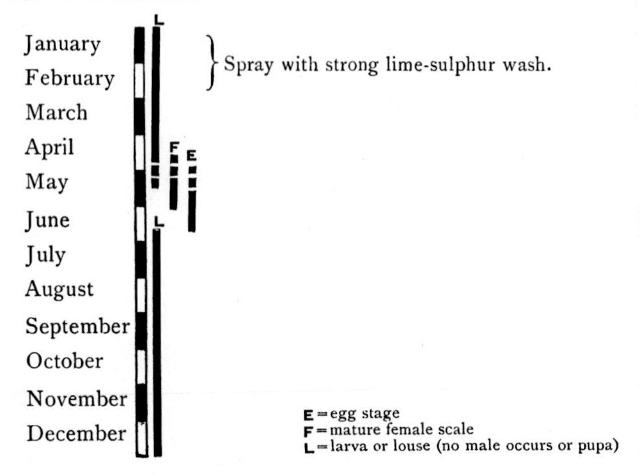


## WOOLLY CURRANT SCALE

This very curious scale occurs occasionally on currant bushes in this country. The appearance during summer is very characteristic; white woolly pads and hairs cover the bushes and extend from branch to branch.

The larvæ hatch out in June and are minute orange-coloured lice, very active. They disperse over the plant and select the young shoots and here anchor and become hardened. A cushion of "wool" appears in early spring and lifts one end of the scale up. Loose strands blow about among the branches. On this wool the eggs are laid and the female then dies and shrivels up. The eggs become distributed by wind and by birds. A large amount of sticky "honey-dew" is produced in the spring. (See fig. 203, page 363.)

Treatment etc. as for Brown Scale (page 362).

# MITES AND SPIDERS

(*ARACHNOIDEA*)

# CHAPTER 22

## Mites

These are not of course insects, but belong to the same order as spiders. Propagation is by egg-laying, but there is nothing corresponding to the *pupal stage* of an insect. The young usually varies slightly from the adult.

The Gall Mites infest buds and leaves, while the so-called "spiders" suck the juices of plants in a similar manner to the aphis, and treatment follows the same lines (see chapter 18).

## **BIG BUD** (Currant gall mite)

Name *Eriophyes ribis* Class *Acarina*
Order *Arachnoidea*

### 1. General

**Description**

**Adult Mite**

| | |
|---|---|
| SIZE. | Very minute (barely $\frac{1}{100}$ inch long). |
| APPEARANCE. | Semi-transparent, glistening: thin cylindrical shape, with few long bristles (two at tail-end). |
| COLOUR. | White or faintly greenish. |
| LEGS. | Four at end of body, around the mouth. |
| MOUTH. | Sucking and chewing (with jaws and sucker). |
| PROGRESSION. | By means of a claw at tail-end the mite can traverse the outside of the buds. |

| | |
|---|---|
| LOCATION. | (1) In great numbers in the buds apparently all the year round. |
| | (2) During April and May, occur also on outside of buds or on stems and flowers. |
| | (3) Possibly some enter the soil for a time. |
| APPEARS IN | Most of the year in the buds. |
| DURATION. | Mites apparently die in the dead buds, but their actual length of life is undetermined. |

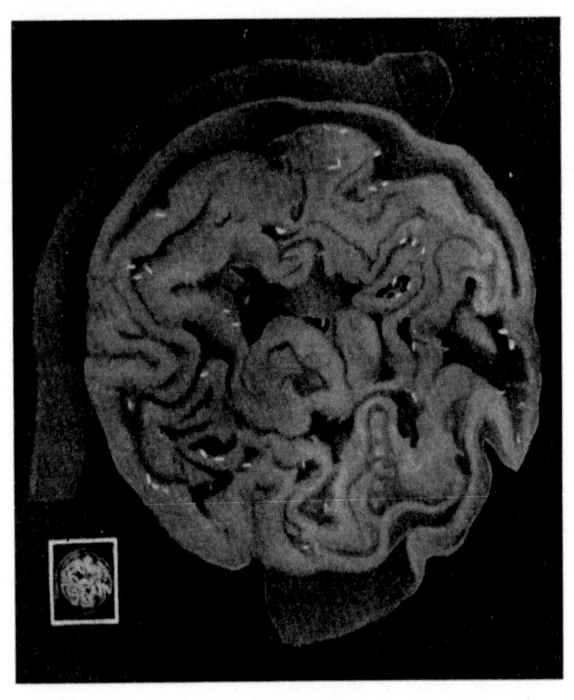

Fig. 204. Cut through a bud (radial section) showing mites. Magnified and natural size (inset).

**Ova (Eggs)**

| | |
|---|---|
| SIZE. | Occur as a whitish powder in the buds. |
| APPEARANCE. | When magnified, spherical, pale green, transparent. |

| | |
|---|---|
| LOCATION. | In thousands inside the swollen buds. |
| APPEAR | All the year round. |
| DURATION. | Very variable. |

### Young

These vary in size and appearance and several moults appear to take place before the adult stage is reached.

## Distribution

Widespread in England.

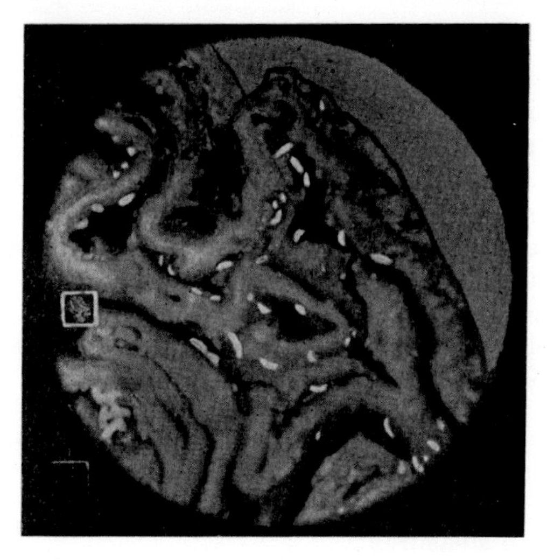

Fig. 205. As fig. 204, but more highly magnified.

## Life-history

The full life-history of this mite has not yet been traced.

The mites occur in the swollen buds apparently at most times of the year, together with many eggs.

Owing to the fact that they still appear in diseased buds when the plant has been entirely cut down and the roots cleansed, it is probable that some of the mites also enter the ground and re-infest the plant.

In April and May, when some of the buds burst, mites crawl over the plant by means of the claws at their tail-end.

The disease is distributed by the agency of wind, insects, clothing, and especially by cuttings from diseased stock.

**Remarks**

Healthy bushes in a plantation are infected from neighbouring plants by the migration of the mites in March and April, and once infected, the bush will be completely covered with "big buds" in the following season.

As regards "nettle-head" or "reversion" in currants, see Appendix, page 704.

## 2. Economic

**Trees Attacked**

Currants.

**Susceptible Varieties**

Many "immune" varieties have been produced, only to become attacked after longer or shorter intervals[1].

**Frequency of Pest**

Common in most parts of the country.

**Nature of Attack**

The buds, instead of appearing conical and healthy, swell up and become bloated and mealy in character.

If the attacked buds open, no fruit is as a rule produced.

Generally the buds turn brown and die during the summer.

**Degree of Damage**

Very serious once infection has taken place.

**Preventive Measures**

1. Insist upon healthy stock from the nurseryman.
2. Immerse cuttings in water at 115° F. for ten minutes before planting, or in cold water for some days.
3. Plant *strong growing varieties*.

**Remedies**

1. Hand pick swollen buds on their first appearance. It is advisable also to remove the buds *above and below* the infested ones, as these usually contain mites[2].

[1] Seabrook's black variety is however apparently more resistant. This appears to be not because it is immune from attack, but because the attacked bud dies and so the mites soon perish.

[2] Theobald.

The plantation should be gone carefully through twice a year, and the infested buds placed in baskets and burnt.

2. Spray in the winter with CARBOLIC ACID[1] 5 per cent. and SOFT SOAP[2] 10 per cent. This destroys the big buds by penetration and does not affect the healthy ones[3].

Fig. 206. Healthy (left) and attacked (right) stems of black currant, showing the "big buds."

3. Spray[3] when first leaves are as big as a sixpence (see No. 6, fig. 246, page 620) with extra strong LIME-SULPHUR (1 in 12 of 1·3 S.G.)[4].

[1] See page 410. [2] See page 452. [3] Lees. [4] See page 612.

## LEAF BLISTER MITE

Name *Eriophyes pyri* Class *Acarina* Order *Arachnoidea*

### 1. General

**Description**

**Adult Mite**

| | |
|---|---|
| SIZE. | Minute (about $\frac{1}{120}$ inch long). |
| APPEARANCE. | Cylindrical, with ring markings round body. |
| COLOUR. | White or pinkish. |
| LEGS. MOUTH. PROGRESSION. | Similar to the Currant *Big Bud* mite. |
| LOCATION. | Winter—under bud scales. Spring—entering leaves through stomata (breathing pores). Summer—in galls of leaves. Autumn—exposed on the trees and entering bud scales. |

**Ova (Eggs)**

| | |
|---|---|
| SIZE. | Minute. |
| APPEARANCE. | Oval, translucent white. |
| LOCATION. | In the tissue of the leaves. |
| HATCH IN | About a week. |

**Young**

These resemble the adult: they lie in a curved position inside the galls on the leaves.

**Distribution**

World-wide. Not common in Britain till fairly recently.

**Life-history**

The winter stage is passed by the mites beneath the bud scales in groups of a score or so. In spring the young leaves of the bud are entered through the stomata (breathing pores).

Each mite produces a minute red blister, gradually increasing in size, and here eggs are deposited.

Later the newly hatched mites crawl out and enter fresh leaves continuously throughout the summer, the old galls darkening in colour.

In the autumn the mites leave the leaves and enter the buds under the scales.

## 2. Economic

### Trees Attacked

Pear, apple.

### Frequency of Pest

Becoming commoner.

### Symptoms of Attack

The young opening leaves are observed to have minute red blisters (galls) which increase in size. The buds are also attacked. On the larger leaves the blisters may be very minute or reach ¼ inch in diameter, and may fall off.

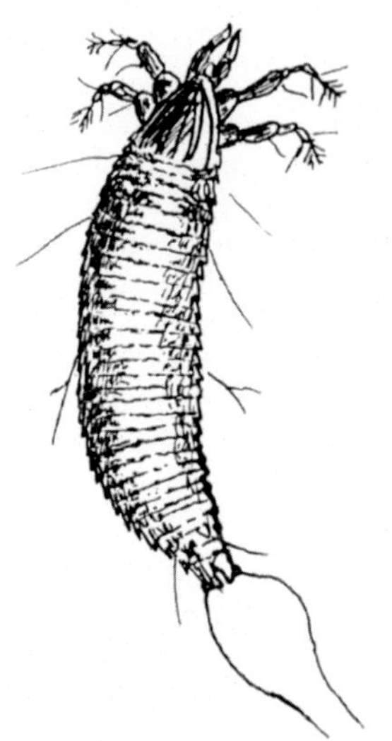

Fig. 207. Diagram of Leaf Blister Mite. Greatly enlarged.

### Preventive Measure

The fumigation of all nursery stock with HYDROCYANIC ACID GAS[1] kills any mites in the buds.

### Remedies

1. Spray with strong LIME-SULPHUR[2] wash either in late autumn or early spring.
2. In small plots, hand pick and destroy affected leaves as early as possible.

[1] See page 423.

[2] See page 612.

# CHAPTER 23

## Red Spiders

In general, the name applies to any small, red, spidery looking mite seen upon trees. Few of these however appear to be harmful. Thus the red spider (*Tenuipalpis glaber*) found in immense numbers on plum trees, which is familiar in the winter stage sometimes covering the shoots with the small round red eggs, is apparently quite harmless.

The Gooseberry Red Spider is another kind altogether (*Bryobia*) and is well known for its attack on the leaves of this plant. Various species of *Tetranychus* occur, and the one causing the most damage is the spider on the hops (*T. malvæ*). Treatment follows the same lines in all cases of spider attack.

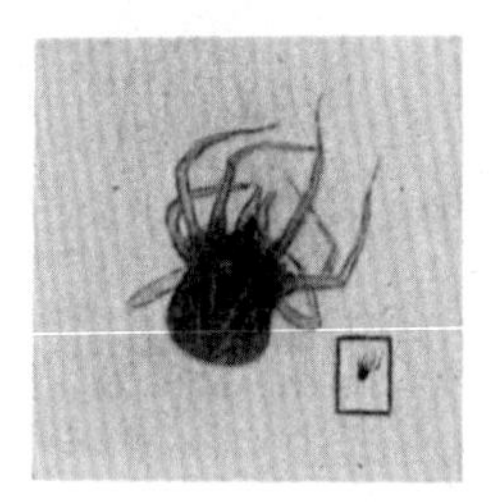

Fig. 208. A harmless variety of Red Spider. Enlarged and natural size (inset).

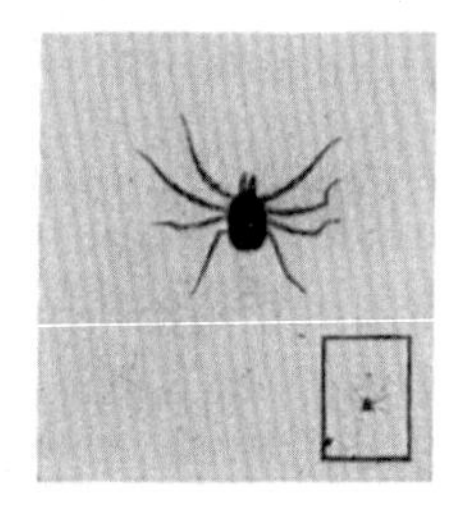

Fig. 209. Another common, and apparently harmless, variety of Red Spider. Enlarged and natural size (inset).

## HOP RED SPIDER

Name *Tetranychus malvæ* Class *Acarina* Order *Arachnoidea*

### 1. General

### Description

#### Adult Mite ("spider")

| | |
|---|---|
| SIZE. | Female about $\frac{1}{25}$ inch long: male about half this. |
| APPEARANCE. | Oval, with four pairs of legs, armed with claws. Legs all about *equal in length.* |
| COLOUR. | Variable, reddish to greenish brown. |
| APPEARS IN | Spring. |
| LOCATION. | On undersurfaces of leaves amongst webs. |
| DURATION. | Till autumn. |

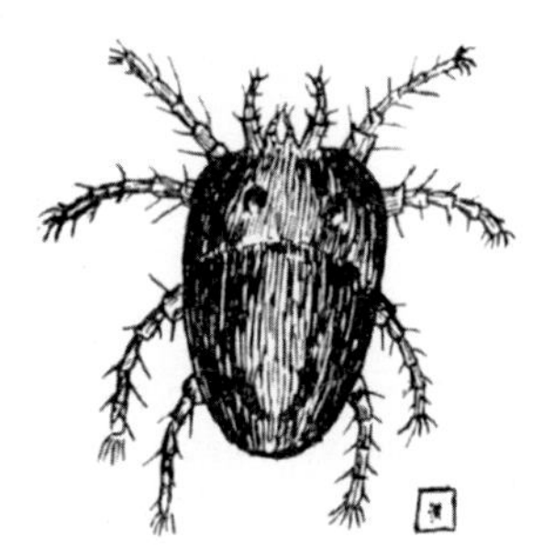

Fig. 210. Hop Red Spider. Enlarged and natural size.

#### Ova (Eggs)

| | |
|---|---|
| APPEARANCE. | Globular, pale yellow. |
| LOCATION. | Amongst and attached to webs on under surface of leaf. |
| APPEAR | Throughout summer. |
| DURATION. | Few days. Eggs laid in autumn probably last throughout winter. |

#### Young

These are distinguished by having only six legs. They rapidly develop into adults.

### Distribution

Widespread in the hop districts.

**Life-history**

The "spinning mites" (which are not true spiders) appear in spring, probably from eggs, though some may winter in crannies or under the ground.

They commence to spin fine webs on the under surfaces of the leaves, attaching these to the leaf hairs.

Eggs are laid and secured to the web, and these rapidly hatch out and the young spiders quickly become adult. This accounts for the rapid increase of the pest in suitable weather conditions. The spiders disappear in the late summer.

**Remarks**

Distinguished from the Gooseberry Red Spider by its front legs, and from the harmless variety on the plum by the fact of its web. (See pages 374, 377, 378.)

## 2. Economic

**Plants attacked**

Hops principally.

**Frequency of Pest**

Not infrequent.

**Nature of Attack**

The Hop Red Spider feeds upon the leaves, and causes these to appear greyish or marbled above and shiny beneath owing to the fine web. In severe attacks the hops are greatly weakened, and have a marked droop.

**Conditions favouring attack**

Dry sunny weather appears most favourable to attack, and the spider reproduces very rapidly under such conditions.

**Degree of Damage**

Occasionally serious.

**Remedies**

1. SULPHURING[1] the hops, as is carried out for mould, is a good remedy.
2. Spraying with SOFT SOAP[2] and either LIVER OF SULPHUR[3] or NICOTINE[4] or both will kill the spider. Plenty of soft soap must be used in order to penetrate the web.

[1] See page 626. [2] See page 452. [3] See pages 431, 624. [4] See page 436.

**Remarks on Remedies**

Care must be taken in using liver of sulphur on hops, especially in hot sunny weather, as it is very liable to scorch. (See page 625.)

## GOOSEBERRY RED SPIDER

Name *Bryobia pretiosa* Class *Acarina* Order *Arachnoidea*

### 1. General

**Description**

**Adult Spider**

| | |
|---|---|
| SIZE. | Small ($\frac{1}{30}$ inch adults) but larger than Hop Red Spider (*Tetranychus*). |
| COLOUR. | Usually bright red; but may be greyish or greenish. |
| APPEARANCE. | *Four pairs* of legs, front pair *very long*. |
| OCCURS IN | Early spring and throughout summer. |
| LOCATION. | On under surfaces of leaves, in early spring in crevices on the stem. |
| REMARKS. | This spider does not spin any web on the leaves. |

**Ova (Eggs)**

| | |
|---|---|
| SIZE. | Very minute (almost microscopic). |
| COLOUR. | Shiny red, with a few white hairs. |
| LOCATION. | On twigs and base of thorns. |
| APPEAR | From early spring onwards (till about June). |
| HATCH IN | 4 to 5 days normally. Some eggs probably remain over winter. |

**Young Stage**

| | |
|---|---|
| APPEARANCE. | Small, semi-transparent, with *six legs*. |
| LOCATION. | On under surfaces of leaves. |
| MOULTS. | One; the mite is then an adult spider. |
| APPEARS | Throughout spring and early summer. |
| DURATION. | 4 to 5 days. |
| REMARKS. | They are apparently able to reproduce in under 3 weeks from birth. |

**Distribution**

Occurs wherever the gooseberry is grown.

**Life-history**

This spider probably passes the winter in the egg stage, and hatches out in early spring. Throughout spring and early summer eggs are laid which take only a few days to hatch and less than a month to produce the mature spider. The young hatching out from this egg have only *six legs*, while the adult has *eight legs*, the *front pair* being *very long*.

A remarkable fact about this spider is the extremely rapid rate at which increase can take place at times in suitable weather. Proably in these cases the immature stage is much shorter than normal.

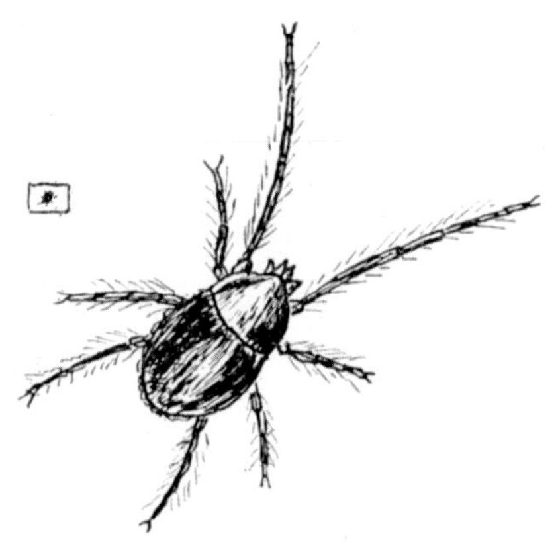

Fig. 211. Diagram of Gooseberry Red Spider. Enlarged and natural size.

**Remarks**

Distinguished from the red spider of hops and other fruit by its larger size, its long pair of front legs, and the absence of a web.

## 2. Economic

**Trees Attacked**

Gooseberries.

**Frequency of Pest**

Common.

### Symptoms of Attack

The older leaves become marbled grey, or silvery in appearance, and the younger leaves are stunted in growth. In bud attacks, both leaves and blossoms and even the young fruits fall off.

### Nature of Attack

The sap is drawn from the leaves by the suckers with which the spider is provided.

### Duration of Attack

Variable: the spiders suddenly increase under suitable conditions, which appear to be warm, dry, sunny weather. They also are more active and do more harm under these weather conditions.

### Degree of Damage

Serious, if in sufficient numbers.

### Preventive Measure

Spraying with CAUSTIC SODA[1] during the winter has been found very beneficial.

### Remedies

There are three substances of recognised value for Gooseberry Red Spider: NICOTINE, SULPHUR COMPOUNDS, PARAFFIN.

At times, one or other of these has failed in effect. It is therefore good practice to spray with all three, using a compound wash, made up as under, or bought specially prepared.

The best time to spray is early March, when a strong wash can be used. If sprayed later, the weaker mixture must be employed, and if there is *any doubt* as to the efficiency of the emulsion, it is best to leave out the paraffin altogether. This will however do no harm in February and early March.

### Remarks on Remedies

1. *February and early March Spraying*: soft soap 15 lbs., paraffin oil 5 gallons, liver of sulphur 4 lbs., nicotine 4 ozs., per 100 gallons wash.
2. *Summer Spraying*: soft soap 15 lbs., paraffin oil 2 gallons, liver of sulphur 2 lbs., nicotine 5 ozs., per 100 gallons wash.

For directions in emulsifying the paraffin, see page 418.

### Calendar of Treatment

The egg, young and adult stages overlap so much that it is useless to construct a calendar.

[1] See page 413.

# EELWORM

(*VERMES*)

# SNAILS AND SLUGS

(*MOLLUSCA*)

# CHAPTER 24

## Worms, Snails and Slugs

## EELWORM

Name *Tylenchus devastatrix* Order *Vermes*

### 1. General

**Description**

**Adult Worm**

| | |
|---|---|
| SIZE. | From $\frac{1}{30}$ to $\frac{1}{15}$ inch long. |
| APPEARANCE. | Very long, thread-like, and much twisted. |
| COLOUR. | Creamy white. |
| PROGRESSION. | By twisting movements of the body. |
| LOCATION. | In the roots of the infested plants, or in soil, manure, etc. |

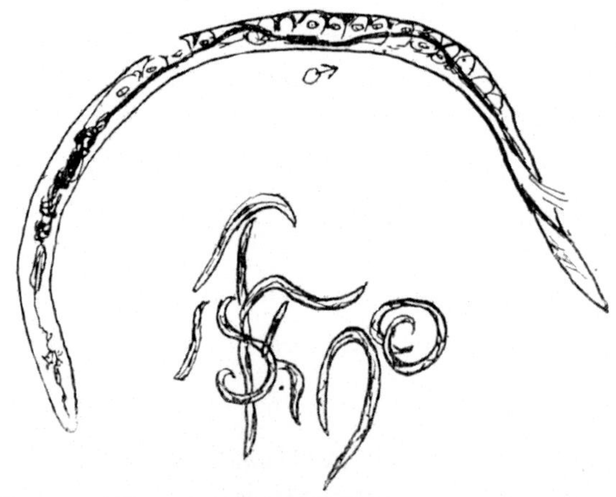

Fig. 212. Diagram of Strawberry Eelworm. Greatly enlarged. Also bunch of worms.

**Ova (Eggs)**

| | |
|---|---|
| SIZE. | Minute, oval. |
| COLOUR. | White. |
| LOCATION. | In the infested roots. |

### Distribution

In most districts of England.

### Life-history

The worms enter the roots from the soil, and cause them gradually to decay. They breed inside the root tissue, and on its death they enter the soil and attack neighbouring plants.

## 2. Economic

### Plants Attacked

Many grasses and roots : amongst fruit, mainly strawberries.

### Symptoms of Attack

The plants gradually fail and finally wither. No marked symptoms appear, and the roots must be carefully examined with a hand magnifier to detect the presence of the worm.

### Remedies

The usual SOIL INSECTICIDES[1] should be watered into the ground surrounding the plants, and if these are badly attacked, they should be taken up and burnt.

# SNAILS AND SLUGS

Name *Helix, various species, Limacidæ* Order *Mollusca*

## 1. General

### Description

#### Adult

| | |
|---|---|
| APPEARANCE, ETC. | Too well known to need description. |
| APPEAR | All the year. |
| DURATION. | Probably some years. |
| LOCATION. | Under stones, etc. in daytime: on the leaves of plants at night. |
| REMARKS. | Active after rain. |

[1] See page 433.

**Ova (Eggs)**

| | |
|---|---|
| ARRANGEMENT. | In heaps of 10 to 50 or more. |
| LOCATION. | In the soil. |
| APPEAR IN | Usually August to November. |
| HATCHING PERIOD. | 2 to 3 weeks. |

**Distribution**

Found universally in temperate climates.

**Life-history**

The slugs and snails are not all, or entirely, destructive to plants, as many feed upon worms and other slugs.

They are particularly active after rain. During the winter they hide in a dormant state under stones, debris, etc., living in some cases many years.

## 2. Economic

**Plants Attacked**

Strawberries are the chief sufferers, though most small plants are attacked.

**Nature of Attack**

Both the fruit and the leaves are eaten.

**Natural Enemies**

The thrush is the most beneficial bird, devouring many hundreds of snails and slugs.

**Remedies**

1. Pen ducks amongst the plants: these devour the slugs greedily.
2. Dress the ground with soot and lime.
3. Sprinkle SOIL INSECTICIDES[1] over the surface.

[1] See pages 392, 433.

# SECTION V

## INSECTICIDES

# SECTION V

## INSECTICIDES

## CHAPTER 25

### Introduction

Insecticides may be divided into five main classes:

1. STOMACH POISONS,
2. CONTACT POISONS,
3. CORROSIVE POISONS,
4. FUMIGANTS,
5. PREVENTIVES.

1. **Stomach poisons** are employed against EATING INSECTS (i.e. those which have jaws to chew their food) principally CATERPILLARS and BEETLES. The poison is then taken with the food and acts through absorption in the stomach and intestinal tract of the insect.

Only those insects which feed upon the surface of plants can be attacked in this manner. Caterpillars, beetles, and sawfly larvæ which live inside the tissues of plants, as the caterpillars of the codling, blister moth, pith moth, goat, clearwing, wood leopard, etc. cannot be reached by this means.

Since most of the stomach poisons available are also poisonous to man, care must be exercised throughout their preparation and employment. Thus, trees having fruit in a mature state cannot be sprayed without danger of poisoning the latter, and the grass in orchards where such poisons are sprayed will be rendered poisonous to stock for some period after the spraying of the trees.

The chief stomach poison employed is:

**Arsenate of Lead.**

In the past the following have been used to a considerable extent but are now almost given up, on account of their liability to cause damage to the foliage:

PARIS GREEN,
LONDON PURPLE.

Although trials with it have not been favourable in all cases, damage to the leaf having been reported, the following compound is being increasingly employed on account of its cheapness:

ARSENATE OF CALCIUM.

All these are compounds of *arsenic*. Two other substances which have been used satisfactorily in particular cases are

POWDERED HELLEBORE,
POWDERED PYRETHRUM.

Another chemical which has met with some success is

Chromate of Lead.

All these substances may be used either in the dry form as powders blown on to the plant, or, as usually and more efficiently employed, in suspension in water (hellebore is not used in this manner) and applied to the plant as a **misty spray**. The *character of the spray* is important, the idea being to just *coat* the surfaces of the leaves, etc. with the substance, but not to wet the leaf sufficiently for the fluid to run off to any extent.

None of these poisons are of any value for *sucking insects*[1].

### 2. Contact Poisons

These are employed against SUCKING INSECTS, i.e. those insects which live upon the plant juices of fruit trees, and possess some form of sucking beak adapted for piercing the tissues of the plant. Such insects are the APHIDES (or "lice"), PLANT BUGS, and SUCKING MITES AND SPIDERS. The poison is taken into the body, probably chiefly through the breathing pores (spiracles) of the insects and also possibly by direct absorption through the tender skin.

The most valuable of this class of insecticides is

**Nicotine** (or Tobacco).

This substance is an extremely active insecticide and has the great advantage of being without action in weak solutions on plant leaves and tissues.

Another popular contact poison is

QUASSIA.

[1] With the exception of pyrethrum.

Quassia appears also to have a beneficial action upon the foliage of the plant, and is thought by some to prevent future attacks. It is however much less active than nicotine and tobacco.

In addition to nicotine and quassia, there are doubtless many other substances obtainable from plant or animal sources which would be satisfactory contact insecticides. One or two are already known and form the basis of proprietary remedies, while others remain to be discovered. As a contact insecticide however, nicotine in its pure form, if reasonably cheap, leaves little to be desired. Of recent years it has unfortunately been difficult to obtain adequate supplies in this country.

CARBOLIC ACID probably comes into this class; its use has however almost died out on account of its scorching tendencies.

PYRIDINE, a substance allied chemically to nicotine, has been shown to have some value, especially against the capsid bug, but it is very much inferior in this respect to nicotine.

### Soft Soap.

This substance is, by itself, an efficient contact insecticide, acting by blocking up the pores of the insect and so suffocating it. It is however in a class to itself, owing to its *spreading action* when used with many other insecticides. It enables such insecticides to have a more certain action, by ensuring a thorough WETTING OF THE INSECT with the spray. This is discussed in detail on page 452.

Other substances having a similar wetting power (though to a less extent) are

SAPONIN,
Starch and Flour Pastes,
Glue, etc.

Only the first of these is used in practice.

Contact poisons require quite a different method of spraying from that used with the other class (stomach poisons). It is necessary here to *thoroughly wet the body of the insect*, and a high pressure of the pump is required together with the use of plenty of the liquid. This is further discussed on pages 632, 661.

## 3. Corrosive Poisons

These substances have a *corroding* or burning action on the skin or exterior surface of insects or their eggs. They are mainly used during the winter, in the dormant state of the tree, as they all tend to injure the leaf unless in a sufficiently dilute state. Their efficiency on insect eggs, even in the strongest forms possible to use, is open to grave doubt, and unreliable statements have been made of their success in this direction. The reason for this has been the natural shrivelling up of unfertile eggs which always occurs to a large extent in the aphis and bug class of insects.

The commonest of the corrosive poisons are

CAUSTIC SODA (or potash),
LIME.

Lime has also a further use which is described below. The action of the following upon insects is a little uncertain. Besides behaving as corrosives, they may also to some extent work as contact poisons (see above).

LIME-SULPHUR,
LIVER OF SULPHUR,
PARAFFIN and other mineral oils.

Liver of sulphur has also a destructive action on *red spiders* and lime-sulphur on *scale insects*. These are also both used largely as *fungicides* (Section VIII).

Paraffin is occasionally used by itself on trees in the dormant state, but is usually *emulsified* with soft soap, or in other ways. It may act as a contact poison to some extent. The nature of emulsions and their action are discussed on page 417.

Corrosive poisons require to be sprayed in a similar manner to the previous class (contact poisons). Plenty of liquid is to be used and a high pressure maintained in the pump.

GAS LIME,
KAINIT,
Soot,
Ashes.

These are used as insecticides for insects under the ground and probably belong to the class of corrosive poisons, though relatively weak in their action.

SALT.

This substance appears to have a pronounced action on insect eggs, especially when used with lime and particularly in the case of

*Psylla* (Apple Sucker). Its action is uncertain, but it probably extracts water from the vital fluid in the interior of the eggs by what is known as *osmotic action.*

4. **Fumigants.**

This class of insecticides acts in the form of vapour or gas and kills by being absorbed through the breathing pores of the insect. Fumigants are used on plants which can readily be enclosed in confined spaces, on nursery stock, and against insects in the soil. For insects above ground the most efficient fumigant is

HYDROCYANIC ACID GAS.

This is prepared from a *cyanide* as described on page 423. It is extremely poisonous to human beings and great care must be exercised in its use. It is very active and insects are poisoned by a few seconds exposure to the pure gas.

NICOTINE or Tobacco.

This is employed as a fumigant in the form of vapour by burning the tobacco leaves or evaporating the nicotine in a confined space.

For insects in the ground the most commonly employed substances are

CARBON BISULPHIDE,
NAPHTHALENE.

The former is a liquid which readily vaporises. It is injected into the ground by means of a syringe and is chiefly used against root forms of aphis (woolly aphis, phylloxera of vine, etc.).

Naphthalene is a solid which also vaporises, and is used in the form of a powder dug into the ground. It forms the basis of some proprietary soil-insecticide compounds.

5. **Preventives.**

Under this class come those substances which stop the insects from reaching the leaves and branches of the trees, and which prevent egg-laying or egg-hatching. The one of the greatest utility is the material known as

BANDING COMPOSITION (or "grease").

This is placed on bands around the trunks of trees to prevent insects from reaching the branches. It is chiefly useful against the *wingless female of the Winter Moth*, which crawls up th trunks to lay its eggs in the branches. Compounds vary very much in their efficiency (see page 405).

LIME.

This, when sprayed in early spring on the trees, prevents the successful hatching of insect eggs by covering them over with an adherent coat. The new born insect is thus imprisoned and killed. Substances which have been used for a similar purpose include

SILICATE OF SODA,
GLUE (gelatine).

# CHAPTER 26

## **Insecticide Materials** (arranged in alphabetical order)

## ARSENATE OF CALCIUM

**See also Arsenate of Lead, etc.**

### 1. General

**Employment**

For destroying caterpillars etc. (as for arsenate of lead, page 397).

**Description**

Calcium arsenate (or "arsenate of lime") is a white powder, similar in appearance to arsenate of lead, but not so heavy. It is not at present manufactured in this country in any quantity, but may be made by the grower himself as described later.

There are very varying opinions as to its liability to scorch the trees, and many authorities condemn its use on this account. In insecticidal value it is about equal to arsenate of lead.

**Preparation**

It is necessary to have lime present in excess of the amount required to form the arsenate, otherwise scorching is *bound to occur*. The following method has been found to give satisfactory results in America:

FOR THE PREPARATION OF 100 GALLONS OF SPRAY SOLUTION.

1. Take **2½ lbs. of good quality[1] quicklime** ("*Buxton*" is the best to use in this country if obtainable) and slake it in a bucket by gradual additions of water. Mix into a thin cream with further water.

2. Dissolve **2 lbs. of commercial, "fused," arsenate of soda[2]** (or 3½ lbs. of crystallised arsenate) in **50 gallons of water** in a large tub of not less than 100 gallons capacity.

[1] The lime must not be "air-slaked" and should be in lump form and white or greyish-white in colour (see page 428).

[2] This is a violent poison and must be handled with very great care. For first aid treatment in case of poisoning see Appendix VII, page 705.

3. Pour the "milk of lime" (obtained as in 1) slowly, with constant stirring, into the arsenate solution. Then add water to make **100 gallons** of spray.

This is now ready for use.

If more than this proportion of lime is used, there may be scorching of the leaves due, not to the arsenate, but to the caustic soda produced by the reaction of the lime on the arsenate.

The ideal way would be to make say three trials, varying the amount of lime slightly each time, and find with the particular materials available which gives the least harmful mixture.

## 2. Method of Use

### Strength of Spray

As above.

### When and how to Spray

As for arsenate of lead (page 402).

### Quantity of Spray Used

As for arsenate of lead (page 402).

### Remarks

As scorching has been found with arsenate of calcium, it is advisable to make a trial first before using on a large scale.

## 3. Scientific

### Chemical Composition

Apparently the tri-basic salt is produced, $Ca_3(AsO_4)_2$ = 57·8 per cent. $As_2O_5$, on precipitation of arsenate of soda solution with milk of lime.

### Properties

Arsenate of calcium is soluble in water about 1 in 200, so is much too soluble for spray purposes.

If excess of lime is employed (about half as much again) the arsenate remains almost insoluble.

### Tests for Purity

Would be similar to those for arsenate of lead (page 403). The commercial product is hardly available at present.

### Insecticidal Value

About equal to arsenate of lead.

# ARSENATE OF LEAD

**See also Arsenate of Calcium, Paris Green, London Purple, Chromate of Lead**

## 1. General

### Employment

For the destruction of caterpillars, beetles, sawfly larvæ and *all leaf-eating insects.*

### Description

Arsenate of lead is a heavy white powder. It dissolves in water to only a very slight extent. It is obtained by adding a solution of a lead salt (such as the acetate or nitrate of lead) to a solution of a soluble arsenate (such as arsenate of soda) or arsenic acid. Although not so poisonous as "white arsenic" or as a soluble arsenate, it is, nevertheless, a powerful poison if taken into the stomach, and must be handled with great care[1]. It is a stable substance, i.e. not liable to change on keeping or exposure to air, and consequently its value as a spraying material does not alter on storing with the exception that, in the form of a *paste*, it is liable to dry and become hard by evaporation of the water (see below).

For spraying purposes it can be made by the grower by adding a solution of acetate of lead ("sugar of lead") to a solution of arsenate of soda. The quantities of each need to be very carefully adjusted or scorching of the leaves will result. Particularly is it necessary to avoid excess of arsenate of soda. For this reason, it is much better for the grower to use the *factory prepared arsenate of lead*, the forms in which this is obtainable being detailed below.

Arsenate of lead is a heavy substance and when used, suspended in water, for spraying tends to settle rapidly. The coarser the particles of arsenate are, the more quickly this settlement takes place. It is therefore very important to obtain it in as *fine a form as possible*, and manufacturers vie with one another in producing it in the finest obtainable condition. Simple tests for the grower to try with his own purchases are described below.

### Preparation

On the large scale, arsenate of lead is prepared by adding a soluble salt of lead to a soluble arsenate in the correct proportions. The

[1] For first aid treatment in case of poisoning see Appendix VII, page 705.

white insoluble arsenate of lead settles out, and is obtained by filtration, and either dried to a powder or left in the form of a paste. If the grower wishes to prepare his own wash (which is not advised except in special cases) he should proceed as follows:

**To make 100 gallons of spraying fluid.**

Dissolve **10 ozs. of "crude" or "dry" arsenate of soda**[1] or **17½ ozs. of "crystallised" arsenate of soda** in **50 gallons of water** in a tub. This may be best done by tying the solid in a piece of sacking and suspending it in the water just under the surface. It may then be left till completely dissolved. Meanwhile, dissolve in a similar manner **35 ozs. (2 lb. 3 ozs.) acetate of lead** in a further 50 gallons of water in another vessel. When both are dissolved, stir the lead solution gradually into the arsenate solution. A white milky liquid is produced. This should now be tested in the following way:

Allow a little of the milky fluid to settle in a pail or can, and to the clear liquid add a little solution of arsenate of soda. An immediate cloudiness should be produced, showing the presence of an excess of lead acetate. If no cloudiness occurs, more lead acetate solution must be added to the main bulk until, when settled, it gives this test.

A more satisfactory, although a more troublesome, way to prepare the wash is to add the acetate of lead solution to the arsenate of soda solution, until about ¾ of the total is used, and then to settle each time before adding more, noticing if cloudiness is produced on further additions of the acetate. When no further cloudiness is seen, the wash is ready for use.

**Commercial Brands**

Arsenate of lead is sold in three forms:

1. Powder,
2. Paste,
3. Cream.

[1] Arsenate of soda occurs on the market in two forms, the "crystallised" and the "dry" or "crude." The latter is much stronger and is generally cheaper to buy. It however varies a good deal in strength and hence the need for testing when a grower makes up his own wash. Although a large excess of lead acetate will also harm (and is very uneconomical), the *important thing to avoid is an excess of arsenate of soda;* as this will scorch severely. The acetate of lead, as purchased, is not liable to vary in strength.

The **powder** is suitable for dry spraying, although it is not a common practice in this country (see page 706). It is however the worst form to use for a wet spray. This is because on drying the white precipitate, it loses much of its fineness, and on remixing with water will be found relatively coarse and quick-settling.

The **paste** form is the one in largest use at the present time. Pastes vary very much in the fineness of the particles and also in their chemical properties. There should be practically no free arsenate of soda present, and on mixing with the right quantity of water it should take a good time to settle (see tests below). The paste should not be too dry owing to evaporation of water, or it will be difficult to mix in the water, and will be relatively coarse.

The **cream** form is the most suitable of all. It mixes immediately on pouring into water, and its particles are generally in the finest possible state of division. Further, it may be *measured* out in making the wash, an improvement on the somewhat inconvenient weighing of the paste form.

A very important point in buying arsenate of lead in any form is to know the *strength in terms of arsenic oxide.* As the poisonous properties of the substance are due to the arsenic present and not to the lead, its efficiency, from a chemical point of view, depends on this figure. Most reputable firms will supply or even guarantee the strength in terms of arsenic oxide, "$As_2O_5$," of their product. While, however, the value of a given arsenate of lead as an insecticide is dependent upon its strength in terms of arsenic—more of the weaker products being required to be used per 100 gallons of wash made up—growers should give preference to those makes which have the finest particles and which therefore remain in suspension in the wash for the longest time. Not only do these prevent uneven spraying by not settling rapidly in the tank of the machine, but they are more efficient when sprayed on the leaf.

## Simple Tests

1. TO ENSURE FREEDOM FROM SCORCHING.

(*a*) Procure some *acetate of lead* from the chemist. Dissolve about a teaspoonful of this in water in an 8 oz. medicine bottle, using distilled, well boiled, or rain water for the purpose. (The acetate is best dissolved in 4 ozs. of water and the bottle then filled up and shaken.)

(*b*) Take about a teaspoonful of the sample of arsenate of lead, and shake it up with about 4 ozs. of rain water (or distilled). Allow to settle.

(*c*) Pour off the clear settled liquor from the top of the arsenate of lead into a clean medicine bottle, and add a little of the clear acetate of lead solution. *Only the slightest trace of cloudiness should be seen.* If a distinct milkiness appears the sample contains free arsenate in solution and there is a danger of scorching foliage. This may be remedied by the addition of sufficient acetate of lead to the made-up wash, but it is safer to refer the matter to the manufacturer.

2. TO TEST THE FINENESS OF THE PARTICLES.

This is only a rough-and-ready test, but it will give useful information.

Place a new sixpenny piece on one pan of a small pair of scales[1] and balance on the other pan the correct quantity of the sample of arsenate of lead. Shake this well up with a small quantity of water in a 16 oz. medicine bottle. When it has completely mixed with the water, fill the bottle ¾ full and shake well. Allow to stand. The liquid will now be up to the 12 oz. mark. The white powder will commence to settle at once, but the rates of settlement with various brands are very different.

Make a note of how long it takes before there is a layer of clear liquid down to the 6 oz. mark (that is, the length of time that it takes for the powder to settle half way down).

In the case of the best samples, this will not occur for many hours. In many of the coarser arsenates, it happens in as many seconds.

As stated above, this is only a rough test of fineness. For instance, although a sample may contain a great many coarse particles which settle rapidly, these will not be shown if there are also enough finer particles of arsenate to give the test. It is however an excellent sorting-out test for a grower to use.

[1] If no small scales are handy it will be well to make up a larger quantity. For this purpose, take a pound of the sample, and mix it up well with half a pailful of water in a tub. Then add a further 3½ pails of water. Stir well and at once fill a medicine bottle of this for trial.

3. TO FIND THE PERCENTAGE OF ARSENIC PRESENT.

This gives the actual strength of the sample from the insecticidal (or "killing") point of view. It is not possible for the grower to ascertain this. He should get an undertaking[1] from the manufacturer or send a sample to an analyst, asking for percentage of arsenic present *as arsenic oxide* ($As_2O_5$). At the present time samples vary from about 12 to 20 per cent. $As_2O_5$ as pastes, and about 25 to 33 per cent. reckoned on the dried paste, or as powders. About 33 per cent. is, at present, the highest possible percentage of arsenic as arsenic oxide which can exist in a genuine arsenate of lead in the dry form.

### Action on Insects

Arsenate of lead is for use with biting or chewing insects only. It is of no use for aphis ("green fly"), etc.

Caterpillars are killed in periods varying from 1 to 12 days, according to the dose of arsenic taken before they go "off feed." Thus a caterpillar may devour a whole leaf before feeling sufficiently ill to stop feeding. In most cases however caterpillars go "off feed" in a short time on leaves sprayed with arsenate of lead, and although they do not die at once, there is no further damage done.

## 2. Method of Use

### Strength of Spray

This varies with the strength (in terms of arsenic) of the material and the manufacturer's directions should be consulted. Assuming an average strength of 15 per cent. of arsenic as arsenic oxide in the pastes, an efficient strength is **3 to 5 lbs. per 100 gallons of water.**

The strength of the spray however is obviously very dependent on the character of the jet used in the machine, and the amount put on the tree. Thus some growers prefer to use up to 7—10 lbs. of paste per 100 gallons of water, and to use only about a half the usual quantity of wash on each tree, employing a very fine spray jet. The same amount of arsenate of lead actually reaches the leaves in each case (see *How to Spray* below). The strength of home-made sprays is given on page 398.

[1] Legislation will probably soon be introduced to compel the manufacturer to state the strength of arsenic present (as arsenic oxide) in his product.

### Preparation of Spray

Directions for preparing one's own spray from the original materials are given on page 398.

In the case of arsenate paste or powder, the correct amount of material is weighed and well mixed, first with a small quantity of the water in a pail. No lumps should remain. This is then poured off into the main bulk of the water and stirred well.

The liquid must be well stirred every time any is taken out for spraying, unless the machine is capable of taking the whole amount at one time.

### When to Spray

A great deal is saved by spraying *as soon as possible after the pest is first noticed.* Much less spray is used when the leafage is younger, and the small caterpillars are more quickly killed. Further, they are destroyed before the chief damage, i.e. to the young foliage and buds, is done.

Directions as to the time of the year to spray will be found at the end of each description in Section IV.

### How to Spray

A fine nozzle giving a MISTY SPRAY should be used, see figs. 252, 253, page 635. The idea is to envelop the tree in a fine "fog." If too much spray is used, or if the "mist" is too coarse, the liquid will drip off the leaves carrying the arsenate of lead with it, causing both loss of material and inefficiency.

### Quantity of Spray Used

It is only possible to give approximate figures, as the amount necessarily varies so much with the state of the foliage.

The following table has been found useful:

| Spread of trees, feet | Amount of diluted spray per 100 trees | Weight of arsenate of lead paste per 100 trees (using 4 lbs. per 100 gallons) |
|---|---|---|
| 7 | 200 gallons | 8 lbs. |
| 10 | 270 ,, | 11 ,, |
| 15 | 460 ,, | 18½ ,, |
| 20 | 720 ,, | 29 ,, |
| 25 | 1120 ,, | 45 ,, |
| 30 | 1500 ,, | 60 ,, |
| 35 | 2200 ,, | 88 ,, |
| 40 | 2500 ,, | 100 ,, |

To ascertain the amount which will be required per acre, the above figures should be multiplied by the following factors, according to the distance apart of the trees:

6 ft, 12·1; 10 ft, 4·35; 12 ft, 3·02; 15 ft, 1·93; 18 ft, 1·34; 20 ft, 1·08; 24 ft, 0·75; 30 ft, 0·48; 40 ft, 0·28.

## 3. Scientific

### Chemical Composition

There are at least two arsenates of lead in commercial pastes, and most pastes are mixtures of these two in varying proportions. These arsenates have the following composition:

$Pb_3(AsO_4)_2$ containing 25·5 per cent. of $As_2O_5$ (tri-plumbic arsenate).
$Pb_2H_2(AsO_4)_2$ ,, 33·1 ,, ,, (di-plumbic ,, ).

Pastes contain varying proportions of water, but usually about ½ of the weight of a commercial paste is water (50 per cent.).

Arsenate of lead is manufactured by precipitating a solution of arsenate of soda ($Na_2HAsO_4$) or arsenic acid ($H_3AsO_4$) with lead acetate

$$[Pb(C_2H_3O_2)_2 3H_2O]$$

or lead nitrate $[Pb(NO_3)_2]$ according to the following equation (for the acetate):

$$\underset{\text{arsenate of soda}}{2Na_2HAsO_4} + \underset{\text{acetate of lead}}{3Pb(C_2H_3O_2)_2}$$
$$372 = 1 \qquad 1137 = 3{\cdot}06$$

$$= \underset{\text{arsenate of lead}}{Pb_3(AsO_4)_2} + \underset{\text{acetic acid}}{2HC_2H_3O_2} + \underset{\text{acetate of soda}}{4NaC_2H_3O_2} + \underset{\text{water}}{9H_2O}.$$
$$899 = 2{\cdot}42$$

The di-plumbic arsenate is also produced in varying proportions.

### Tests for Purity

1. Water per cent. (by drying at 100° C.).
2. Arsenic as $As_2O_5$.
3. Insoluble powders other than arsenate of lead: by examination of insoluble residue after treatment with nitric acid: Ca, Ba, Zn, etc. by usual chemical methods.
4. Soluble arsenate and arsenic acid. In filtered liquor as above by usual methods.

### Insecticidal Value

The toxic dose for full grown caterpillars varies a good deal.

Careful investigations by the author have shown that on an average an amount of the paste (50 per cent.) equal to $\frac{1}{20}$ per cent. of the body weight of a caterpillar will stop it feeding in an hour or two and kill it in a few days.

Even if poisoned caterpillars appear to recover, they usually fail to pupate successfully.

### Remarks

Approximately 1 ton of arsenate of lead paste (50 per cent. water) is produced from 12½ cwt. of acetate, equivalent to 7½ cwt. of litharge (PbO).

## BANDING COMPOSITION

### 1. General

### Employment

Used for smearing round the trunks of fruit trees to catch insects crawling up to the branches.

### Description

The practice of "banding" fruit trees is by no means a novel one. It dates from the time of the general recognition of the damage done by the female of the WINTER MOTH (see page 176). This insect is practically wingless, and is obliged to crawl up the trunks of trees in order to lay her eggs on the branches. The caterpillars of the moth are by far the most destructive to fruit of any in this country, and as each moth is capable of laying some hundreds of eggs, the attack can be very greatly reduced by this means.

More than one experienced grower is of the opinion that the attack cannot be *absolutely eradicated* by banding the trees, even when the most elaborate precautions are used to prevent the insect passing the band (as for instance, applying the composition both on the surface of the band, and underneath it also, and using two bands, one over the other, so as to catch any on the second one which may successfully pass the first). The only explanation of this appears to be that in some rare cases the male may carry the female to the branches while in the act of pairing[1].

[1] The author understands that this has been proved to occur in some plantations abroad.

In spite of this there is no question whatever of the immense benefit resulting from careful and systematic banding. If any grower doubts this, it is a good plan to leave a few trees unbanded in an orchard and then notice the difference the following summer between these and the banded ones.

Even if the results were less successful than they usually are, it would still be an excellent practice, since it gives the grower time to get his arsenate of lead, or other poison spray, applied before the damage to the young leaves and buds is too great.

## Properties

Previous to about the year 1910, the only materials available for banding were various kinds of GREASES. These were mainly of the cart-grease or wagon-grease order, and prepared by mixing various proportions of mineral and vegetable oils and fats with or without rosin and treating the mixture with milk of lime or magnesia, etc. The disadvantage of these greases was that, sooner or later, they dried up, even when applied in a thick coat to the trunk. This necessitated renewal once or twice during the winter in order to remain efficient in their action. Further, the grease was injurious to the bark of the tree, and it had to be applied upon grease-proof paper, which was tied with string around the trunk. About ten years ago a new kind of composition for the purpose appeared from America. This was very much stickier and more adhesive than greases (and proportionately more efficient) and it remained sticky throughout the whole season, showing no tendency at all to "skin over" or "dry up."

Although this composition (called "Tanglefoot") may be placed directly upon the bark of trees without injuring them, the use of a band of paper is generally preferred, as it is better to keep the trunk of the tree clean, and the bands can be completely removed each year.

Several similar compounds of German origin, termed "moth glues," also appeared about the same time, but have no sale in this country. Since the introduction of "Tanglefoot" other similar brands have appeared, competing in price and quality with the original article. In all of these, the use of a large band, such as was advisable in the case of the old greases, is no longer necessary. This compensates for the increased cost of the material which is considerably over double the price of the greases formerly used.

### Requirements and Tests

The only tests of any value are practical ones, based upon its behaviour under the actual climatic conditions when in use. These may be summarised as follows:

1. The composition obtained should be, and should remain, sufficiently adhesive to catch both the smallest and largest insects.
2. It should show no tendency to run down the trunk of the tree, even if liberally applied and exposed to the hot sun.
3. It should remain quite sticky throughout the winter and spring and not require renewal.

The only exception to this is in the case of windy and dusty weather, when a film of dust may form over the surface of the band. In such cases, it is only necessary to pass the hand or a stick over the band, and so expose a fresh surface of the material.

4. It should not harden under frost.
5. It should not be so stiff as to render application to the band difficult.

## 2. Method of Use

### Preparation

It is not advised for the grower to attempt to prepare his own banding composition, as this is best done by the use of special plant under factory conditions and control.

### Procedure

For large plantations the following methods will be found very expeditious.

1. The paper, which should be of good, strong, grease-proof variety, is supplied in rolls and sheets. Rolls are best for small and mixed plantations. If trees are of uniform age and size, the sheets may be used and cut to strips of the required size. A width of about 5 inches is generally sufficient, and the paper should overlap about an inch when tied round the trunk.
2. One or more labourers (women are largely employed on this work) carry the paper in strips or rolls and paste it on the trees with ordinary flour paste[1]. They are followed by more who tie the string round. Two lengths of string are used one at the top and one at the bottom margin of the band, and it is

[1] If the bark of the tree is very rough and uneven it should be scraped to a smoother surface before applying the paste.

Fig. 213. Photograph of part of band on an apple tree showing the males and females of the Winter Moth caught by the composition. Natural size.

tied as tightly as possible to avoid spaces between the paper and the bark of the tree. (If the insects can crawl under, it is obviously useless to safeguard the surface.)

3. One or two workmen now follow with pails or boxes of the composition, and wooden sticks with broad ends or other suitable implements for placing the composition on the bands. This is an operation requiring some skill and experience. An even coat, not too thick, should be used, otherwise there is considerable wastage of material.

**Important.** If there is a tree stake, this *must also be banded.*

The half-closed palm of the hand is sometimes used to apply the composition, the workmen or women walking backwards round the tree, and keeping the hand in contact with the band.

## When to Band

Banding should take place *as early in the autumn as possible.* If left too late, there is a chance of the Winter Moth appearing early and the damage is then done. It is well to apply sufficiently early to ensure that every tree is banded BEFORE THE END OF SEPTEMBER.

It is useful, though not necessary, to keep bands sticky and in efficient operation until spring spraying commences, say till the beginning of May. Not only are other moths caught, such as the March, Mottled Umber, etc., but numbers of Apple-Blossom Weevil, which frequently crawl up the trunks, may be prevented from reaching the buds. In any case, the bands should be kept operative *till the end of March* to catch the *March Moth.* A periodic inspection of the bands from October onwards should be made.

It is well to remove the old bands in the early summer, and destroy them, so that there is no chance of the eggs, of which there are usually quite a number, hatching. It is always worth while to examine the bands closely and many growers keep a record of the number and variety of pests caught.

## Quantity of material required

The following figures are based upon much experience and can be relied upon as a close guide. They allow for a fairly liberal application of the banding composition.

A. PAPER. *Assuming a width of 5 inches and allowing an overlap of one inch.*

| Diameter of trees (inches) | Length of paper for 100 trees (feet) | Average weight (ozs.) | Length of paper *per acre* for distances of trees apart as under (feet) | | | | |
|---|---|---|---|---|---|---|---|
| | | | 6 | 10 | 15 | 20 | 25 |
| 1 | 34½ | 2½ | 397 | 137 | 62 | 35 | 21 |
| 2 | 60 | 4¼ | 690 | 240 | 108 | 60 | 36 |
| 3 | 87 | 6¼ | 1050 | 348 | 157 | 87 | 52 |
| 4 | 113 | 8 | 1300 | 452 | 203 | 113 | 68 |
| 5 | 140 | 10 | 1610 | 560 | 252 | 140 | 84 |
| 6 | 165 | 11¾ | 1900 | 660 | 297 | 165 | 99 |
| 7 | 191 | 13½ | — | 764 | 344 | 191 | 115 |
| 8 | 218 | 15½ | — | 872 | 392 | 218 | 131 |
| 9 | 244 | 17½ | — | 976 | 440 | 244 | 146 |
| 10 | 270 | 19¼ | — | — | 486 | 270 | 162 |

The string required is roughly double these lengths.

B. COMPOSITION. *For paper of 5 inches width, bigger bands in proportion.*

| Diameter of trees (inches) | Weight of "grease" for 100 trees (lbs.) | Weight of "grease" (in lbs.) *per acre* for distances of trees apart as under (in feet) | | | | |
|---|---|---|---|---|---|---|
| | | 6 | 10 | 15 | 20 | 25 |
| 1 | 1½ | 17¼ | 6 | 2¾ | 1½ | 1 |
| 2 | 3 | 34½ | 12 | 5½ | 3 | 2 |
| 3 | 4½ | 53 | 18 | 8 | 4½ | 3 |
| 4 | 6½ | 75 | 26 | 11¾ | 6½ | 4 |
| 5 | 7¾ | 89 | 31 | 14 | 7¾ | 4¾ |
| 6 | 9 | 103 | 36 | 16¼ | 9 | 5½ |
| 7 | 10 | 115 | 40 | 18 | 10 | 6 |
| 8 | 13 | 150 | 52 | 23½ | 13 | 7¾ |
| 9 | 14½ | 166 | 58 | 26 | 14½ | 8¾ |
| 10 | 15½ | 178 | 62 | 28 | 15½ | 9¼ |

## CARBOLIC ACID

### Employment

Carbolic acid or "Phenol" has, in the past, been largely used both as an insecticide and fungicide. It is however relatively weak and is liable to cause injury to the leaves. For this reason it has been superseded by more powerful and suitable substances.

### Description

Carbolic acid is obtained from coal tar. When pure it is a solid substance, crystallising in long needles. It however liquefies with the slightest trace of water. It is soluble in water to the extent of 5 per cent. and has powerful burning properties. It is also known as one of the best antiseptics (disease-germ destroyers).

### Application

It cannot be used at a higher strength than **10 lbs.** to **100 gallons** of water, as injury to the plant then occurs. At this concentration it is a weak insecticide for aphis, psylla (sucker) and plant bugs. Solutions containing 5 lbs. of soft soap and 5 lbs. of carbolic acid per 100 gallons have been employed for spraying against apple-blossom weevil, and are recommended by *Whitehead* for this purpose.

Carbolic acid has also been advised by *Lees* for destroying those buds of currants which are infested by the big bud mite (see page 371).

## CARBONATE OF SODA

**Employed for softening water. See page 461.**

## CARBON BISULPHIDE

### 1. General

### Employment

Used as a soil insecticide and fumigant for destroying underground pests.

### Description

Carbon bisulphide, or bisulphide of carbon (more correctly *disulphide*) is a colourless liquid, heavier than water. The commercial article has an unpleasant smell, recalling rotten cabbage. It will not mix with water.

IT IS EXTREMELY INFLAMMABLE and must be handled with great care on this account.

It readily vaporises, and its vapour is very poisonous to all insect life, also to human beings if inhaled in any great quantity.

It is used in the form of injections into the ground, principally to kill the root form of woolly aphis (see page 332).

### Simple Tests

The commercial substance may be adulterated with paraffin oil. This may be detected by means of a *hydrometer* (see page 615).

Fill a tall glass cylinder with the liquid, and place the hydrometer in so that it floats (if it sinks to the bottom, a lighter one must be used, and if it floats too high, a heavier one).

Now read the mark on the scale at the level of the liquid. This should not be below 1·26. (On Twaddell's hydrometer = 52° or Beaumé = 30°.)

Another simple test is to place a little of the liquid in a watch-glass or other glass dish, and allow it slowly to evaporate in a room. (*Keep away from flames.*) After a few hours the liquid should have evaporated off, leaving the surface of the glass clear and dry.

If these tests seem to show adulteration, the sample should be sent to an analyst for examination.

### Commercial Data

Usually sold in iron drums or taper cans with screw top. Smaller quantities are retailed by the druggists.

## 2. Method of Use

### Procedure

It is best to use a special form of syringe or *injector*. The carbon bisulphide is placed in this, and the nozzle of the injector is then forced into the soil to a depth of about 6—8 inches. Four or more injections are made at a distance of about 2 to 2½ feet from the trunk, and at equal distances apart. Care should be taken that the roots themselves are not wet with the liquid, and if a root is struck with the injector, a slightly different position should be selected. The vapour is quite harmless to the roots.

### Amount to use

From about 1½ to 5 ozs. according to the size of the tree.

### When to apply

For woolly aphis attack, it is best to apply it in the late winter, but in any case before April. March is a very suitable month.
A day should be selected at the end of a spell of dry weather, as it is much more effective in dry soil.

### Remarks

Besides killing the root form of the woolly aphis, the vapour will destroy weevils, chafers, sawfly maggots, caterpillars, pear midge, etc. It does not appear to kill the pupæ.

## 3. Scientific

### Chemical Composition

It consists of one atom of carbon, combined with two atoms of sulphur ($CS_2$).

### Properties

Colourless, mobile, highly refractive liquid, sweetish odour when pure.
Specific gravity 1·29232 at 0° C.
Boils at 46° C., solidifies at − 110° C.
Very inflammable.
Only very slightly soluble in water (0·17 per cent. at 15° C.).
Miscible with oils.
Prepared commercially by passing the vapour of sulphur over red-hot charcoal.

## CARBON TETRACHLORIDE

### Employment

Used as a soil insecticide, or fumigant for stored produce.

### Description

It is a very heavy liquid, not soluble in water. Has a sweet odour resembling chloroform.

It is used in some cases to replace carbon bisulphide but is much less effective. It has, however, the advantage of being non-inflammable and it is used where there may be danger of fire.

### Method of Use

As a fumigant, about 2—3 lbs. for every 100 cubic feet is required. For treatment of soil, it should be used in the same manner as for carbon bisulphide, but 2—3 times the amount is required. Its employment is not advised except in special cases.

### Chemical Composition

It consists of a combination of one atom of carbon with four of chlorine ($CCl_4$) and is closely allied to chloroform ($CHCl_3$).

## CAUSTIC SODA

**See also Lime, Lime-sulphur**

### 1. General

### Employment

As a strong winter wash for fruit trees, both alone and in the form of emulsion with paraffin oil.

### Description

Caustic soda must be distinguished from carbonate of soda or "soda" which is practically useless as a cleansing agent.

Caustic soda should be kept in *tight packages*, away from the air, as it soon loses its caustic properties, and is converted into carbonate of soda by the action of the "carbonic acid" in the atmosphere.

Although very effective alone, especially at the higher strength of 2½ per cent., the eggs of insects are more certainly destroyed by using an emulsion of paraffin with caustic soda added (for directions see page 420). It may also be used with sulphate of copper (see page 611).

## Properties and Simple Tests

Caustic soda has a *very powerful burning action* on the skin, and must be carefully handled. During spraying operations the faces and particularly the eyes of the men should be protected by masks and they should wear old clothes, as these will be spoilt by the spray.

India-rubber gloves are the best to protect the hands and it is always as well to have a pailful of water with about a pint of vinegar stirred into it for the men to bathe any part affected.

The spray solution should be used as soon as possible, or kept away from the air, as it loses its efficiency if left exposed.

## Commercial Brands

Caustic soda may be obtained in powder form in sealed tins. This is very high strength, usually 98 per cent. purity, and it is the best form in which to obtain it.

It can also be bought in lumps, but these usually contain more moisture and impurities.

A guarantee of strength should always be insisted upon, and if weaker than say 95 per cent. proportionately more must be used in the spray.

## Action on Plants

Caustic soda must only be used on trees *during the dormant season.* It has a powerful burning action on the foliage, and it destroys all forms of low vegetable growths, such as moss and lichen on the bark.

## Action on Insects

Caustic soda, at the strength employed for winter washing, will kill most insects, and it destroys a good percentage of eggs when these are properly wet by the liquid. The chief benefit to be derived from the spray is the destruction and removal of dead bark, moss, lichen, etc., in which harbour many insects, eggs, and fungus spores. For the treatment of the latter, however, *lime-sulphur* is preferable (see page 612).

## 2. Method of Use

### Strength of Spray

**Per 100 gallons of wash.**

Use **20** to **25 lbs**. of pure (98 per cent.) fresh caustic soda.

### Preparation of Spray

Place the caustic soda in an iron or wood vessel (iron is not affected by it, but with galvanised vessels the zinc lining is rapidly attacked) and add about 5 gallons of water.

Stir occasionally till dissolved (heat is generally developed). Then add the bulk of the water and stir well.

### When to Spray

During the winter months only. It is advisable to choose a spell of dry weather, as a shower falling after the application of the spray will wash the material off the trees.

### How to Spray

Use plenty of liquid and a powerful, somewhat coarse spray. The object is to force the liquid between the crevices of the rough bark and under the moss and lichenous growths.

### Quantity of Spray Used

The following figures represent an average of quantities used in practical work.

| Spread of trees | Quantity of liquid per 100 trees | Weight of caustic soda (98 per cent. purity) per 100 trees |
|---|---|---|
| ft. | galls. | lbs. |
| 7 | 160 | 3½ |
| 10 | 220 | 4½ |
| 15 | 380 | 7½ |
| 20 | 600 | 13 |
| 25 | 920 | 20 |
| 30 | 1300 | 30 |
| 35 | 1850 | 42 |
| 40 | 2400 | 55 |

To obtain the quantity required per acre, multiply the figure for 100 trees by the following factors, according to the distance apart in feet: 6 feet 12·1, 10 feet 4·3, 12 feet 3, 15 feet 2, 18 feet 1·3, 20 feet 1, 24 feet 0·75, 30 feet 0·48, 40 feet 0·28.

### 3. Scientific

#### Chemical Composition

NaOH=40.

#### Properties.

Powerfully caustic and alkaline.

Combines with acids to form salts of sodium. (Thus it combines with hydrochloric acid or "spirits of salts" to form sodium chloride or "common salt.")

Rapidly absorbs $CO_2$ from air, becoming "carbonated."

Deliquescent (absorbs moisture from the air).

## CHROMATE OF LEAD

**See also Arsenate of Lead, etc.**

#### Employment

As a caterpillar poison.

#### Description

Chromate of lead is produced by adding a solution of acetate of lead to bichromate of soda. The chromate of lead is thrown down as a fine yellow precipitate known as "chrome yellow." Although not so poisonous as arsenate of lead, it has been reputed to adhere to the leaves for a longer time, and so to remain effective for a greater period.

#### Strength of Spray

If the dry chromate is used, a suitable quantity is from **3 to 5 lbs. per 100 gallons of water.** If the chromate is prepared, the quantities to be used are as under.

#### Preparation of Spray

Take 4 lbs. of crystallised lead acetate, and dissolve in 50 gallons of water in a large vat holding 100 gallons. Take a large pailful of this solution and place on one side. Dissolve separately about 2 lbs. of bichromate of soda crystals in 20 gallons of water. Add this to the lead acetate until further additions produce no further cloudiness to the settled liquor. Now add the pailful of lead acetate solution, fill up to the 100 gallon mark and stir well. This gives 100 gallons of spraying solution.

## COAL TAR

### Employment

Coal tar is not used as an insecticide. It is useful however to paint over the exposed wounds of trees both after pruning and especially after cutting out canker growths, etc.

Stockholm tar is preferred by some growers for this purpose, but coal tar is cheaper and appears to answer equally as well. Lead paint or carpenters' knotting also fulfils the same purpose, but tar is preferable in cases of fungus disease, as it prevents successful germination of the spores.

### Description

Coal tar is too well known to need description. It is a very complex mixture of substances and is obtained in the distillation of coal in the production of coal gas. It is readily obtained from any gas works.

## EMULSIONS (Paraffin)

### 1. General

### Employment

1. With soap for obtaining better penetrating or "wetting" power.
2. With soap, as an insecticide, principally for aphis.
3. With other substances such as iron and copper solutions (Woburn emulsions) as insecticides.
4. With caustic soda as a winter wash for destroying eggs.

### Description

Practically all oils, and many other liquids insoluble in water, possess the property of producing *emulsions* when shaken up with solutions of certain substances.

In these cases the oil is broken up into very minute globules each of which is surrounded by a film of the watery solution. A thick milky fluid is produced, and in some instances the emulsion becomes almost or quite solid.

A familiar instance of an emulsion is *milk* which consists of an emulsion of about 3 to 4 per cent. of butter fat in a solution of casein.

Emulsions may last only a short time or they may be practically permanent. Thus, if oil is shaken with water alone, the emulsion produced rapidly separates, the oil floating to the top of the water almost immediately.

Since paraffin oil has a very pronounced burning action on foliage it follows that only emulsions which are practically permanent are safe to use for spraying.

The most widely used and the most satisfactory emulsions are those in which paraffin oil is emulsified with solutions of soft soap. Much investigation has been carried out to find the best proportion to use of each so that an efficient wetting of the leaf and insect is obtained and consequent killing action, with at the same time safety to the plant.

Since the preparation of the emulsions for spraying must often be left to semiskilled and unskilled people there is always *some* risk in the use of paraffin as an insecticide. If however emulsions of the right strength are properly prepared and intelligently used there is no question of either their efficiency or safety.

In addition to the emulsions as ordinarily prepared, paraffin oil is also employed as a more or less "soluble paraffin soap." This is produced by boiling soft soap and paraffin oil together in various proportions. Under these conditions the oil dissolves in the soap. On dilution a certain amount of the oil is thrown out of solution but is in the form of an emulsion. Such "soluble paraffin" and "paraffin soap" is often used, but it has been shown not to be anything like so effective as the freshly prepared emulsion made by the grower himself.

## 2. Method of Use

### Preparation

1. *Emulsion with soft soap.*

   The soap is first boiled with a proportion of the water and the paraffin oil then added and thoroughly churned up by charging and discharging a pump fitted with a *rose nozzle.* Any other kind of agitation is unsuitable for the purpose. (A garden syringe is very efficient.) The remainder of the water is then added and the whole well stirred up.

2. *Emulsions with iron and copper solutions.*

The sulphates of iron or copper may be used to produce emulsions. To a solution of one of these is added the requisite quantity of lime and the paraffin oil is then churned up with it in the usual manner.

3. *With caustic soda.*

This is for winter washing. The emulsion with soft soap is made in the ordinary way and the soda then added.

Some separation may occur but this does not appear to affect the efficiency of the spray. The iron and copper solutions may also be used with caustic soda[1].

## Action on Insects

The action of paraffin emulsion on insects is not fully known. It appears to have some corrosive action on the skin of caterpillars and also acts as a direct contact poison. Emulsions form an excellent vehicle for applying other insecticides such as liver of sulphur and nicotine.

## Strength of Spray

A. USING SOFT SOAP.

1. *Summer Spray.*

**Per 100 gallons of wash.**

Soft soap 10 lbs.
Paraffin oil 1 gallon.

2. *Early Spring or Autumn Spray.*

In this case either the buds are not fully opened or the leaves are old and not very susceptible to the spray.

**Per 100 gallons of wash.**

Soft soap 20 lbs.
Paraffin oil 2 gallons.

These quantities are for water of medium hardness (e.g. 15 or thereabouts[2]). There should be in the first case at least 6 lbs. and in the second case 15 lbs. of **free soap** present. Consequently, if harder water than this is used, correspondingly more soap must be employed to give permanent and safe emulsion.

[1] See Pickering, *Woburn Reports.*

[2] See page 458 on hardness of water.

B. USING COPPER OR IRON SULPHATES.

**Per 100 gallons of wash.**

Copper sulphate 6 lbs.
Lime water 85 lbs.
Paraffin 1½ gallons.

The copper sulphate is first dissolved in a little of the water and the settled clear lime water run into it. The paraffin is then added and the whole churned up with water to the 100 gallons. In the case of iron sulphate, the same quantity is used as of the copper but milk of lime may be employed in place of water. The quantity necessary is about 3 lbs. of quicklime, slaked and made up into a cream with water (for further particulars see Pickering, *Woburn Reports*).

C. USING CAUSTIC SODA.

The caustic soda is dissolved in a little water and added to the emulsion prepared either with soap, copper, or iron sulphates.

**Per 100 gallons of wash.**

Emulsion as above, 100 gallons.
Caustic soda 20 to 30 lbs.[1]

### How to Spray

As for Soft Soap, see p. 455.

### Quantity of Spray Used

As for Soft Soap, see page 456.

## GAS LIME

### Employment

As a soil dressing for killing all forms of insect life. It is specially fatal to snails and slugs.

### Description

Gas lime is the waste slack lime after it has been employed for the purification of coal gas, and it contains a great many substances besides lime. When fresh, it is very destructive to vegetation. If it is mixed with other soil, it soon changes by absorption of oxygen from the air. If used fresh, it should not be

[1] According to purity and strength required (see page 415).

placed too close to the tree or plant, and should be spread upon the surface of the ground till the air has had time to act upon it.

### Chemical Composition

The following is an analysis of fresh gas lime:

| | |
|---|---|
| Calcium hydrate | 15·1 % |
| ,, carbonate | 24·2 |
| ,, thiosulphate | 11·8 |
| ,, sulphide | 6·9 |
| ,, oxysulphide | 3·2 |
| ,, sulphate | 0·2 |
| ,, sulphite | 0·5 |
| ,, cyanide | 0·2 |
| Iron sulphide | 0·6 |
| Sulphur | 4·3 |
| Silica | 1·8 |
| Alumina | 0·7 |
| Tar | 0·2 |
| Water | 30·3 |
| | 100·0 |

## HELLEBORE

**See also Pyrethrum**

## 1. General

### Use

Hellebore is poisonous to all leaf-eating insects and is employed to a considerable extent in dealing with SAWFLY (including slug worm) attack on small bush fruit, such as currants and gooseberries. It is practically never used on a large scale, or for caterpillar on fruit trees.

### Description

Hellebore powder is obtained by grinding up the roots of the White Hellebore (*Veratrum album*) and the Green Hellebore (*Veratrum viride*). The former plant is common in some parts of Europe but not in Britain, and the latter occurs fairly commonly but is a native of America.

Both plants owe their action mainly to the poisonous principle termed *Veratrine*, which is employed medicinally.

### Properties and Simple Tests

Hellebore powder should consist of the dried ground root. It should be finely powdered and contain no coarse particles. It is liable to adulteration with flour or starch and this can be detected by examination under a microscope.

### Action on Insects

Hellebore is a poison to leaf-eating insects. The powdered root is only from one-half to one-third as poisonous to caterpillars as arsenate of lead.

## 2. Method of Use

### Application

The powdered root is used alone or mixed with flour. It may then be blown on to the leaves of the plant by means of a sulphurator or, as usual in a smaller way, with a pair of bellows, or simply by placing in a muslin bag and "dusting" this over the plants by shaking the bag. If preferred, the powdered root may be stirred into water in the proportion of **2—4 lbs.** to **100 gallons** and this at once sprayed on the plants. Unless a very fine spray is used, however, much of the actual poison, being dissolved by the water, runs off the leaf, and the other method is preferable.

### Time to Apply

It is best applied to the plants at dusk so that the dew may assist it in adhering to the leaves, and render it more active.

### Remarks

In the case of gooseberries and currants, and small bush fruit, it is preferable to use Hellebore in place of arsenate of lead, since there is risk of poisoning the fruit with the latter. Even with Hellebore care must be taken, as this is also a poison, but it is much weaker than arsenate of lead and will probably be washed or blown off the fruit before this is picked.

## 3. Scientific

The alkaloid *Veratrine*, to which Hellebore owes its action, is obtained in white crystals, melting at 202° C. It is soluble in 1000 parts of water and in 3 of alcohol. It forms soluble salts with acids.

### Tests

It dissolves in concentrated sulphuric acid giving a yellow coloration which *gradually changes to blood-red.*

# HYDROCYANIC ACID GAS

## 1. General

### Employment

For the fumigation of plants to kill all insect life. Chiefly for nursery stock and plants under glass.

### Description

This gas is produced when cyanides are treated with mineral acids, such as sulphuric acid (oil of vitriol). When dissolved in water it is called *Prussic Acid.* Both the cyanide and the gas are EXTREMELY POISONOUS, and must be handled or produced with the greatest care.

Although chiefly used for nursery stock and plants under glass, it is often employed, especially abroad, for young trees in the open. In such cases, an impervious kind of canvas sheet or tent is erected or hoisted over the tree and the gas produced inside this cover. In greenhouses, fumigation with this gas is used for the destruction of mussel scale, aphides, mealy bug, thrips, red spider, and weevils. Fumigation of nursery stock kills all kinds of insect life in the young trees, and also a proportion of the eggs. It cannot however be relied upon to destroy all eggs, even when the action is prolonged and the gas used at high strength.

### Materials employed

1. POTASSIUM CYANIDE.

   This is sold in two forms:

   *a.* Lump cyanide, of 98—100 per cent. purity.

   *b.* Stick cyanide, of about 40 per cent. purity.

   It is better to purchase the former quality and to get a guarantee of strength.

   It occurs as white hard lumps.

2. SODIUM CYANIDE.

   This is cheaper to buy and just as effective for fumigation.

   It is sold as "130 per cent. strength." This is because it liberates 30 per cent. more gas, weight for weight, than the potassium cyanide. It is about 98—99 per cent. actual purity.

It dissolves more readily than the other salt and is, on this account, preferable when cheaper.

Both these cyanides are extremely poisonous and should be kept in tight packages, prominently labelled "POISON."

3. SULPHURIC ACID (oil of vitriol).

The "pure concentrated acid" is the best to purchase. It should test not under "1·83 specific gravity" though this is not a reliable test of strength, and the purity should consequently be guaranteed. Strong sulphuric acid has a violent burning action on the skin and should be used with great caution.

## 2. Method of Use

### Quantities required

For every 100 cubic feet of space use:

1. FOR HARDY PLANTS.

½ oz. of sodium cyanide ("130 per cent.").
⅔ oz. of potassium cyanide (98—100 per cent.).

or

1 oz. of sodium cyanide for 200 cubic feet of space, and
1 oz. of potassium cyanide for 154 cubic feet of space.

2. FOR TENDER PLANTS.

0·154 oz. of sodium cyanide ("130 per cent.").
⅕ oz. of potassium cyanide (98—100 per cent.).

or

1 oz. of sodium cyanide for 650 cubic feet of space, and
1 oz. of potassium cyanide for 500 cubic feet of space.

For every 1 oz. of potassium cyanide used take 1 fluid ounce of sulphuric acid and 3¾ fluid ounces of water.

For every 1 oz. of sodium cyanide used, take 1½ fluid ounces of sulphuric acid and 5 fluid ounces of water.

### Procedure

The following precautions should be observed to the letter and the operation carried out by *a skilled person* to avoid risk of accidents.

1. See that every chink and crevice is closed up and the house made as air-tight as possible.
2. Place the correct amount of water in an earthenware jar and stir the strong acid, previously weighed, into it; allow to cool.

3. Weigh out the calculated quantity of cyanide and wrap it in a piece of gauze or close-mesh netting. Tie this round with a long piece of string and pass the end of the string through a hole in the wall or roof of the house, arranging so that it is suspended just over the jar of acid and can be lowered into it from the outside.
4. Close the door of the house and block up all the crevices at the sides, bottom and top.
5. From the outside, lower the packet of cyanide into the acid.
6. Allow an hour to elapse, then open the door wide *from outside*, but do not enter for *several hours*.

In cases of poisoning by the gas, a douche of cold water on the face, or the use of sal volatile is advised. Prompt action is essential.

### When to Fumigate

The evening is the best time. There must be no direct sunlight. The temperature of the house should not be over 60° F. and the plants, as well as the surface of the ground, should be as dry as possible.

Since eggs are not killed by this strength of the gas, a second fumigation should be made after an interval of about two weeks.

### Nursery Stock

An airtight box or tent may be used for this purpose.

A higher strength of gas should be employed. It is advised to use 1 oz. of sodium cyanide or $1\frac{1}{3}$ ozs. of potassium cyanide for every 100 cubic feet of space[1]. There is a good chance of eggs being killed by this strength, though reliance must not be placed on it. The stock should be *dry*, and the box must be opened so that the fumes blow away from the operator.

## 3. Scientific

### Chemical Composition

Hydrocyanic acid is the cyanide of hydrogen, HCN, *i.e.* carbon combined with one atom each of nitrogen and hydrogen. Liberated with effervescence from all cyanides on treatment with mineral acids. (The *ferrocyanides* are quite different chemically.)

[1] *Report on Economic Zoology.* Wye, 1908, p. 82.

### Properties

Colourless gas, lighter than air, having an odour resembling almonds.
*Extremely poisonous.*
Readily soluble in water (the solution is known as prussic acid).

## KAINIT

**See also Naphthalene, Gas Lime, Bisulphide of Carbon**

## 1. General

### Employment

Kainit has been found of benefit as a top dressing for soil around fruit trees. It is said to kill or injure the pupæ of various insects, and also those larvæ which live in the soil, such as the cockchafer grub and swift caterpillar. For complete list of insects which spend the winter in the pupal state in the ground and of larvæ which live in the soil, see under "Naphthalene" (page 433).
As it is also an excellent fertiliser its use is recommended even if more powerful insecticides, such as Naphthalene, are also employed.
It is of proved value in cases of injury by slugs and snails.

### Description

Kainit is a natural mineral deposit, and is a mixture of salts of potash and magnesia. It also contains variable amounts of common salt. The potash is the most valuable constituent from the manurial point of view, and the amount present varies from 12 to 20 per cent.

### How to Buy

The percentage of potash should be guaranteed and the price charged should be in relation to the amount present. An average grade contains about **13 per cent. of potash** present (as $K_2O$).

## 2. Method of Use

### When to Use

It is best to dress the ground

1. When the caterpillar is nearly mature.

   On falling to the ground to pupate it is then most easily killed.

2. When the pupal state is nearly over.

   This is early for the Winter Moth[1], December Moth, March Moth and Mottled Umber, but for the others the best month is probably early March.

3. For beetle grubs, when these are young, in early autumn.

### How to Use

For mature caterpillars, spread the Kainit over the surface of the ground. In other cases work it into the surface soil.

### Quantity Used

The amount employed, of course, varies with the size of the trees and their distance apart. For an average orchard of fair-sized trees, about 5—6 cwt. per acre is required, and this is placed around the trees to correspond with the spread of the branches.

## 3. Scientific

### Chemical Composition

Pure Kainit corresponds to the formula:

$$K_2SO_4 + MgSO_4 + MgCl_2 + H_2O.$$

An average analysis of the commercial material gives:

$K_2SO_4$ 21; KCl 2; $MgSO_4$ 14; $MgCl_2$ 12; NaCl 35; $CaSO_4$ 2; Insoluble 0·8; Water 13·2.

Total Potash as $K_2O = 12 \cdot 7$ per cent.

### Tests for Adulteration

The value depends upon the potash present which may be determined by the Platinic chloride or Perchlorate methods.

[1] Banding is of course the best remedy for Winter Moth and the treatment of the ground may then be safely neglected (see page 404).

# LIME

**See also Gas Lime**

## 1. General

### Employment

1. As a spring "cover wash" to prevent the hatching of aphis, psylla (sucker) and other insect eggs.
2. In the preparation of "Lime-Sulphur" (see page 613).
3. In the preparation of "Bordeaux mixture" (see page 606).
4. In the preparation of "Arsenate of Calcium" (Calcium Arsenate) (see page 395).
5. As a water softener (page 461).

### Description

Lime is prepared by burning limestone and chalk in a kiln. It varies a great deal in purity according to the quality of limestone used.

Lime when freshly burnt is "caustic" and unites vigorously with water, much heat being developed. If the lime is kept exposed to the air it rapidly crumbles and is called "air-slaked." Such lime is quite useless for insecticidal purposes.

The best quality of commercial lime is that prepared in Derbyshire from the fine grade lime-stone, and termed "BUXTON LIME." This can be obtained in truck loads delivered to any station in Great Britain and will often pay to use even at a slightly higher cost.

### Slaking

Freshly burnt, or *Quick*, lime is necessary for all insecticidal purposes and no other kind will serve.

It should be in lumps with as little powder as possible. The powder in commercial samples of lime is due to the air slaking of the lumps on exposure, and consists largely of chalk or carbonate of lime which has little or no efficiency, especially if used for the preparation of Lime-Sulphur, Bordeaux, Arsenate of Calcium, or for water-softening purposes.

The first operation in the use of lime is to slake it. On treatment with water, lime combines chemically with about one-third its weight of water forming slaked lime or "calcium hydroxide." In order to thoroughly slake lime in a short time, a minimum of water must be used so that the heat formed by the chemical action keeps the whole mass at a high temperature. The lime is placed in a tub or suitable vessel and a little water added. When heat commences to be developed small additions of water may be made. The lumps then rapidly crumble and on stirring a thick cream is produced—"milk of lime." The remainder of the water may now be added, stirring well with a wooden paddle. It should then be strained through sacking before use, to get rid of any coarse or gritty particles.

If the mixture is allowed to settle the clear solution of lime in water contains about 0·22 per cent. of dissolved lime (as quicklime).

### Action on Plants

Lime, if freshly prepared and sprayed on to young leaves, has a considerable scorching action. It may however be safely applied to the shoots before the buds are fully opened in early spring.

### Action on Insects

Lime is employed either alone or with the addition of salt, silicate of soda, or other adherent for sealing up eggs of insects so as to prevent their hatching out, and also to protect the buds against insect attacks such as weevil etc.

## 2. Method of Use

### Strength of Spray

1. LIME ONLY.

**Per 100 gallons of spray.**

*2 cwt. lime.*

Slake about four to six hours before using. After slaking, add 30 gallons of water at once and allow to stand. (It will not be too long if it is allowed to remain overnight but it should be well covered up.) When spraying, one pailful of the above paste may be taken to two pailfuls of water.

If possible it is best to use a strainer of about 16 to 20 meshes to the inch. In any case strain through sacking.

2. LIME AND SALT.

**Per 100 gallons of spray.**

*2 cwt. lime.*

Slake as above, add 30 gallons of water and then 25—40 lbs. of salt. Stir till dissolved and use as above.

3. LIME AND SILICATE.

As above with the addition of 10—12 lbs. of silicate of soda (water-glass) per 100 gallons of wash.

## When to Spray

As late as possible in the dormant season. It is still safe to apply when the buds have commenced to open (see figures 244—247). An added advantage of spraying with lime is that it delays the opening of the buds for a few days and thus protects them against late frosts.

## How to Spray

A powerful machine is required, preferably fitted with a special spray nozzle (see fig. 254, page 636). The whole of the tree should be sprayed, the branches and terminal buds having a good coating all over.

# 3. Scientific

## Chemical Composition

Lime, when pure, consists of the oxide of calcium, $CaO$.

Varying proportions of silica and other metals are usually present.

## Properties

Hard lumps; caustic; rapidly absorbs carbon dioxide from the air, forming calcium carbonate (chalk), $CaCO_3$.

On treatment with water, combines vigorously with one molecule, forming $Ca(OH)_2$.

# LIME-SULPHUR

## Employment

In addition to its use as a fungicide (see page 612) several authorities are agreed that it lessens aphis attack if used in the late winter. Actual trials have not shown it to have any reliable killing action on eggs of insects, and the solution of its apparent beneficial action is still to be discovered. In addition it has been found fatal to the *scale insect* both upon apples and other fruit trees, and on the brown scale of currants. Trees are sprayed for this purpose during the dormant season.

### Strength of Spray

As for fungus diseases, see page 614.

## LIVER OF SULPHUR

### Description

For a full description, see under "Fungicides," page 624. Liver of Sulphur has been found useful against RED SPIDER.

It must be used with soft soap, sufficient being present to neutralise the hardness of the water and about 3—4 lbs. to 100 gallons *extra* ("free soap").

Great caution must be used in spraying with Liver of Sulphur as it varies so widely in composition (see page 624). It is liable to be very caustic, and in this condition will burn the leaves of tender plants if used in strong solutions. Growers should insist on having the POTASH salt, as a good deal of so-called Liver of Sulphur has been the *sodium* compound. This is likely to be much more injurious to the plant than the former.

### Method of Use

For red spider (on gooseberries, hops, etc.) in the spring or summer, use 2—4 lbs. of Liver of Sulphur to 100 gallons of water and 3—4 lbs. of *free soft soap* as above (see also page 454).

It is better to employ a compound wash of paraffin emulsion with Liver of Sulphur and Nicotine as given under the description on page 379.

## LONDON PURPLE

**See also Paris Green, Arsenate of Lead, etc.**

### 1. General

### Employment

As for arsenate of lead (for destroying all leaf-eating insects).

### Description

"London Purple" is a waste product from dye works, and is an impure form of calcium arsen**i**te (*not* arsen**a**te). It contains about

75 per cent. of this substance and also soluble arsenic to a considerable extent.

Even with pure calcium arsenite the solubility is sufficiently marked to produce severe scorching of the leaves under unfavourable conditions. For this reason, although this is a cheap poison and to that extent attractive, it is strongly advised to use the safer and equally efficacious arsenate of lead whenever this is obtainable.

### Properties

All the arsenites and "white arsenic" are **very poisonous**, much more so than the arsenates, and they must be very cautiously handled in consequence[1].

### Preparation

Calcium arsenite may be prepared by treating **2½ lbs. of quicklime** with sufficient water to slake it and then adding 3 gallons of water and **1¼ lbs. of white arsenic** (arsenious oxide, $As_2O_3$) and boiling for about 30 minutes. The rest of the water is then added, and the mixture is ready for spraying. Using the waste product, London Purple, it should be made up with an equal amount of quicklime before spraying, but the mixture need not be boiled.

## 2. Method of Use

### Strength of Spray

London Purple ½ lb. }<br>Quicklime ½ lb. } Water 100 gallons.

Slake the lime with sufficient water in a pail, mix in the London Purple and pour into the bulk of the water, stirring well. For the preparation of pure calcium arsenite spray, see above.

### When and How to Spray

See remarks under arsenate of lead, page 402.

### Quantity of Spray Used

As for arsenate of lead, page 402.

[1] For treatment in cases of poisoning see Appendix VII, page 705.

## NAPHTHALENE

**See also Kainit, Bisulphide of Carbon, Soot and Gas Lime**

### 1. General

#### Employment

Naphthalene is the most successful substance yet found as a **soil insecticide** or fumigant for killing all harmful insects in the ground. Such insects may be in the form of larvæ (grubs, maggots, caterpillars), pupæ or adult insects. The following is a fairly complete list:

#### Insects [injurious to fruit] which live in the soil during the winter:—

1. LARVÆ.

   Caterpillars: Swift.

   Grubs: Leaf Weevils, Nut Weevil, Raspberry Weevil, Red-Legged Weevil, Ground Beetles, Raspberry Beetle (often), Cockchafer, Rose chafer.

   Maggots: Apple Sawfly, Gooseberry and Currant Sawfly, Slug worm, Nut Sawfly, Plum Fruit Sawfly, Plum Leaf Sawfly, Social Pear Sawfly, Pear Midge.

   Aphides: Woolly Aphis.

2. PUPÆ.

   Of Moths: Buff Tip, Clouded Drab, December, Dot, Hawk, March, Mottled Umber, Peppered, Winter.

   Of Weevils and Beetles: Twig-cutting weevil. All the above described under "grubs" occur.

   Of Flies and Sawflies. All the above described under "maggots" occur.

3. ADULT INSECTS.

   Ground Beetles, Woolly Aphis, Big Bud Mite (?), Eelworm, Slugs (often).

#### Description

Napthalene is a production of the distillation of coal. When pure, it occurs as a glistening white solid, either in flakes or lumps, or moulded into cakes, sticks or balls. It possesses a somewhat pleasant characteristic smell.

It forms the basis of most, if not all, of the advertised soil insecticides, which consist of naphthalene ground up with ashes or clinker or other gritty materials.

Its action is due to the fact that it slowly vaporises ("sublimes") at the ordinary temperature and this vapour has a stupefying action on insects, resulting, under prolonged treatment, in death.

In addition to its use as a soil insecticide it is largely employed as a preventive of moths in clothes, and as a "disinfectant" in lavatories, etc.

### Simple Tests

The best test for the grower to use is to place a small amount in a warm place and see if any residue is left which will not vaporise off. It is liable to contain tar oils through not being sufficiently refined, but these should not exceed 2—3 per cent. Paraffin wax may be fraudulently added, and this is an objectionable impurity, as it prevents the free vaporisation of the naphthalene.

### Commercial Brands

It is best to buy a fairly pure product and to insist on the percentage of naphthalene present being stated.

It should be as finely ground as possible and not contain lumps.

## 2. Method of Use

### When to Apply

Any time during the winter months is a suitable time. For the pupæ of the winter moth, the application must be made in the autumn.

If the trees are banded in the autumn, it is unnecessary to use naphthalene for this purpose.

### How to Apply

The powdered naphthalene is best mixed before application with about double its weight of fine soil or ashes. It should then be dug fairly deeply in around the trunks of the trees as far over as the spread of the branches.

Fairly deep digging-in is necessary, as the vapour of the naphthalene will not penetrate downwards to any extent.

### Quantity of Material Used

1. For general application to land, the amount to use for efficiency is about 2½ to 3 cwts. per acre.

2. For application to trees, as under:

| Spread of trees, feet | Weight of naphthalene per tree, lbs. | Cwts. per acre. Distance apart in feet | | | | | | | |
|---|---|---|---|---|---|---|---|---|---|
| | | 10 | 12 | 15 | 18 | 20 | 24 | 30 | 40 |
| 7 | ½ | 2 | 1½ | 1 | ⅔ | ½ | ⅓ | ¼ | ⅛ |
| 10 | ¾ | | 2¼ | 1½ | 1 | ¾ | ½ | ⅓ | ⅛ |
| 15 | 1½ | | | 3 | 2 | 1½ | 1 | ⅔ | ⅓ |
| 20 | 2½ | | | | 3 | 2½ | 1½ | 1 | ½ |
| 25 | 3¾ | | | | | 3¾ | 2½ | 1¾ | ¾ |
| 30 | 5 | | | | | | 3 | 2 | 1 |
| 35 | 7 | | | | | | | 3 | 1¼ |
| 40 | 9 | | | | | | | 4 | 2 |

## 3. Scientific

### Chemical Composition

Naphthalene is a hydrocarbon of the aromatic series $C_{10}H_8$.

Its structure is as follows:

```
        H       H
        C       C
   HC⫽    ╲ C ╱   ╲⫽CH
   |        ||      |
   HC╲    ╱ C ╲   ╱⫽CH
        C       C
        H       H
```

### Properties

Commercial naphthalene should melt at 79° C., boil at 217°—218° C. and volatilise without leaving any residue.

### Tests for Adulteration

1. When heated with concentrated sulphuric acid at 170°—200° C. the colour should not become a darker shade than grey or faintly purple.
2. It should volatilise without residue (or in commercial samples not over 20 per cent. residue).

### Insecticidal Value

The author has found that pure naphthalene, at normal temperatures, loses 0·18 per cent. of its bulk in 24 hours by vaporisation.

In terms of the surface area exposed to air, it loses 0·01 gramme per square centimetre, or ⅓ oz. per square foot every 24 hours.

# NICOTINE

## 1. General

### Employment

It is of the greatest value for destroying sucking insects, as aphis ("lice"), psylla ("sucker"), capsid bug, red spider, etc. In addition, leaf-eating insects, especially if soft-bodied, are killed if stronger solutions are used.

### Description

Nicotine belongs to the class of substances known as *alkaloids*. These are very active poisons, obtained largely from plants. It occurs in tobacco in amounts varying in different species of plants from ½ to 7 or 8 per cent.

### Properties and Simple Tests

Nicotine, as it comes into the grower's hands, is a yellow to dark brown fluid with a very pungent odour. It is a MOST VIOLENT POISON. Thus a single drop placed on the tongue of a dog is said to be fatal. The greatest care should be taken not to inhale the fumes, especially in pouring it out of the cans or drums in hot weather[1]. There are no simple tests for nicotine which the grower can apply.

The hydrometer is of no service, as the specific gravity of nicotine is very close to that of water, which is one of the likely adulterants. Notes on the analysis are given on page 441.

### Preparation

Nicotine is manufactured by three principal firms in this country. To cope with the increasing demand for it, others are now installing plant for this purpose. These remarks apply to the so-called PURE NICOTINE 95—98 PER CENT.

[1] In cases of poisoning by nicotine fumes, a draught of strong coffee, or brandy, followed by deep breathing in the open air are good. Camphor is also good as an antidote. If the nicotine is swallowed, an emetic of mustard and water should be given at once, and the patient kept lying down. In severe cases, the hypodermic injection of $\frac{1}{20}$ grain of strychnine is advised.

In America a cruder form is produced by a special patented process. This is termed BLACK LEAF 40, and is guaranteed to contain an equivalent of 40 per cent. of nicotine.

### Pure nicotine

A guarantee of strength should be insisted upon, and, if any quantity is bought, it will be well to have a sample analysed.

The most likely adulterant is water, but there may be also ammonia, or pyridine added fraudulently. The latter is especially difficult to detect, and the analyst should be prepared to guarantee a sample *free from pyridine*[1].

### Black leaf 40

In this compound the nicotine is not in a free condition, but combined with sulphuric acid. It is therefore liable to remain on fruit sprayed with it instead of rapidly evaporating off as pure nicotine does. A certain amount of crude material from the tobacco is also present. It is however cheaper than the pure nicotine and efficient for many purposes. An analyst should be asked to specify the *percentage of nicotine present.*

### Action on Plants

Nicotine, being itself a "vegetable substance," has little or no action on plants at the strength required to kill insects, and may be safely employed even on the most delicate foliage.

### Action on Insects

Nicotine is perhaps the IDEAL INSECTICIDE for all sucking insects. It is extremely efficient, a rapid and uniform kill being assured at effective strengths, especially in the presence of soft soap as a carrier.

Caterpillars are also killed by using an increased proportion of soft soap and nicotine, but for this purpose they *must be hit by the spray*. The nicotine, unlike the arsenates, will not remain as a poison on the leaf, but rapidly evaporates off.

[1] The author has worked out a reliable method for determining the amount of pyridine in nicotine (see *Analyst*, 1919, **44**, 363).

## 2. Method of Use

### Strength of Spray

**4 to 5 ozs.**[1] of pure nicotine **per 100 gallons** of wash is quite efficient for APHIS ("lice," "greenfly," etc.) and PSYLLA (apple sucker). There should also be present 2 *to* 4 *lbs. of free soft soap* in solution (i.e. soap over and above that required to "kill" the hardness of the water (see page 458)).

For the CAPSID BUG, a rather higher strength of nicotine is required; **6 to 7 ozs.** with 4 *lbs. of free soft soap* (page 454) has been found to give very efficient results.

For CATERPILLARS, to ensure a good effect, it is safest to use from **10 to 12 ozs.** of nicotine with 5 *to* 6 *lbs. of free soft soap.*

Many growers use nicotine along with arsenate of lead and without soap. In the author's opinion this is only advisable if labour difficulties prevent a separate spraying with nicotine and soft soap. This is for two reasons:

1. The "fog" type of spray used for applying arsenate of lead is unsuitable for applying nicotine.
2. The effect of nicotine is increased enormously when used with soft soap.

### Procedure

The soft soap is prepared by boiling in the usual manner. The hardness of the water should be known, so that the correct amount of "free soap" is obtained in the final wash. The nicotine is then carefully measured out in an 8 *oz. measuring glass* (procurable at a druggist's—as used in photography), added to the diluted soap solution, and well stirred.

### How to Spray

With a high pressure pump and medium[2] nozzle, using plenty of solution, and giving a powerful spray.

### Quantity of Spray Used

The following figures (page 440) are compiled from practical routine spraying and are a good average to work upon. They are based upon a normal, early summer leafage on the trees.

[1] This quantity can be measured as fluid ounces, since 1 fluid ounce of nicotine weighs roughly 1 oz.

[2] Adjustable nozzles with collar screwed up (see page 634, figs 250, 251).

Fig. 214. Capsid bugs (from Cambridge) magnified; inset, natural size.

| Spread of trees (feet) | Dilute wash required per 100 trees (galls.) | Nicotine 95—98 per cent. required per 100 trees | | | | |
|---|---|---|---|---|---|---|
| | | Strength as under per 100 gallons of wash | | | | |
| | | 4 oz. | 6 oz. | 8 oz. | 10 oz. | 12 oz. |
| 7 | 260 | 10½ oz. | 15½ oz. | 1 lb. 5 oz. | 1 lb. 10 oz. | 1 lb. 15 oz. |
| 10 | 350 | 14 oz. | 1 lb. 5 oz. | 1 lb. 12 oz. | 2 lb. 3 oz. | 2 lb. 10 oz. |
| 15 | 610 | 1 lb. 8½ oz. | 2 lb. 2½ oz. | 3 lb. 1 oz. | 3 lb. 13 oz. | 4 lb. 5 oz. |
| 20 | 960 | 2 lb. 7 oz. | 3 lb. 10½ oz. | 4 lb. 14 oz. | 6 lb. 1½ oz. | 7 lb. 5 oz. |
| 25 | 1470 | 3 lb. 11 oz. | 5 lb. 8½ oz. | 7 lb. 6 oz. | 9 lb. 3½ oz. | 11 lb. 1 oz. |
| 30 | 2100 | 5 lb. 4 oz. | 7 lb. 14 oz. | 10 lb. 8 oz. | 13 lb. 2 oz. | 15 lb. 12 oz. |
| 35 | 3000 | 7½ lb. | 11¼ lb. | 15 lb. | 18¾ lb. | 22½ lb. |
| 40 | 3800 | 9½ lb. | 14¼ lb. | 19 lb. | 23¾ lb. | 28½ lb. |

To ascertain the amount of wash and of nicotine required PER ACRE the above quantities must *be multiplied by the following factors:*

| Distance of trees apart in feet | Factor |
|---|---|
| 6 | 12·1 |
| 10 | 4·35 |
| 12 | 3·02 |
| 15 | 1·93 |
| 18 | 1·34 |
| 20 | 1·08 |
| 24 | ·75 |
| 30 | ·48 |
| 40 | ·28 |

## 3. Scientific

### Chemical Composition

Nicotine has the following composition[1]:

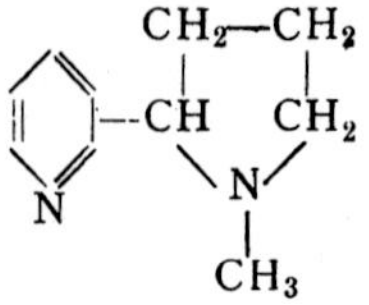

$CH_2$—$CH_2$
| |
(pyridine ring, N)—CH $CH_2$ or n. methyl pyridyl pyrrolidine.
\ /
N
|
$CH_3$

[1] Synthesised by Pictet in 1904.

### Properties

Nicotine is a strongly basic liquid, boiling at 247° C. and of specific gravity 1·011 at 15·5° C.

It is very soluble in water, and also in most oil solvents.

Has a pungent odour and is a violent poison.

### Tests for Adulteration

It is liable to adulteration with water, alkalies (including ammonia) and especially pyridine. Adulteration with the latter is difficult to detect. The refractive index gives a good indication of adulteration (see Fryer, *Analyst*, 1919, **44**, 363 for methods of analysis and determination of pyridine). A table of refractive indices of pure nicotine calculated by the author is given on page 702.

## PARAFFIN

### Employment

Chiefly as an **emulsion** (see page 417) or as *soluble paraffin* or *paraffin jelly* for aphis ("green fly"), including woolly aphis, and for caterpillars. It is also used with caustic alkali for the destruction of insect eggs during the dormant season.

### Description

Paraffin oil is one of the fractions from the distillation of crude petroleum or from shale oil. The lighter fractions of petroleum are known as "naphtha," including "petrol" and other naphthas. The fractions heavier than paraffin oil are used as lubricating oils. Between paraffin oil and lubricating oil there is a fraction which is too heavy for the one and too light for the other. This is called "Solar Oil" or "Solar Distillate" and has been found especially suitable for emulsions for spraying purposes [1].

Further particulars will be found under "Emulsions," page 417.

[1] See Pickering, *Woburn Reports*.

# PARIS GREEN

(Also termed "Emerald green," "Schweinfurt green," "Imperial green")

**See also Arsenate of Lead, Arsenate of Calcium, London Purple, Hellebore, Pyrethrum, Chromate of Lead**

## 1. General

### Employment

As for arsenate of lead, for the destruction of all leaf-eating insects (caterpillars, etc.).

### Description

Paris green, or aceto-arsenite of copper, is a fine emerald green dye. It is sold in powder and in paste form, the latter containing about 75 per cent. of the dry substance. It is a powerful caterpillar poison, the latter containing about 24·2 per cent. of arsenic (i.e. 37 per cent. in terms of $As_2O_5$ as compared with arsenate of lead paste, 12—20 per cent.). It has, however, been proved to have dangerous scorching tendencies[1] on foliage. The injury is lessened but not removed by the use of lime with Paris green, and it is best to always use lime when this spray is employed. When it can be obtained, however, it is preferable to use arsenate of lead.

## 2. Method of Use

### Strength of Spray

1. POWDER.

   ¼ to ½ lb. Paris green.
   ¾ lb. quicklime.
   100 gallons of water.

Slake the lime with a small quantity of the water, stir in the Paris green and add to the bulk of the water. Stir well.

2. PASTE.

   5 to 10 ozs. Paris green.
   10 ozs. quicklime.
   100 gallons of water.

Proceed as before.

[1] Thus Pickering found 75 per cent. of apple trees decidedly scorched, using 8 ozs. of the dry substance to 100 gallons of water.

### When and How to Spray

Spray as soon as caterpillars and other leaf-eating pests are seen. The spray must be "fogged on" as with arsenate of lead (see page 402), a high pressure and fine nozzle being used and excess of spray avoided.

### Quantity of Spray Used

The following represent average figures:

| Spread of trees, feet | Amount of diluted spray per 100 trees, gallons | Weight of Paris Green, per 100 trees | |
|---|---|---|---|
| | | powder, lbs. | paste, lbs. |
| 7 | 200 | ¾ | 1 |
| 10 | 270 | 1 | 1¼ |
| 15 | 460 | 1¾ | 2¼ |
| 20 | 720 | 2¾ | 3½ |
| 25 | 1120 | 4 | 5¼ |
| 30 | 1500 | 5½ | 7¼ |
| 35 | 2200 | 8 | 11 |
| 40 | 2800 | 10½ | 14 |

## 3. Scientific

### Chemical Composition

Paris Green is a double arsenite and acetate of copper:

$$3CuOAs_2O_3Cu(C_2H_3O_2)_2 = 618.$$

Cu = 41·1 per cent.

As = 24·2 per cent.

### Properties

A nearly insoluble, emerald green powder.

Very poisonous[1].

### Tests for Adulteration

The best test for insecticidal purposes is the percentage of arsenic present, both in a soluble and insoluble form.

[1] For first aid treatment, see page 705.

# PYRETHRUM

## 1. General

### Employment

Has a limited use against soft-bodied insects. It acts as a fumigant, giving off a poisonous vapour.

### Description

Pyrethrum is cultivated in Japan and other parts of Asia. Two plants are mainly used for the preparation of the powder.

*Pyrethrum cinerariæ folium.*

*Chrysanthemum coccineum.*

The flower heads are the only parts of the plant containing the poisonous principle. They are collected, dried and powdered, and the powder is either used alone, or mixed with three times its weight of flour, or with flowers of sulphur.

### Properties

The powder soon loses its potency if exposed, and it should be freshly ground or kept in sealed, air-tight packages.

## 2. Method of Use

### Strength of Spray

The powder is generally used, either alone, or in admixture with flour. Extracts may also be employed using soft soap solution or alcohol in which to macerate the flowers.

### Preparation of Spray

The soap extract is prepared as follows:

**Per 100 gallons of wash.**

Soft soap 10—15 lbs. Boil with 4—5 gallons of water until dissolved, and add 15 lbs. Pyrethrum powder (fresh). Stir and make up to 100 gallons with cold water.

### Quantity of Spray Used

As for Soap (page 456).

### Remarks

The soap formula is an excellent one and would be in great favour if the flowers could be relied upon for freshness and the price were sufficiently low to compete with other insecticides.

# PYRIDINE

## 1. General

### Employment

It has been used occasionally as a contact insecticide, especially for *capsid bug.*

### Description

Pyridine is a colourless liquid, dissolving in water, and having an extremely disagreeable and pungent odour. It rapidly causes headache when inhaled. It is only a weak insecticide and although quite cheap it tends to injure the foliage when used at the strength required to kill aphis or capsid. As far back as 1905, the author made a large number of tests with this substance, since, resembling nicotine as it does in many ways, it appeared a promising insecticide. These have been repeated more recently, but the results have in all cases proved disappointing[1].

Pyridine has been used as *an adulterant of nicotine,* and as it is very much cheaper and extremely difficult to detect, it offers a great inducement to unprincipled people for this purpose (see page 441).

## 2. Method of Use

### How to Spray

As for nicotine.

### Quantity of Material Used

Pyridine, in order to have any effect, must be used at a strength of **15—16 ozs. per 100 gallons** of wash, along with 3—4 lbs. of *free* soap (see page 454). At this strength it is liable to injure the foliage, especially when weather conditions are unfavourable.

## 3. Scientific

### Chemical Composition $C_6H_5N = 91$.

Pyridine consists of benzene in which one of the CH groups is replaced by Nitrogen.

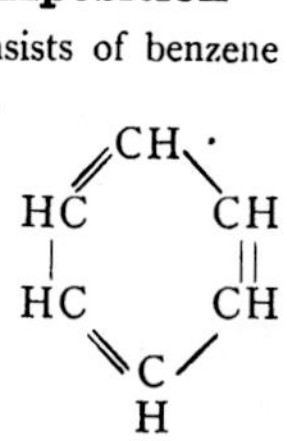

```
     CH                 CH
   //  \              //  \
 HC     CH          HC     CH
  |     ||           |     ||
 HC     CH          HC     CH
   \\  /              \\  /
     C                  N
     H

  Benzene            Pyridine
```

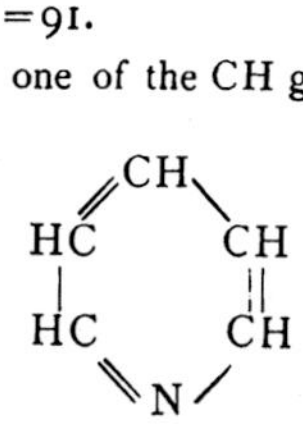

[1] One series of tests with pyridine on capsid bug in Cambridge appeared fairly satisfactory, but the results were not confirmed and there is risk of injury to the plant if used above a definite strength.

### Properties

Colourless, pungent liquid, soluble in water, and very stable.
Specific gravity 1·003 (0° C.). B.P. 114·8° C.

### Tests for Adulteration

May be tested for water by titration with standard acid, using methyl orange as indicator. Each c.c. of normal acid = 0·091 gramme pyridine.
The refractive index is also a good test for purity (see table by author, page 703, and also *Analyst*, 1919, **44**, 363).

# QUASSIA

## 1. General

### Employment

For destroying *sucking insects* such as aphis (green fly) and psylla (sucker) in a solution of soap.

### Description

Quassia occurs in the form of billets or chips of wood from the Quassia tree. There are two varieties of wood:

1. JAMAICA QUASSIA (*Picræna excelsa*): yellowish white, coarse texture, billets usually over 5 inches in diameter.
2. SURINAM QUASSIA (*Quassia amara*): deeper colour than above, harder and heavier than Jamaica, billets generally under 3 inches in diameter.

The wood is intensely bitter owing to its containing the bitter principle, known as *quassin*, which is poisonous to aphis and other insects. The wood is more efficiently extracted by hot than cold water, and the extraction is more thorough if a small quantity of an alkali (such as carbonate of soda) is present.

### Commercial Data

Quassia occurs on the market in two forms:

1. Natural.
2. Kiln dried.

It is always better and safer to buy the kiln dried, as the drying prevents fermentation and deterioration of the wood. As the quassia wood loses a good deal of moisture in the process of drying, the cost is correspondingly higher. Since however it contains much more, weight for weight, of the active principle, *quassin*, it is not in reality more expensive.

Chips should not be kept too long before use, and should be stored in a dry building. It is better to buy the billets, and chip with a machine as required.

Quassia *extract* occurs on the market as a treacly, black fluid. It is usually supposed to be about 100 times as strong as the wood, but in practice it is subject to great variation and growers should make a trial of a sample before buying any large bulks.

### Action on Insects

Quassia is a distinct poison to aphis, psylla and other soft-bodied sucking insects. Its action is not so rapid as that of nicotine, which kills in a few hours at the lower strengths (page 437). It appears to have the property of causing the lice to relax their hold on the leaves and fall off the plant, thus leaving the leaves cleaner than other insecticides. This is probably because a certain amount of quassia is absorbed by the plant and renders the sap distasteful. Quassia is also credited by many growers with preventing further attacks of fly for some days after spraying.

Not only is quassia quite harmless to plants, but it appears even to have a distinct stimulating action, especially in the case of hops.

## 2. Method of Use

### Strength of Spray

Very varied strengths have been recommended. This is probably due to the different condition and quality of the chips used.

If fresh kiln dried chips are used, and the method of extraction given below is followed, 5 lbs. per 100 gallons of diluted wash should be a sufficient quantity. An adequate amount of *soft soap* must also be employed. The amount depends upon the hardness of the water. From 3 to 4 lbs. of "free soap" should be present (see page 454).

### Preparation of Spray

It is quite certain that many growers fail to extract anything like all the "goodness" from the quassia chips. The author has often found rejected chips which contain a high percentage of *quassin* left in them. In order to extract in the most efficient way, boiling water should be used, and a double treatment should be given. For this purpose two tanks are employed for treatment of the chips and the chips should be boiled about two hours in the first tank. The liquor is then strained and added to the soft soap solution.

When the water has boiled for about 10 minutes, a small quantity of carbonate of soda is added (at the rate of about a handful to 100 gallons or roughly ·05 per cent.). This aids the dissolving out of the quassin, but more than this must not be used, as the foliage may suffer. After the chips have been once treated and the liquor strained off, they should be boiled a second time with an equal quantity of water. This time no carbonate of soda is added. At the end of about an hour, the liquor is strained, and added to a fresh lot of chips in the other tank, and the boiling continued as at first for two hours with an addition of carbonate of soda. The original chips are now thrown away. In this way, each lot of chips is twice extracted before it is rejected.

### When to Spray

Both for hops and fruit, but particularly for the former, it is best to spray directly any signs of aphis ("fly" or "lice") are seen on the leaves, as by so doing a serious attack is often prevented.

### How to Spray

Plenty of solution must be used, and a high pressure so that the aphides are "hit." On the other hand the spray should be fine enough to reach leaves which are not directly in the "line of fire" but hidden behind other leaves and branches (see page 632).

### Quantity of Spray used

As for Soft Soap (page 456).

## 3. Scientific

### Chemical Composition

Not fully investigated.

The bitter principle from Jamaica quassia has been termed *picrasmin*.

There are apparently two forms:

α. Picrasmin $C_{35}H_{46}O_{10}$ melting at 204° C.

β. ,, $C_{36}H_{48}O_{10}$ ,, ,, 209° C.

The percentage in the chips appears to vary from 0·75 to 1·0 per cent.

### Properties

The *picrasmin* or *quassin*, the active principle of quassia, is a white crystalline powder, sparingly soluble in water, more soluble in hot and alkaline solutions. Readily soluble in *chloroform*, by which it can be extracted from acid solutions.

It gives a brown coloration with weak ferric chloride solution.

Its solutions are intensely bitter.

### Tests for Strength

The best method is to extract the chips with a dilute boiling alkali solution, acidifying with sulphuric acid, and shaking out with chloroform. Separate off the chloroform and evaporate to dryness. Weigh the residue so obtained.

A less satisfactory method is to precipitate with a solution of Tannic Acid, and then to filter and weigh the precipitate.

### Insecticidal Value

The insecticidal value of pure *picrasmin* (or *quassin*) appears to be about a half to a quarter that of pure nicotine.

It however takes a much longer time to kill than the latter.

### Remarks

The bitter principle, *picrasmin*, has not yet been extracted on the commercial scale.

If a cheap and suitable method of extraction could be found, the pure substance would be extremely convenient to use in place of the present tedious and often wasteful method of chip-boiling.

## SAPONIN

**See also Soap, Water**

### 1. General

### Employment

As a wetting and spreading agent for causing sprays to wet the leaves and insects, and so bring insecticides into contact with the surfaces of the insect and fungus pests.

### Description

Saponin occurs when pure as a white powder. It varies in its chemical composition and the various kinds are obtained from several sources. Thus the *Quillaia* or *soap bark*, and the root of the *soapwort*, both contain varying amounts of Saponin. Several other plants are also known to contain it.

Saponin is especially of use where soap could not be used, owing to chemical action. It has recently been very successfully employed with LIME-SULPHUR to ensure better contact and penetrating power when spraying for fungus attacks. (See page 619.) In this case soap would be useless, as it immediately curdles in solutions of lime-sulphur, forming an insoluble lime soap, and stopping up the sprayer.

In every case where soft soap can be employed it is preferable to saponin for spraying, because, in addition to being an efficient spreading and wetting agent, it has also a definite killing action on insects.

For the above reason, saponin is of great use when testing the action of insecticides since one can be certain that the mortality is due entirely to the insecticide and not partially to the soap.

In the case of very hard waters, where water softeners cannot be employed, it is preferable to use saponin, as the hardness of the water has no effect on its spreading power.

### Commercial Brands

Pure saponin is obtainable from chemists as a fine white powder. It is however much too expensive for agricultural use, and the grower is advised to use a commercial preparation by a well-known firm[1]. This is in the form of a liquid containing 10 per cent. of pure saponin. In such a form saponin is only slightly dearer to use than soft soap.

### Action on Insects

Saponin has little or no action on insects in dilute solutions.

## 2. Method of Use

### Strength of Spray

**Per 100 gallons of wash.**

Use **2 to 5 lbs. saponin** (pure) or **2 to 5 gallons** of **10 per cent solution**[2].

[1] The Yalding Manufacturing Co., Ltd., Yalding, Kent.

[2] See note above.

To this solution, before spraying, the insecticide must be added, and, in order to kill efficiently, more must be used than in the case of soap.

Thus, in the case of a very hard water, 6—7 ozs. of nicotine should be used per 100 gallons of wash with say 3 to 4 gallons of 10 per cent. saponin solution. For use with lime-sulphur solutions, see page 619.

### Preparation of Spray

Stir in the correct amount of solution into the bulk of the water, together with the insecticide.

In the case of the pure powder, dissolve first in about 5 gallons of water and stir this into the bulk of the water.

### When and How to Spray

As for soap, page 455.

### Quantity of Spray Used

As for soap, see page 456.

## 3. Scientific

### Chemical Composition

The saponins are all glucosides of composition $C_nH_{2n-8}O_{10}$. The exact composition is not yet established, but upon decomposition they yield glucose and sapogenins.

### Properties

Some of the saponins are very poisonous, others are almost harmless.

Bitter acrid taste.

Soluble in water, forming frothy solutions.

The pure substance is obtained by precipitating infusions of the bark and roots of plants with neutral and then basic lead acetate. The precipitates are decomposed and the saponin solutions evaporated and purified by solution in chloroform and precipitation by ether.

# SOAP

**See also Saponin, Water, Carbonate of Soda**

## 1. General

### Employment

Soap is an efficient, although a slow acting insecticide when used in sufficient quantity. It apparently acts by blocking the breathing pores of the insect and so causing suffocation.

Its main use however is to cause solutions of insecticides to "wet" the insects and penetrate into the breathing apertures.

### Description

If water is sprayed upon the surface of a leaf it will be found that the small particles of the spray run together, forming large drops which do not wet the leaf surface and readily run off altogether. Some leaves show this effect to a more marked extent than others.

It is found that certain substances, when added to the water, enable it to spread over the leaf surface[1]. The most effective of these is SOFT SOAP. Hard soap is also a good "spreader" but it is more liable to injure the plant.

Quite a small quantity of soft soap is effective for this purpose; thus using distilled or pure water 2 to 3 lbs. of soft soap per 100 gallons of the water is all that is necessary.

Unfortunately the water available for spraying purposes always contains some impurities—even rain water, which is the purest form of naturally occurring water. Many of these impurities destroy a proportion of the soap, forming a chemical compound with it which is thrown out of solution as a "curd." It is therefore necessary to use sufficient soap to remove these impurities, and "kill the hardness" before the soap can be reckoned on as having any effect.

The degree of impurities in solution, or the HARDNESS of the water available, is therefore a very important question in spraying, and it is discussed in detail below (page 458).

[1] The scientific explanation of this is that it is due to the reduction of the surface tension of the liquid.

**Properties and Simple Tests**

Soft soap is manufactured by boiling together CAUSTIC POTASH and OIL.

Almost any sort of vegetable, animal or fish oil may be used, but a mineral oil is in a different class altogether and quite unsuitable. Oils commonly employed are Linseed Oil, Bean Oil, and Cotton-seed Oil for the better grades, and Whale, Cod liver and Japan Fish or Herring Oil for the lower grades.

The boiling process is by no means an easy one and requires skilled supervision. It is thus not possible for a grower to make his own soap.

The better grades of soft soap are clear and transparent, of a light colour and almost without odour. They are firm in consistence without being stiff or leathery. In cold weather, or after being kept for some time, whitish specks, streaks, or small lumps appear in the soap. This is due to so-called FIGGING and the particles are the harder soaps crystallising out. It is no detriment to the soap in any way.

The lower qualities are of a darker colour and have often a pronounced disagreeable "fishy" smell. Such soaps often contain proportions of free caustic soda or carbonate of soda, which is liable to injure the tender foliage, e.g. that of hops.

All soft soaps, whether of high or low grade, should be clear and transparent.

Lower grade soaps often contain portions of ROSIN, and although this is not an adulterant in the sense that it is of no value, it reduces the quality and the soap is especially likely to be objectionable when used with hard water. In such cases a more or less sticky "curd" is produced which is likely to cause trouble in the nozzles of the spraying machine.

When soap is carefully made, it can be produced quite "neutral" or free from excess of potash. Such soaps are not so suitable for spraying purposes as those containing a certain amount of "*extra*" or "*free*" *potash.* The amount however must be carefully controlled and must not be allowed to exceed certain limits or there is a risk of scorching to the plant.

Soap containing free or *uncombined oil* is highly objectionable, being liable to clog the sprayers, but it is seldom met with in good makes.

HARD soap is made by boiling together various fats, such as tallow, palm, coconut, etc. and CAUSTIC SODA. It is more difficult to dissolve, forms a harder curd with water and is not so good for the plant nor as efficient as soft soap. Any excess of caustic soda is especially liable to cause injury to the plant.

During the great war, a substitute for soft soap was largely produced. This was manufactured with caustic soda, but specially selected oils were used and special precautions were taken to prevent any excess of caustic soda. These soaps proved fairly satisfactory but were nowhere equal in efficiency to the true potash soft soaps, and many insisted on using the latter even at the advanced prices. (The reason of the scarcity was, of course, the cessation of supplies of potash from Germany, only small quantities from other sources being available.)

### Strength of Soap

Just as safety in spraying depends upon the high grade and quality of the soap, so its strength depends upon the percentage of fatty matter in the soap.

Since in the manufacture of soap the oil is split up, giving *fatty acids* which combine with the potash, it is usual to speak of the percentage of fatty acids in the soap.

Soft soaps may contain from 33 per cent. of fatty acids upwards. Good grades contain from 38 to 42 per cent., or even higher.

When over about 42 per cent. strength, soaps tend to become stiff and leathery in consistency and are very difficult to dissolve, even in boiling water. If rosin is present, this should be stated separately by the manufacturer and not given as fatty acids.

## 2. Method of Use

### Strength of Spray

The amount of soap to use depends upon

1. The strength of the soap.
2. The hardness of the water.

A. Assuming a high grade of soap of say 40 per cent. fatty acid strength, and a soft water, the amount of soap necessary is

**2—3 lbs. of free[1] soap per 100 gallons.**

If about 3 lbs. of soap are necessary to "kill the hardness" this means

**5—6 lbs. total soap per 100 gallons.**

[1] See page 458 on hardness of waters.

B. With the same grade of soap and a **medium hard water,** more *free soap* is necessary, because the curd formed will require more soap to keep it in suspension and to prevent it sticking together and clogging the nozzles of the machine,

**3—4 lbs. free soap per 100 gallons.**

If about 5 lbs. of soap is required to "kill the hardness" this gives

**8—9 lbs. total soap per 100 gallons.**

C. Where **hard waters** are in question, the amount of free soap must still be further increased, say

**4—5 lbs. free soap per 100 gallons.**

To this must be added the quantity of soap required to *kill the whole of the hardness.* This may amount to a very large figure so that such waters take

**10—15 lbs. total soap per 100 gallons,**

or even more than this.

In such cases much can often be done by water softening and a great saving in soap accomplished. Of course it is always preferable to use a softer source of supply if this is in any way obtainable. For further particulars of water and the question of softening, see section on WATER (page 458).

## How to Spray

Soap solutions, both with and without insecticides, are sprayed with considerable force[1], using a high pressure on the pump and plenty of liquid. The object is to reach and hit each insect and to coat it over with a complete film of fluid. Only by so doing can one be certain of covering the spiracles or breathing pores.

## Quantity of Spray Used

The table overleaf is from the author's practical routine figures:

To ascertain the amount of soap required per acre, multiply the figures on page 456 by the following factors, according to the distance apart in feet.

| Distance apart | Factor | Distance apart | Factor |
|---|---|---|---|
| 6 | 12·1 | 20 | 1·08 |
| 10 | 4·35 | 24 | ·75 |
| 12 | 3·02 | 30 | ·48 |
| 15 | 1·93 | 40 | ·28 |
| 18 | 1·34 | | |

[1] The adjustments for spray nozzles of the universal type is as on figs. 250, 251, page 634.

| Spread of trees, feet | Dilute wash required per 100 trees, gallons | Soft soap (in lbs.) required per 100 trees (at strengths in lbs., as under, per 100 gallons) | | | | | | | | | | | | |
|---|---|---|---|---|---|---|---|---|---|---|---|---|---|---|
| | | 3 | 4 | 5 | 6 | 7 | 8 | 9 | 10 | 11 | 12 | 13 | 14 | 15 |
| 7 | 260 | 7¾ | 10½ | 13 | 15½ | 18¼ | 21 | 23½ | 26 | 28½ | 29 | 34 | 37 | 39 |
| 10 | 350 | 10½ | 14 | 17½ | 21 | 24½ | 28 | 31½ | 35 | 38½ | 42 | 45½ | 49 | 53 |
| 15 | 610 | 18½ | 24½ | 30½ | 36½ | 43 | 49 | 55 | 61 | 67 | 73 | 79 | 86 | 92 |
| 20 | 960 | 29 | 38 | 48 | 58 | 67 | 77 | 87 | 96 | 106 | 115 | 126 | 134 | 144 |
| 25 | 1470 | 44 | 59 | 73 | 88 | 104 | 117 | 132 | 147 | 162 | 175 | 191 | 208 | 220 |
| 30 | 2100 | 63 | 84 | 105 | 126 | 149 | 168 | 189 | 210 | 231 | 252 | 273 | 294 | 315 |
| 35 | 3000 | 90 | 120 | 150 | 180 | 210 | 240 | 270 | 300 | 330 | 360 | 390 | 420 | 450 |
| 40 | 3800 | 114 | 152 | 190 | 228 | 246 | 304 | 342 | 380 | 418 | 456 | 494 | 532 | 570 |

### 3. Scientific

#### Chemical Composition

Soft soap consists of potassium oleate with a little stearate or palmitate, and varying quantities of linoleate, linolenate and clupanodonate. The latter acid predominates in soap made from fish oils.

It also contains small quantities of caustic potash free, and carbonate of potash.

Soda may be present in varying amounts. The percentage of combined potash varies from 7—8·5 per cent.

#### Properties

Soft, transparent, solid, yellow to dark brown in colour, and a sharp taste due to the free alkali.

Readily soluble in hot water, in which it should give a clear solution. Smell—odourless, or nearly so, to "fishy" or offensive.

#### Tests

1. The percentage of fatty acids yielded on solution and acidification with mineral acid.
2. Free and combined alkali.
3. Amount of soda present (by difference after subtraction of potash).
4. Silicate, starch or other filling material.
5. The character of the fats used, determined by an examination of the fatty acids obtained as above.

## SOOT

#### Employment

When a better insecticide is not available, soot may be used as a surface dressing on ground round young fruit trees and bushes. In addition to being obnoxious to caterpillars and beetle larvæ, particularly when fresh, it serves to keep away snails and slugs.

#### Description

Soot varies a good deal in its chemical composition according to the kind of coal from which the smoke is obtained. It contains, as a rule, about 2–3 per cent. of sulphur, from 15 to 35 per cent. of carbon, and may also have a high proportion of tarry matter. Some kinds of soot contain over 6 per cent. of nitrogen and are therefore of manurial value.

## WATER

### Employment

Water is of itself an insecticide, as it is capable of drowning insects by flooding the breathing tubes. It is not however possible to achieve much good by spraying with water alone, as it will not adhere so as to wet the leaf surface and the bodies of insects. For this purpose a "spreader" is required, such as soft soap, saponin, etc.

Most insecticides are applied in solution or suspension in water. For this purpose, use has to be made by the grower of any source of supply which is convenient to hand. Such water always contains impurities in the form of more or less *mineral salts.*

On the amount of these present depends the HARDNESS of the water.

### Hardness of Waters

For spraying with SOAP, or washes containing soap, a hard water is unsuitable, as the mineral salts destroy a proportion of the soap and more must be used to produce the desired effect. After the *hardness* of the water (which is due to certain lime and magnesia salts) is *killed* the soap is no longer destroyed, and a "lather" or "head" is soon formed on the top of the water. This is an evidence that there is FREE SOAP present. Without this free soap, it is useless to spray and expect any "spreading power."

### Degrees of Hardness

It is very unfortunate, and apt to be confusing, that there are two figures commonly used for indicating the hardness of waters.

The old way of expressing the hardness is by "degrees" on **Clark's Scale**. This is nominally equivalent to grains of carbonate of lime (calcium carbonate) per gallon of water or *parts per* 70,000.

The modern way is to express hardness in PARTS OF CARBONATE OF LIME PER 100,000 and, as most analysts now express their results in this form, it has been adopted in this book.

The word DEGREES has been left out as misleading and the number alone is given. Thus, if a given sample of water shows mineral salts present equivalent to 15 parts of carbonate of lime per 100,000, it is spoken of as **having a hardness of 15.**

To convert the hardness numbers to the old "degrees" of Clark's Scale, it is fairly correct to multiply by 7 and divide by 10.

Thus a water of "hardness 15" should give a result of 10·5 degrees on Clark's Scale.

## Variation in Waters

Not only do waters from different sources vary a great deal in hardness, but they are liable to differ considerably from time to time owing to evaporation by the sun, additions of rain water, feeding from springs, drainage from surrounding lands, etc.

It is therefore advisable for each grower to either have his water supply analysed[1] or to do a simple test himself before making up his spraying liquid.

## Sources of Supply

The following are the chief sources of supply open to growers:

1. POND OR DITCH WATER.

   This class of water is liable to very great variations in hardness. Some pond waters, especially after heavy rains, are very soft, and others may be extremely hard, especially in chalky districts. Thus variations of from 3 to 30 are continually met with.

2. RIVER WATER.

   This is usually the very best kind of supply for the grower (excluding rain water, which is seldom available). It is subject only to slight variations and is usually soft, especially in the cases of the larger rivers.

   Thus the Medway tests on an average about 12 to 15 and the Thames 19 to 20. Tributaries of these rivers may be a little harder.

3. COMPANY'S WATER.

   As this is usually derived from springs, it is generally of medium hardness, although the services in different districts vary greatly. The individual supplies are not subject to variation to any great extent and an analysis can be obtained from the supply company concerned. There is also pressure in the mains and this saves the labour of pumping or carrying.

   The Mid Kent Co.'s water has a normal hardness of about 17 to 18.

[1] One well-known soap manufacturer does this free of charge.

4. SPRING, STREAM AND WELL WATER.
   This varies from medium to very hard, the latter especially in Kent, the water having passed through beds of chalk or chalky soil before issuing. They often exceed 50 hardness. It is not advisable to use such waters for spraying purposes unless no other source of supply is available.

5. SEA WATER.
   This is so heavily charged with mineral salts as to be quite unusable. Also the large amount of salt present makes it injurious to plants.

## Soap required for softening different waters.

The following gives a table showing the amount of soap required to *kill the hardness* in water of various degrees of hardness:

| Hardness of Water | | Soap[2] (high quality) required to kill hardness, lbs. per 100 gallons |
|---|---|---|
| Number[1] (parts per 100,000) | Degrees on Clark's Scale | |
| 0 | 0 | 0 |
| 1·2 | ·8 | ¼ |
| 2·4 | 1·7 | ½ |
| 3·5 | 2·4 | ¾ |
| 4·7 | 3·3 | 1 |
| 5·9 | 4·1 | 1¼ |
| 7·1 | 5·0 | 1½ |
| 8·3 | 5·8 | 1¾ |
| 9·5 | 6·6 | 2 |
| 11·8 | 8·2 | 2½ |
| 14·2 | 10·0 | 3 |
| 16·6 | 11·6 | 3½ |
| 19·0 | 13·1 | 4 |
| 21·3 | 15·0 | 4½ |
| 23·7 | 16·6 | 5 |
| 26·1 | 18·2 | 5½ |
| 28·4 | 20·0 | 6 |
| 30·8 | 21·5 | 6½ |
| 33·2 | 23·3 | 7 |
| 35·5 | 25·0 | 7½ |
| 37·9 | 26·5 | 8 |
| 40·3 | 28·2 | 8½ |
| 42·6 | 30·0 | 9 |
| 45·0 | 31·5 | 9½ |
| 47·5 | 33·2 | 10 |

[1] See page 458.

[2] Of say 40 per cent. fatty acids (see page 454).

### Softening water for Spraying

It is very desirable for the grower to use any means he can to soften his water before the addition of soap or soap washes, as by so doing a great saving can be made, particularly in the case of hard waters. The ordinary process of water softening is however not usually feasible, as it involves the use of tanks and filters, and time is required for the water to settle before use.

In particular cases it may be possible that the requisite plant is available, and the water can then be thoroughly softened. The usual method is to add a proportion of milk of lime or lime water, then to settle or filter, and afterwards to add a measured amount of caustic soda or carbonate of soda and filter. The amounts of each must be carefully adjusted in the case of each water supply. Full information can be obtained from the many text-books on the subject.

A modern and extremely efficient softening process is by means of a substance called "Permutit" which removes all hardness from water on merely pumping the latter through a layer of the material. It would probably pay large growers to install a small plant.

It is however possible to effect a considerable softening with some samples of water by the use of a solution of carbonate of soda ("washing soda"). This is a very simple and cheap method and should be used wherever the water is hard. It is however very important not to exceed the required amount as, by so doing, severe scorching to the foliage has in the past resulted. The amount to use in each case may be gauged fairly accurately if the hardness of the water is known, as follows:

CARBONATE OF SODA REQUIRED FOR SOFTENING WATER.

Rule: Divide the hardness number[1] by 2, and call this figure ozs. of carbonate of soda crystals. This is the right amount for 100 gallons of the water.

The crystals should be dissolved in a little (preferably hot) water and the solution stirred into the bulk of the water. Allow to stand for at least half an hour before adding the soap for spraying.

*Example:*

Some water from a pond was sent to be analysed, and tested 26 hardness. It therefore required 13 ozs. of crystal carbonate of soda per 100 gallons to soften.

[1] See pages 458, 460.

In many cases, especially where the water consists largely of magnesium salts, carbonate of soda will not effect much softening. In such cases it is preferable to use first a softener consisting of a cheap soapy material, so as not to waste good soap. A patent for this idea has been granted to a Kent firm of insecticide manufacturers[1]. The softener is first stirred into the bulk of the water, and the soap or wash is then added.

In the case of excessively hard waters, where no other source of supply is available, it is best to use **saponin** in place of soap. This will be less expensive than the amount of soap which would be required, and is unaffected by the hardness of the water.

[1] Yalding Manufacturing Co., Ltd. The article is termed "Potassine Water Softener."

# SECTION VI

## BENEFICIAL INSECTS

# SECTION VI

## BENEFICIAL INSECTS

## CHAPTER 27

It might well be thought that the fruit grower owes nothing but a grudge against the whole insect world, for the damage and destruction of his crops. This is however by no means the case. Indeed it is even questionable if a fruit grower would get any crop at all without insect agency.

Of the first importance are those insects which ensure fertilisation of the blossom by conveying pollen from one flower to another. Of

Fig. 215. Winter Moth pupæ and three parasitised larvæ which failed to pupate. Natural size.

these, the **bees** are by far the most valuable, and benefit has often been derived by keeping hives of bees in or near orchards. Many other insects also effect pollination though less systematically.

In spite of much prejudice **wasps** are without doubt beneficial to a considerable extent, some varieties effecting pollination and others living largely upon other insects. They are however a great nuisance when, as frequently happens, they attack ripe fruit.

The remaining classes of beneficial insects all assist the grower by preying upon insect pests. Although, unless in exceptional cases, they cannot be relied upon to clear off an attack, they nevertheless do an extremely important service in diminishing the number of adults, and so keeping the pest within bounds[1]. Thus, in the present year, the author found a very large percentage of the Winter Moth caterpillars attacked by an ICHNEUMON fly, and none of these were able to pupate, being killed just as they reached their full growth (see fig. 215).

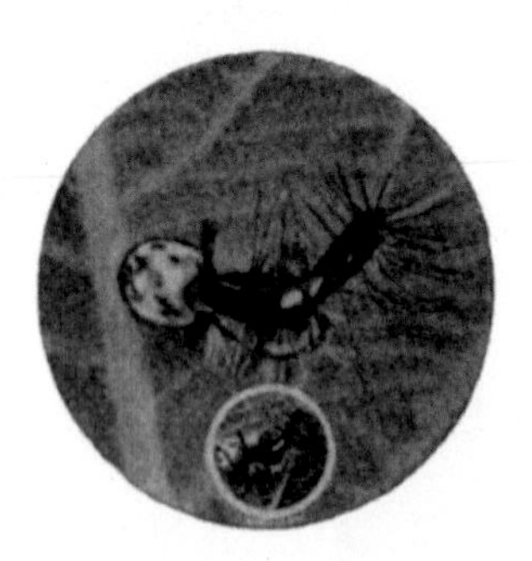

Fig. 216. Cocoon of Ichneumon parasite and skin of victim (Vapourer caterpillar). Magnified ($\times 3$) and natural size.

Fig. 217. Caterpillar of large white butterfly with maggots of Ichneumon just emerging and spinning cocoons. Inset, Ichneumon fly.

In the previous year all the caterpillars examined of a common species of butterfly attacking the cabbage (the *large white butterfly*) were found to be "parasitised" by an Ichneumon fly, not a single healthy one being discovered (see fig. 217). In consequence, the cabbages were, in the present season, practically free from this pest.

Beneficial insects which attack insect pests may be divided into two classes:

1. Those which devour their victims;
2. Those which pass their early stages inside the bodies of their victims, or *parasitise* them.

[1] As regards the propagation and use of beneficial insects for combating various pests, see Appendix IX, page 708.

## Caterpillars

There are several caterpillars which are "cannibals" and prefer to devour others of their own class than to live upon vegetation. Chief in importance of these to the fruit grower is the **Dunbar**, which is often found upon apples and other fruit trees. This caterpillar, which

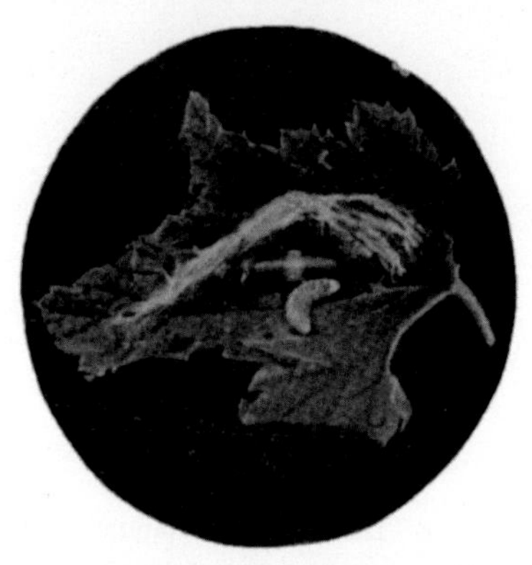

Fig. 218. Ichneumon maggot just emerged from parasitised Dunbar caterpillar, and about to pupate. Also a specimen which has pupated.

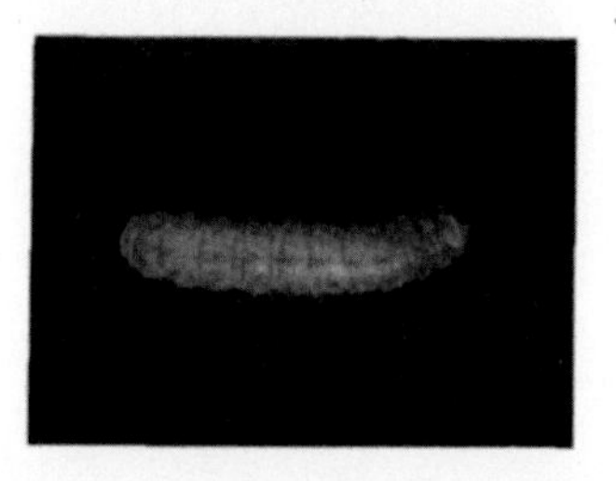

Fig. 219. Dunbar caterpillar. Natural size.

resembles somewhat that of the Winter Moth, being of a light green colour, will devour many destructive caterpillars before it is mature. This caterpillar is also commonly parasitised by an Ichneumon fly, and in this case, since the caterpillar is beneficial, the Ichneumon must be looked upon as an enemy! (see fig. 218). No caterpillars are known to be parasitic in the bodies of other insects.

Fig. 220. Dunbar pupa. Natural size.

Fig. 221. Dunbar moths. Natural size.

**Beetles**

The most important beneficial beetle is the well-known **lady-bird** beetle. Both the adult and its larva or "nigger"—particularly the latter—live upon aphis, and especially upon the hop aphis. The yellow eggs, in regular groups, resembling skittles, are familiar objects on the leaves of the hop (see fig. 222). The "nigger" is a small, black, active grub with long legs, and yellow spots on the body, and the pupæ are formed on the leaf, are quite inactive and have a shield-like appearance. The lady-bird undoubtedly does an immense amount of good and is apparently frequently the means of clearing off an attack upon the hop—at any rate it is often found in large numbers just before the aphis disappears.

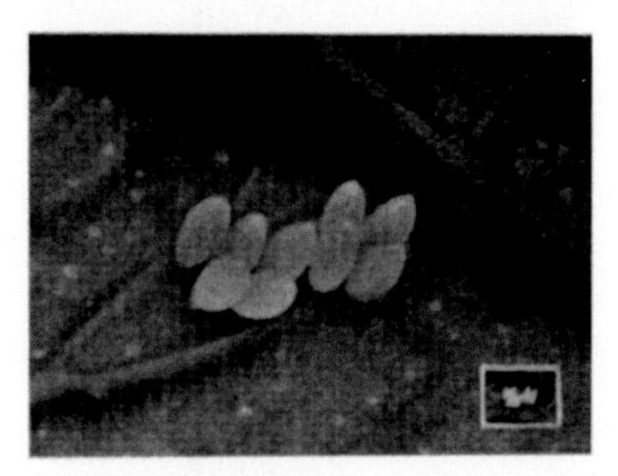

Fig. 222. Eggs of Lady-bird on hop leaf. Magnified ( × 6) and natural size.

There are two other insects, the larvæ (maggots) of which feed upon aphides. These both belong to the two-winged fly family (*Diptera*). The first is the so-called **Wasp-fly** which belongs to the group called HOVER-FLIES, on account of their habit of hovering over the flowers of plants. The Wasp-fly is very conspicuously banded with black and yellow and is commonly mistaken for a wasp, but is quite unlike it in other respects. The larva (maggot) is a curious greyish slimy-looking creature, often found upon fruit leaves, and feeding upon various species of aphis (see fig. 224).

The other beneficial insect of this class which devours other insects is the **Lace-winged fly**[1], the larva of which is frequently found upon hop leaves and is called the **aphis lion** on account of its feeding upon aphis, and its voracious habits (fig. 226).

[1] See illustration of adult insect on page 711.

This is a curious and interesting insect, especially in the manner of laying its eggs, which are placed upon stalks formed by a secretion of the insect's body. The fly first places its tail end on the leaf and then elevates it, exuding the sticky thread which at once hardens. The egg is then laid on the top of the

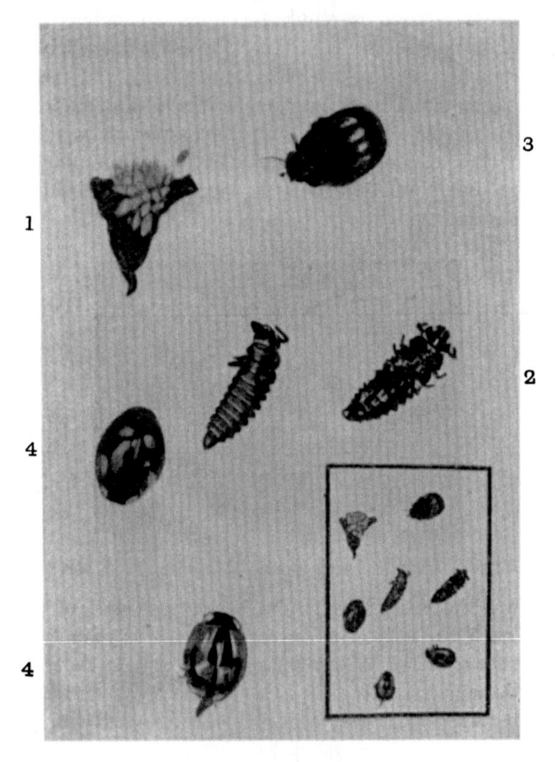


Fig. 223. Various stages of Lady-bird beetle. 1, Eggs. 2, grub. 3, pupa. 4, adult (two varieties). Enlarged (× 3) and natural sizes.

stalk so formed. This procedure is to protect the egg by placing it out of the reach of insects which would devour it. On hatching the young larva crawls down the stalk on to the leaf (see fig. 226).

In the second group, viz. those insects which pass part of their life as parasites in the bodies of others, by far the most important are the **Ichneumon flies**. These form a well-marked group of four-winged flies (*Hymenoptera*[1]). They have mostly slender wasp-like bodies, and long legs and feelers, and gauzy wings. By means of a long and slender tube at the end of the body (the *ovipositor*) they pierce the skin of caterpillars and other insects, and insert their eggs within the body. They attack almost every species of insect known, either in its larval or pupal stage. Thus fig. 227 shows aphides on nettles parasitised by an ichneumon fly. The eggs, of which there may

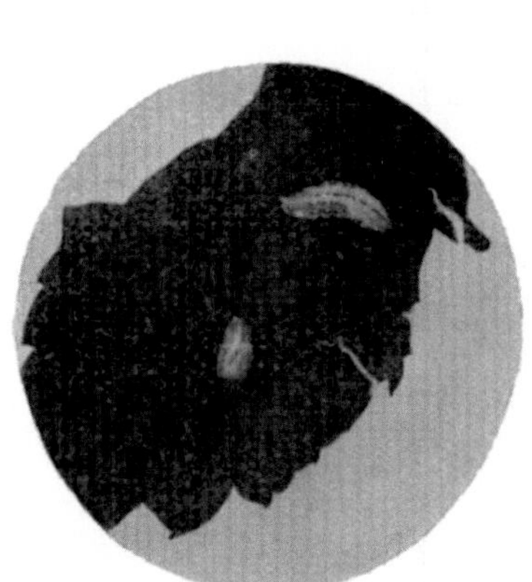

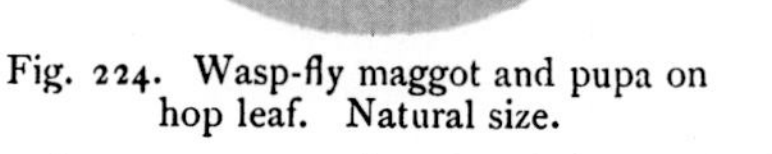

Fig. 224. Wasp-fly maggot and pupa on hop leaf. Natural size.

Fig. 225. Wasp-fly. Magnified (× 3) and natural size.

be a great number, hatch in the body of the victim, and the footless maggots live and grow upon its vital juices. Great care is however taken not to attack any vital organs so that the host is not killed until it is mature. When this happens, a hole is bored from the interior in the insect's skin, and the maggots emerge, spinning silky cocoons near by the dead body of their victim, which soon shrivels up (see figs. 217, 229). In many cases, the pupa is attacked and is finally killed in the same way (see fig. 228).

Another class of insects, very similar to the Ichneumons in their

[1] See page 233.

habits (and often referred to as *Pseudo-ichneumons*) belong to the two-winged flies (*Diptera*[1]). These are known as **Tachina flies**. They closely resemble the common bluebottle in appearance. They lay their eggs in the bodies of other insects, especially caterpillars, just as do the Ichneumons (see figs. 230—232).

The **Chalcid flies** are another important group. These are small, two-winged flies which puncture the bodies of their victims in

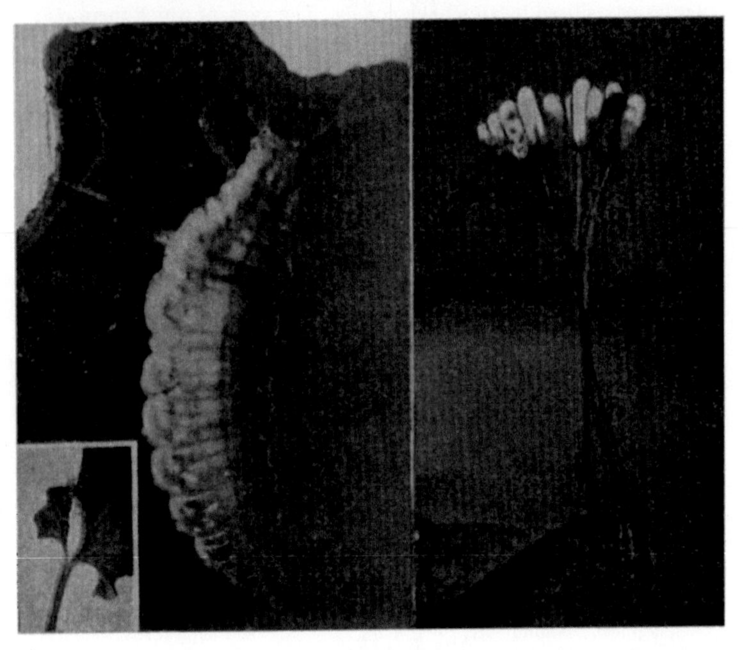

Fig. 226. Larva of Lace-winged fly ("aphis lion") magnified ( × 7). Inset, natural size. Also eggs of same, showing cluster on stalks. Highly magnified.

a similar manner to the two preceding groups. They may be told by their bodies, which are commonly of a brilliant metallic tint. Some of them are very minute, and their larvæ (maggots) live, not in the larvæ, but in the eggs of other insects. Still smaller flies are known as parasites which infest, not only insect pests, but Chalcid and Ichneumon flies. To this extent they are not "beneficial" but destructive, since they prey upon insects which are friendly to the grower.

[1] See page 234.

Fig. 227. Parasitised nettle aphides (dead). Inset, healthy aphis and Ichneumon fly. Magnified (× 3) and natural size.

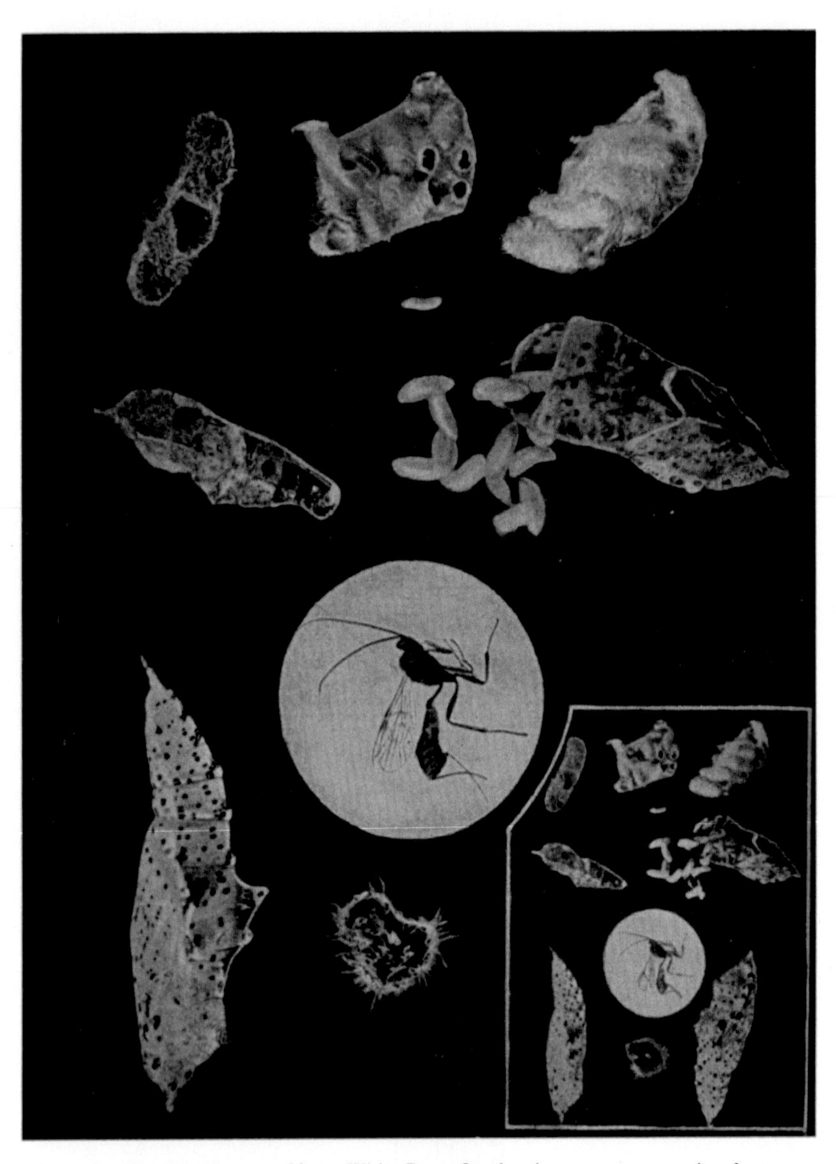

Fig. 228. Parasitised pupæ of large White Butterfly, showing maggots emerging from pupa case, cocoons of parasite, healthy pupa and Ichneumon fly. Magnified (× 2½) and natural size.

Fig. 229. Ichneumon flies and their cocoons, showing also shrivelled skin of parasitised caterpillar. Magnified ( × 3) and natural size.

Fig. 230. Tachina fly infesting caterpillars. Magnified (× 3) and natural size.

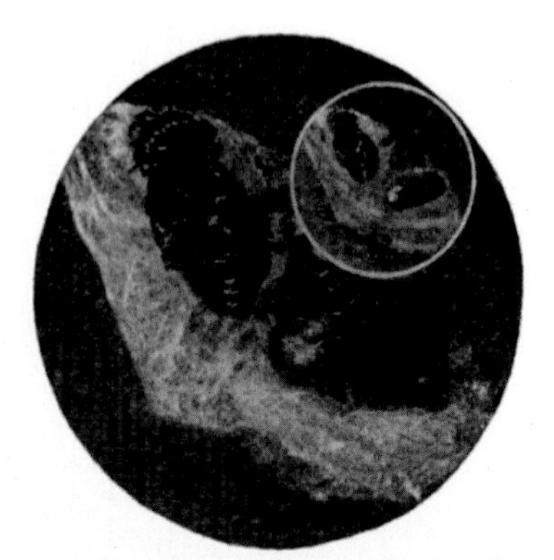

Fig. 231. Pupa of Tachina fly, showing also old skin of caterpillar (Gold-tail Moth). Magnified (× 3) and natural size.

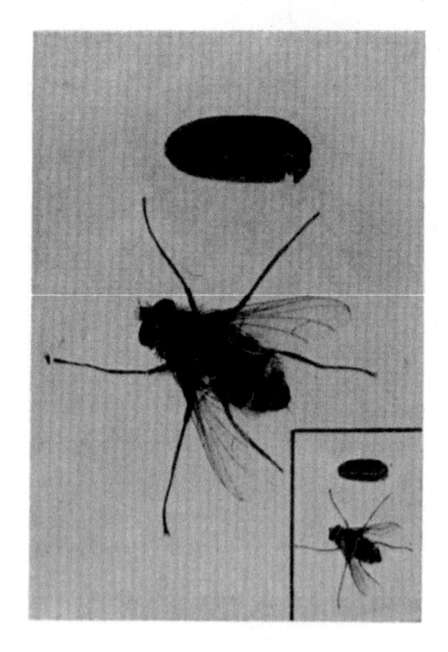

Fig. 232. Tachina fly and pupa case. Parasitic on Tortrix caterpillars. Magnified (× 3) and natural size.

# PART II

## FUNGUS DISEASES OF FRUIT AND THEIR CONTROL

# SECTION VII

## FUNGUS DISEASES

# SECTION VII

## FUNGUS DISEASES

## CHAPTER 28

### About Fungi

In addition to insect pests the fruit grower has a large number of fungus diseases to contend with. Some of these are very serious and extremely difficult to control, and almost all cause serious losses in the fruit crop (see Chapter 1).

Fungi belong to a low division of plant life, and differ from the higher plants in having no flowers and in producing no green colouring matter (chlorophyll).

A great many fungi are quite harmless to plants, growing upon decaying animal or vegetable matter. Others are unable to grow except upon the living tissues of plants, and are thus always *parasitic* in habit. Many, again, are sometimes parasitic and at other times not so.

Although fungi do not produce flowers, they bear "fruit" and "seed" of a special kind, called SPORES, which, under suitable conditions, are able to germinate and produce fresh growths of the fungus. Spores may be of two kinds, known as SUMMER and WINTER (or scientifically *conidial* and *perithecial*), which will be referred to again later.

The body of the fungus is made up of long, very fine threads, or filaments (called *hyphæ*). These usually form an interlacing, mat-like tissue, called the spawn, or *mycelium*. It is this mycelium which is responsible for the damage to fruit trees, since the tissues of the plant are broken down and finally killed, after supporting the fungus by means of the sap from the cells of which they are composed.

The threads and the spores of the fungi which attack fruit are much too minute to be distinguished with the naked eye. It is only therefore by the aid of a powerful microscope that their structure can be investigated, and in describing the various forms of fungus disease an account is given of the general appearances of the fungus growths as seen by the unaided sight, or with the help of a small hand-glass. By this means, and by the effects produced upon the diseased parts of the plant, it is usually possible readily to identify the particular disease in question. In a few cases little or no sign of the fungus itself is to be seen at the time when it is most harmful (e.g. in the Apple Blossom Wilt Disease, page 494). In such instances the disease must be identified by the *symptoms* it produces, such as the wilting and death of the blossom in the instance given.

For the sake of completeness, a summary is given of the microscopical characteristics of the various fungus diseases described in this section of the book. This is placed in small type at the end of each account and may be neglected by the grower.

In the interests of students, a brief statement may be given of the scientific aspect of the subject.

## The Mycelium

The hyphæ consist of transparent tubes, the walls of which usually contain more or less *chitin*, cellulose being generally absent.

These tubes are more or less filled with a colourless *cytoplasm*; and minute globules of oil may be present, but no starch grains, as in higher plants. In the higher fungi the hyphæ are divided into larger or shorter cells by partitions, or *septa*, and are spoken of as *septate*; but lower forms are not thus divided, the tube of the filaments being continuous.

The hyphæ usually branch freely, and the tip of each branch grows continuously, and in this manner an interlacing *mycelium* is produced. The interweaving of the hyphæ in the mycelium is often so close that a flat sheet of delicate threads is produced, resembling felt or, in some cases, tough leather. In other cases, notably in the higher fungi, the mycelium becomes hardened and dark in colour, forming compact masses of varying shape. Growth then ceases, and in this form, termed in the plural *sclerotia*, the fungus is able to remain dormant throughout the winter season, becoming active in the following spring or summer.

The mycelium is usually embedded in the plant tissue, but it may be entirely outside, as in the case of the POWDERY MILDEWS, and in such cases, branches, termed *haustoria*, are sent into the tissues of the leaf or stem which form suckers or roots to supply the fungus with nutriment. In such cases it is obvious that the fungus is more open to treatment and control by chemical agents than when it is enveloped by the plant cells themselves.

## The Spores

The spore-bearing portions of the fungi answer to the reproductive organs of the higher plants. Each spore consists of a single cell, produced and set free by the parent plant, and capable of giving rise to a new growth of the same kind.

Great variations exist in the size, shape and character of the spores, and of the method in which they are produced. Spores are commonly spherical or oval in shape, but they may be spindle-shaped, oblong, club-shaped or long and slender.

Spores may also be produced by a sexual process, termed "sexually-produced spores," or without, termed "asexual." The latter are by far the commonest, and sexually-produced spores are only found in the lower forms, if we except the *Ascomycetes* (which is a debatable point).

Sexually-produced spores may be of two kinds, termed *zygospores* and *oospores*.

The asexually-produced spores are also divided as follows:

1. Endospores.
2. Conidia.
3. Oidia.

**Endospores** are produced within a *sporangium*, or spore case, by a division of the protoplasm. The sporangium is carried on a *sporangiophore*, by means of which it is attached to the mycelium. Some endospores are naked and are able to swim freely about by means of movable hairs (or *cilia*), and others have a definite cell-wall.

The ASCUS, which is a modified sporangium, contains a limited number of *ascospores* and is characteristic of a large class of fungi, many of which are important parasites, giving rise to plant diseases.

**Conidia** are very simple spores, borne naked upon the top of a branch hypha termed a *conidiophore*. They are single cells produced by a partition in the top of the latter, and then set free. They are of very varied sizes and shapes. A special variety of conidiophore, termed a *basidium*, is characteristic of another large group of fungi, the *Basidiomycetes*.

**Oidia** are spores formed by the mycelium becoming divided into a number of short segments, which may remain united in chains or become free from each other. They often, on germination, produce short hyphæ, bearing conidia or sporangia.

Many fungi produce more than one variety of spore at different times from the same mycelium. Thus the members of the important family of the ASCOMYCETES produce both conidia ("summer spores") and ascospores (a form of endospore), or so-called winter spores.

The discovery of different kinds of spores produced by the same fungus has led to great confusion in the classification, and many of the fungi as now known are undoubtedly capable of existing in other forms than those at present described.

## Germination

Some spores, usually the endospores and conidia, germinate in a few hours in the presence of moisture and a suitable temperature. Others, termed *resting spores*, require a certain lapse of time before they will germinate.

On germination, a delicate *germ-tube* is emitted by the spore, which, in the presence of appropriate nutriment, produces hyphæ and grows into a mycelium. Spores often possess a double cell-wall, and in this case the outer wall is punctured and the inner wall is prolonged into the wall of the hypha.

Fungus spores are disseminated by wind, air currents, flies, bees and other insects, and snails.

Thus areas widely separated from infected trees may become diseased by means of these agencies.

The disease is **carried over** from season to season on the same plant in different ways. Often there is a winter fruit of the fungus producing *winter spores* which infest the young leaves or flowers the following spring or summer.

This fruit may be produced on the branches or on leaves or fruit which have remained on the tree throughout the winter.

In other cases, portions of the mycelium (spawn) winter in the tissues of the branches, or in the young buds and become active again in the following year.

Plants are **infected** with fungus disease by means of spores which are carried to the plant by various agencies. One of the most curious and important features of fungus attack is that plants are only capable of infection by certain kinds of spores. Thus, the spores of gooseberry mildew falling upon an apple leaf are incapable of germinating and so causing infection on the apple. This is considered to be due to various different chemical substances which exist in the cells of all plants, and which are capable of attracting certain fungus spores only to the exclusion of others. In the case of the *powdery mildews*[1] this feature is carried to greater lengths, different "strains" of what is apparently the same species of fungus being capable of growing on certain plants only, and incapable of infecting others. It must not therefore be thought that a fungus producing say leaf spot on apples will be a source of danger to cherries, or that, because the strawberries are badly mildewed, there is danger of infecting a crop of gooseberries near by.

[1] As Professor Salmon has shown.

# CHAPTER 29

## Causes and Prevention of Fungus Diseases

As to the **causes** of fungus attacks this is still to some extent an unsettled question. It is quite certain that the STATE OF THE WEATHER has a great deal to do with liability of a plant to fungus attack. The conditions favourable to fungus attack may be summarised as:

1. Absence of direct sunshine.
2. Cold in late spring and summer.
3. Excessive moisture due to lack of air currents.

Such conditions are usually obtained in "close" and "stuffy" weather, and experienced growers have an instinctive knowledge when to expect fungus attacks by observing the weather conditions. Our forefathers believed that disease was caused directly by unfavourable weather, and although we now know that it is only a secondary cause, providing conditions suitable for the growth of the fungus, it is nevertheless a fact that in bright sunny weather infection with fungus diseases is rare, and even badly infected plants are able to resist the attack.

Weather conditions are however by no means the only predisposing causes of fungus attack, or even the most important. Thus many fungi, especially those attacking fruit, are unable to gain an entrance to the plant except through **wounds,** and are thus termed *wound-fungi.* Wounds may be caused by many agencies, but the three commonest are:

1. Insect punctures.
2. Wind.
3. Hail.

Since the fungus spores are extremely minute, it is evident that the tiniest rupture of the surface of the skin of a plant is sufficient to adinit the disease. Insects, especially aphis, are undoubtedly the most usual instruments of infection. Thus we know that the *canker fungus* invariably follows upon attacks of woolly aphis, and no doubt there are numbers of instances of the same kind. Wind and hail cause bruises and rupture of the skin, especially of fruit and leaves, and so lay the plant open to the entrance of fungus spores.

The foregoing are natural causes which, except to some extent in the case of insect attacks, cannot be prevented by growers. In addition to these, however, there are many **preventable causes** of fungus attack. Of these the most important is BAD CULTIVATION, resulting in:

Fig. 233. Typical injury caused to apple by hail. Natural size.

1. An insufficiency of air supply to the roots of the plant.
2. Inadequate drainage, causing a water-logged condition of the soil.
3. Insufficient protection against extremes of heat and cold.

Further, a very common preventable cause of fungus infection is neglect to remove fallen and rotten fruit and leaves, on which the fungus bears fresh crops of spores in the winter, or leaving the diseased fruit or dead and decayed leaves and branches on the tree to carry on the attack from one season to another. A very great deal may thus be accomplished by clean and careful cultivation, general hygienic methods, and attention to the trees during the dormant season.

The neglect to adequately protect **cut and exposed surfaces** of trees after pruning or injury is responsible for much fungus infection. In such cases the wounds should be carefully coated over with tar, shellac varnish, or wax (the first named is the safest), so as to give the spores no loophole to effect an entrance.

Fungus diseases will never adequately be dealt with until united and simultaneous action is ensured by all growers, of whatever size their orchards may be. In order to secure this, it is practically essential that the Government have compulsory powers in regard to spraying and treating diseased plantations. Otherwise, as previously pointed out (page 7), an enlightened and painstaking grower will always be open to attack from the neglected orchards of his neighbour.

In addition, there is the question of wild varieties of fruit, which may serve as breeding grounds for the fungus. Although much may be accomplished by united action, it is on the whole improbable that diseases can ever be completely stamped out. It is specially necessary to keep a rigid watch on imported stock, and in suspected cases to keep stock in quarantine until examination has been made.

## Authorities

Our present knowledge of fungus diseases is, as in the case of insect pests, the accumulation of the work of many observers. It is somewhat invidious to select from the names of so many eminent mycologists, but it might be mentioned that in recent years we are indebted to Professor Salmon and Dr Wormald of Wye for special light on the powdery mildews and on fruit rot respectively.

The following is a short list of some past and present workers in this field:

| | | | |
|---|---|---|---|
| Aderhold | Duggar | Pole-Evans | Stoneman |
| Bailey | Eriksson | Schrenk | Wiltshire |
| Barker | Gimingham | Scott | Ward |
| Brooke | Goethe | Smolak | Wormald |
| Cooke | Hartig | Somerville | Woronin |
| Cotton | Massee | Southworth | |
| De Bary | Percival | Spaulding | |

# DISEASES OF THE APPLE

PLATE XIII

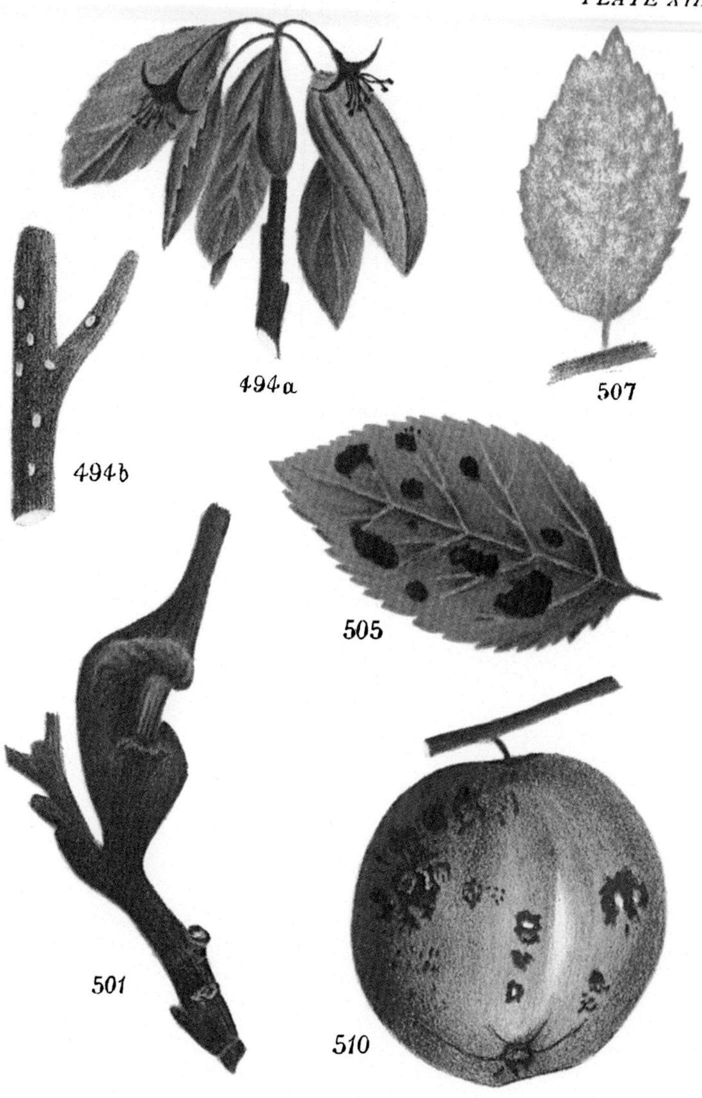


494 Apple Blossom Wilt *(a)* on blossom, *(b)* on branch
501 Apple Canker 505 Apple Leaf Spot
507 Apple Mildew 510 Apple Scab

chromo-lith Cambridge University Press

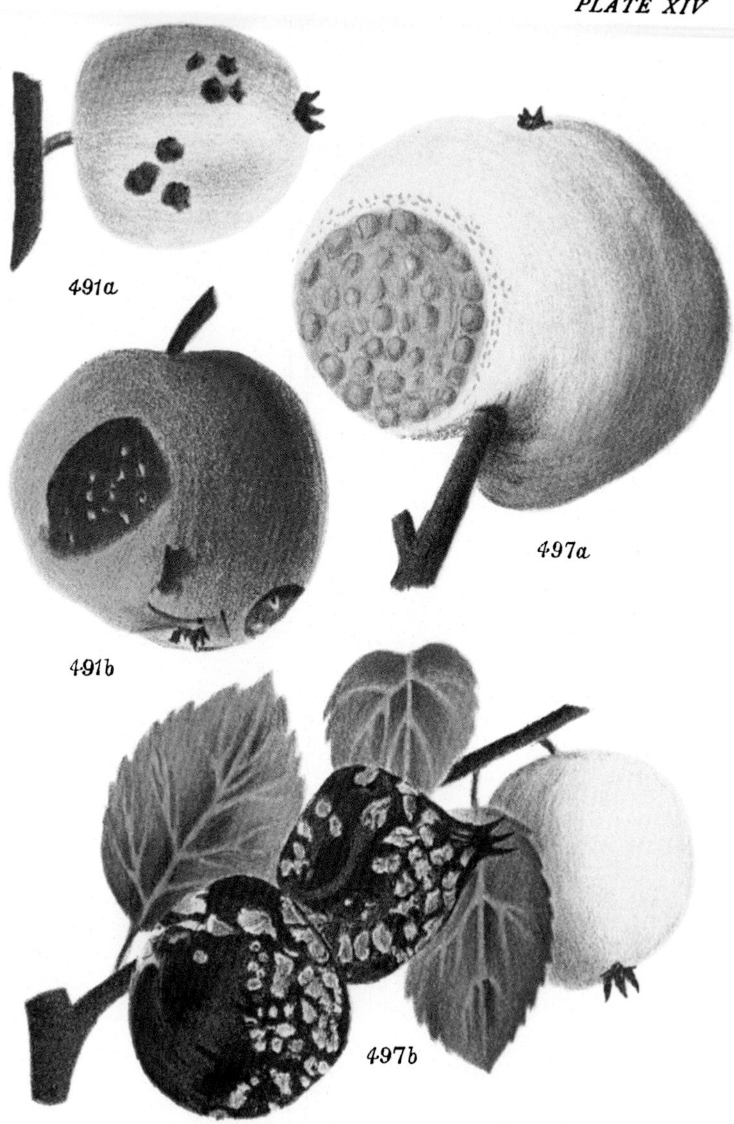


491 Apple Bitter Rot (*a*) early stage, (*b*) later stage
497 Apple Brown Rot (*a*) early stage, (*b*) later stage

chromo-lith. Cambridge University Press

# CHAPTER 30

## Diseases of the Apple

## APPLE BITTER ROT

Name *Glomerella rufo-maculans* Order *Ascomycetes*

### Plants Attacked

Apples: less commonly pears, peaches, grapes.

### Related Diseases

Apple scab, pear scab, apple canker, etc.

### 1. General

This is a very serious disease, and occurs in practically every country where the apple is cultivated. It is estimated that in America the loss of fruit occasioned runs into millions of dollars annually. The fruit and the branches are both affected, but the damage to the latter often passes unnoticed.

### Occurrence and Symptoms

This first appears under the skin of the apple as a small, brown spot, but there may be two or more visible at the same time on the fruit. The spots usually increase rapidly in size, being more or less round, and having a definite outline.

The affected part next sinks in, the diseased area increases, and small circles of black points appear, one within the other. This is the fruit of the fungus, and pinkish sticky masses of spores (conidia) soon appear, which may be washed off by rain.

This fungus is often associated with a canker disease on the young branches. This appears as burnt-looking, cracked patches on the bark, parts of which may break away leaving open wounds. On these patches, summer (conidial) fruit is developed, the spores from which become mature earlier in the year, and infect the young apples by falling upon them.

It has been found however that this canker stage (Glœosporium) is not necessary to carry on the fungus from year to year, as the spores on the fruit will retain their vitality over the winter.

**Effect on Plant**

The diseased apples often drop off the tree, becoming prematurely ripe, but in certain cases of very rapid development they may hang on the branches in a "mummified" condition till the following year. The diseased part of the flesh of an attacked apple has a very bitter flavour.

**How infection occurs**

1. By the spores of the ascigerous ("winter") fruit upon the diseased apples falling from the cankered branches of mummified fruits.
2. From the conidial (summer) spores, which may also be carried by insects from other trees.

The spores appear to be able to cause infection in apples with perfectly whole skins, the germ-tube penetrating the stomata (breathing pores).

**Conditions favourable to Fungus**

Close, warm, wet weather. Cool, dry, summer weather is a definite check to the disease.

**Frequency of Disease**

Common.

**Distribution**

World-wide.

**Treatment (Control)**

1. Remove and destroy all diseased fruit, both that which has fallen and any "mummified" apples hanging on the branches.
2. Prune off any cankered shoots down below the limit of the diseased portion, and burn these. In the case of the larger branches cut out the cankered areas, and dress with TAR[1].
3. Spray with BORDEAUX MIXTURE[2].
   (1) before buds burst.
   (2) 5 weeks after.
   (3) at intervals of two weeks after, till fruit is ripe.

In case of cool and dry weather, less sprayings will suffice.

[1] See page 417. [2] See page 604.

**Calendar of Treatment**

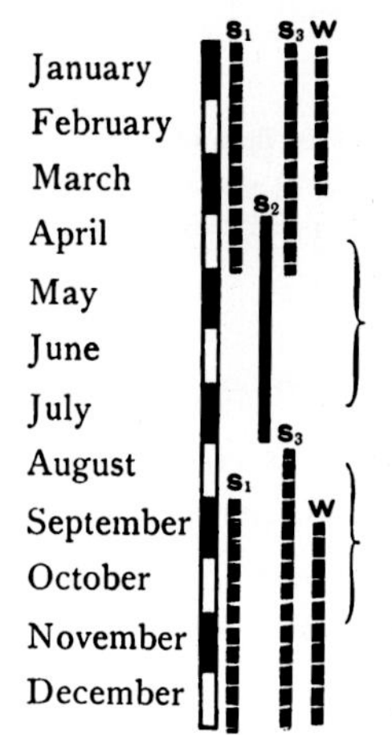


Spray with Bordeaux at intervals of two weeks.

Remove mummified apples on trees.
Prune or cut out canker on branches.

$S_1$ = summer (conidial) stage on branches
$S_2$ = ,, ,, fruit
$S_3$ = ,, ,, mummified fruit
W = winter (ascigerous) stage on decayed fruit

## 2. Scientific

### A. Conidial Stage

| | |
|---|---|
| APPEARANCE. | As brown patches, with concentric rings of fructifications. |
| LOCATION. | Under skin of fruit and in canker spots on branches. |
| APPEARS IN | May or later. |
| DURATION. | Often throughout winter on fallen or mummified apples. |
| MYCELIUM. | Throughout discoloured portions of the apple, and in the cambium of the branches. |
| CONIDIOPHORES. | Erect on stromatic mass of mycelium. |
| CONIDIA. | Embedded in a gelatinous matrix, readily soluble in water. |
| Appearance. | Oblong to ovate hyaline. |
| Size. | $12—16 \times 4—6\mu$. |

### B. Ascigerous Stage

| | |
|---|---|
| APPEARANCE. | Greenish masses. |
| LOCATION. | On decayed apples. |
| APPEARS | At variable times. |

| | |
|---|---|
| DURATION. | The canker spots probably last two years. |
| MYCELIUM. | Dark olive green. |
| PERITHECIA. | |
| Appearance. | Subglobose. |
| Location. | Buried in decayed flesh of apple. |
| ASCI. | Oblong-clavate: 55—70 × 9μ. |
| ASCOSPORES. | 8: resemble conidia: curved 12—22 × 3½ – 5μ. |

## APPLE BLOSSOM WILT[1]

Name *Monilia cinerea* Class *Helotiaceæ*
Order *Discomycetes, Ascomycetes*

This disease is really a form of the BROWN ROT fungus, but usually attacks and kills the flowers, being rarely seen on the fruit.

### Plants Attacked

Apple.

### Related Diseases

Common Brown Rot (*Sclerotinia fructigena*), "wither-tip" of Plums (page 578).

### 1. General

In recent years this disease has become very common, particularly in certain varieties of apples, and especially *Lord Derby*. It has probably escaped notice in the past, because it is very likely that the damage caused by it was attributed to frost.
Although the fungus produces a "brown rot" on the fruit, it does not usually appear in this manner, but is known by the death of the flowers in spring and the subsequent appearance of a canker on the branch, producing pustules of the fungus the following spring, spores from which infect and kill the blossom.

### Occurrence and Symptoms

The first symptom of the disease occurs in the spring (about May) when the leaves and flowers in a blossom truss commence to flag, and then wilt and finally die. In the distance the leaves look greyish owing to the tendency to curl inward and expose the under-surfaces.

[1] See Wormald, *Annals of Applied Biology*, Vol. III. 4, 1917, 159—204.

After the death of the flowers, the fungus grows along the branch and produces a canker, and, in the following year, grey pustules of the fungus are produced at this spot and infect the flowers. The growth of the fungus then ceases at this place, and the injured surface becomes healed over. The pustules occur only rarely on the fruit.

**Distinguishing Features**

Infection takes place THROUGH THE OPEN FLOWERS ONLY, the fruit is rarely attacked.

**Effect on Plant**

Up to 75 per cent. of the blossom is often killed on the trees, and the effect is thus very serious, trees badly attacked yielding no fruit at all year after year.

**How infection occurs**

By the conidia (summer spores) falling from the pustules of the fungus on the old branches on to the open flowers.

**Susceptible Varieties**

Lord Derby very susceptible, also Cox's Orange Pippin, Allington, Worcester, James Greve, Ecklinville, etc.

Charles Ross, Bramley, Blenheim, and Beauty of Bath appear more resistant.

**Frequency of Disease**

Common in recent years.

**Distribution**

Noticed chiefly in Kent and Sussex, and probably occurs widespread on the continent.

**Treatment (Control)**

1. Wherever practicable, remove in spring all trusses showing signs of wilting before the fungus has had time to penetrate into the branch. By this means, canker of the small branches is prevented.
2. Remove all cankered spurs and branches before the flowers open in spring, cutting back to the healthy wood.
3. Growers are advised to try spraying in spring, just before the blossoming of the tree (as No. 2, fig. 244 on page 616), with SOFT

SOAP[1] and AMMONIUM POLYSULPHIDE[2] ("A.P.S." wash). Lime-sulphur does not sufficiently wet the fungus to be of certain benefit.

### Calendar of Treatment

Cut off cankered spurs to healthy wood before buds open.

Spray with soft soap and ammonium polysulphide (A.P.S.).

Remove wilting trusses of flowers.

Remove any rotten apples (rarely found).

C = fungus producing canker on shoots
P = pustules of fungus producing spores on shoots
F = flowers infected and killed from spores from shoots
A = as brown rot on apples (rare)

## 2. Scientific

### A. Conidial Stage

| | |
|---|---|
| APPEARANCE. | *Grey* pustules. |
| LOCATION. | On cankered shoots only. |
| APPEARS IN | Winter and spring. |
| DURATION. | One season. |
| MYCELIUM. | White, in shoot. |
| CONIDIA. | In general, smaller than the brown-rot conidia (*fructigena*). |
| Appearance. | Hyaline, ovoid. |
| Arrangement. | Simple or branched chains: 10—14 × 7—9·5μ. |

### B. Ascigerous Stage

Not yet discovered.

[1] See page 452. [2] See page 600.

## APPLE BROWN ROT

Name *Sclerotinia fructigena* Class *Helotiaceæ*

Order {*Discomycetes* / *Ascomycetes*}

### Plants Attacked

Apples, plums, cherries, peaches, apricots, etc.

### Related Diseases

*Monilia cinerea*, gooseberry collar rot.

### 1. General

The disease termed BROWN ROT is a very general and very destructive fungus, and affects almost all "top fruits." As far as is at present known only the fruit is attacked.

The attack of the allied "*Blossom Wilt*" disease (*Monilia cinerea*) is very similar on the fruit, but in this fungus the flowers are usually killed before the fruit forms.

In the case of the other fruits attacked, the appearance of the disease is very similar, except that the pustules are not arranged in rings or circles as in the case of the apple.

This disease was formerly known as *Monilia fructigena*, this being the conidial stage, and the only one known in this country. Recently the ascigerous stage has been described, and the fungus is found to be the summer stage of the *sclerotinia*.

### Occurrence and Symptoms

The apples are attacked during the summer or early autumn. Small brown discolorations appear which rapidly increase in size and finally involve the whole fruit. Small pimply swellings meanwhile occur under the skin which soon burst through, appearing usually as rings of yellowish pustules with a powdery surface; these ripen and produce spores which spread the disease to neighbouring trees. The attacked fruit rapidly shrinks in size and deep grooves and wrinkles appear on the

Fig. 234. Apple affected with brown-rot fungus: showing also commencement of rot (caused by contact) on a neighbouring healthy apple. Slightly reduced.

skin. Apples may fall or may hang on the trees in a mummified condition throughout the winter. The mycelium or spawn of the fungus ramifies throughout the entire fruit. On fallen apples, fresh pustules of the disease appear, and on the mummified fruits a new crop of spores are produced in the following summer, which infect the young fruit.

The disease also occurs on *stored fruit* after it is picked and may spread amongst many apples if these are packed together. Stored fruit should therefore be periodically examined and any showing signs of disease removed and destroyed.

In certain varieties of apples the fruiting spur is also attacked, a canker is produced round the base, and pustules of the disease appear in the following spring.

The canker is very similar to that produced by the *Blossom Wilt* fungus, but in this case infection takes place through the *flower* only, while in the case of Brown Rot it is the *fruit* and only the fruit which is the channel of infection.

## Distinguishing Features

Distinguished from the allied *Blossom Wilt* (*Monilia cinerea*) by the lack of infection of the flower, which the former usually kills.

## Effect on Plant

The apples attacked become withered, dry and "*mummified*," and either fall off the trees or remain in this condition throughout the winter.

## Degree of Damage

Always serious, and frequently ruining the crop.

## How infection occurs

The spores (conidia) are produced in great numbers from the rings of yellowish pustules on the attacked fruit. The spore probably gains an entrance to the healthy fruit through some slight cut, bruise, or incision caused by wind or insect agency

## Frequency of Disease

Common.

## Distribution

Widespread.

## Treatment (Control)

Since the *mummified* fruit hanging on the trees is a source of fresh infection the following year, it should always be *entirely removed and destroyed.*

Apples which show any signs of the disease should be plucked off in their early stage, where this is possible.

Cut off also any infected or cankered spurs, either in the summer (best time) or during the winter months.

Spray with strong LIME-SULPHUR[1] before the buds burst in the spring.

## Calendar of Treatment

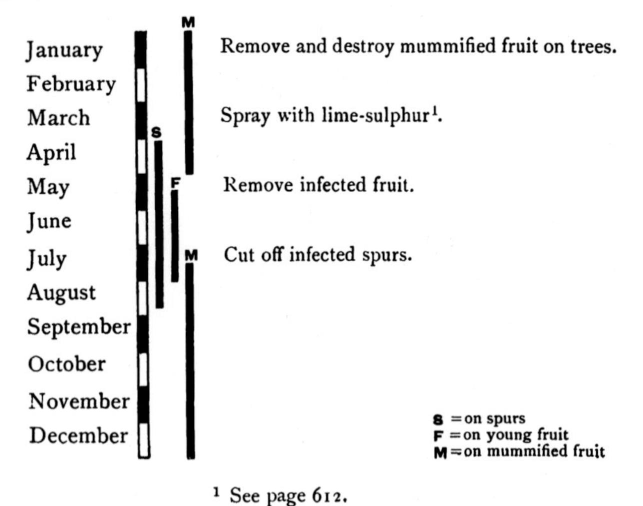


[1] See page 612.

### 2. Scientific

**A. Conidial Stage**

| | |
|---|---|
| APPEARANCE. | In yellowish powdery tufts, forming concentric circles (in apple). |
| LOCATION. | On fruit and on spurs. |
| APPEARS IN | Summer and throughout winter, on decayed fruit. |
| MYCELIUM. | Ramifies throughout fruit. |
| CONIDIA. | |
| Appearance. | Hyaline, ovoid. |
| Arrangement. | Simple or branched chains $21—25 \times 10—12\mu$. |

**B. Ascigerous Stage**

Not known in England.

## APPLE CANKER

Name *Nectria ditissima* Class *Hypocreaceæ*
Order *Ascomycetes*

**Plants Attacked**

Apples, pears, gooseberries.

**Related Diseases**

Apple bitter rot (*Glomerella rufo-maculans*).

### 1. General

This fungus is very common, especially on apple trees and in neglected orchards. It cannot attack healthy bark, but obtains an entrance through wounds, especially those caused by the WOOLLY APHIS. In fact, attacks by this insect are almost invariably followed by the canker. The fungus does not die at the end of a season, but goes on increasing till death of the branch results, and, on the trunk, large cankers are produced.

**Occurrence and Symptoms**

The first sign of the fungus is the cracking of the bark, which usually splits into regular semi-circular grooves. These isolated segments die and drop off.

Fig. 235. Two branches of apple badly attacked by canker fungus.

Later the wood is also attacked, and small branches are completely girdled and the branch destroyed.

More usually, an oval-shaped or irregular callus is formed round the wound.

Two kinds of spores are produced, the first (conidia) being liberated from the fruit, which occurs during the autumn as white pustules. What is generally termed the winter fruit (ascigerous stage) is produced in spring, appearing as dark red clusters on the edges of the wound. From these, the "winter spores" (ascospores) are liberated.

**Effect on Plant**

The yield of fruit is undoubtedly diminished by canker attack, as the vitality of the tree is necessarily weakened. The effect of the canker is cumulative, increasing each year, and finally resulting in the death of the tree.

**Degree of Damage**

In spite of the above, many badly cankered trees continue to yield fair crops of fruit, but they form centres of infection for the other healthy trees.

**How infection occurs**

The fungus spores gain an entrance through wounds of every kind, especially those caused by the punctures of the woolly aphis.

**Frequency of Disease**

Common, especially in neglected orchards.

**Distribution**

Widespread in Europe : less known in America.

**Prevention**

Insect attacks should be kept in check, especially that of the WOOLLY APHIS[1].

An annual treatment of the bark by spraying with the stronger LIME-SULPHUR[2] in the winter, or early spring, will keep it in a healthy condition and make it more resistant to attack.

**Remedies**

1. Remove and burn all badly cankered branches.
2. Cut out cankers where these are slight, or dress wounds with gas-tar, or other protectant[3].

[1] See page 327. [2] See page 612. [3] See page 417.

3. Cut down all trees which are badly attacked, as these form dangerous centres of infection.
4. Spray annually in the winter and spring with winter-strength LIME-SULPHUR[1]. (Spraying is useless unless accompanied by measures 1 and 2.)
5. Cut back the tree and top graft with less susceptible variety such as *Bramley Seedling*.

## Calendar of Treatment[2]

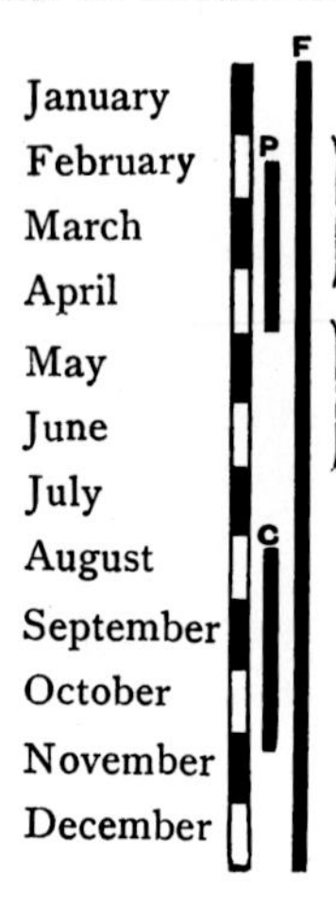


Spray lime-sulphur[1].

Keep down insect attacks.

Spray lime-sulphur.
Remove cankered branches (see above).

F – fungus, or mycelium (perennial in wood of branch)
P – "perithecial," "ascigerous," or "winter stage"
C – conidial, or "summer stage"

## 2. Scientific

### A. Conidial Stage

| | |
|---|---|
| APPEARANCE. | White pustules. |
| LOCATION. | On cankered areas. |
| APPEARS IN | Summer. |
| DURATION. | Few months. |
| MYCELIUM. | Perennial in the wood of branches. |
| CONIDIA. | |
| Appearance. | White; fusiform. |
| Arrangement. | 3—5-septate; $35—60 \times 4{\cdot}5—5\mu$. |

[1] See page 612.

[2] The canker fungus is perennial (occurs year after year).

**B. Ascigerous Stage.**

| | |
|---|---|
| APPEARANCE. | As small dark red clusters. |
| LOCATION. | On margins of cankered areas. |
| APPEARS IN | Autumn and winter. |
| DURATION. | 2 to 3 months. |
| PERITHECIA. | |
| Appearance. | Blood-red; in dense clusters. |
| ASCI. | Cylindrical, 8-spored. |
| ASCOSPORES. | Ovate-oblong, hyaline, 1 septate, $6—8 \times 3—4\mu$. |

## APPLE LEAF SPOT

Name *Sphæropsis malorum* Class *Sphæroidæ*
Order *Deuteromycetes*

**Plants Attacked**

Apples.

**Related Diseases**

Gooseberry shoot spot, pear leaf spot, strawberry leaf blight, currant leaf spot.

### 1. General

**Occurrence and Symptoms**

In the spring, reddish-brown spots appear upon the apple leaves, of variable size and shape. Occasionally these have a purple margin, and later the spots have a greyish appearance. The disease continues developing and involving fresh leaves throughout the summer.

The patches show the limit of the fungus mycelium, and the minute black points on the diseased patches are the pores or openings through which the conidia (spores) escape.

**Effect on Plant**

The leaves droop and fall, and if the disease continues unchecked the tree is seriously weakened, and the fruit suffers in yield and quality.

**How infection occurs**

By means of the conidia (summer spores) scattered by wind and insect agency, etc.

**Conditions favourable to Fungus**

Damp, muggy weather.

**Frequency of Pest**

Not common.

**Distribution**

Widespread—especially noted in America.

**Remarks**

This fungus produces a rot in the apple and a canker on the bark of the tree in America[1].

**Treatment**

Spray the trees with BORDEAUX MIXTURE, summer strength (see page 609), 2 to 3 weeks after the blossom has fallen, and follow by a further spray about one month later.

**Calendar of Treatment**

| Month | Stage | Treatment |
| --- | --- | --- |
| January | | |
| February | | |
| March | s | |
| April | s | |
| May | s | Spray with summer strength Bordeaux. |
| June | s | Apply again. |
| July | s | |
| August | | |
| September | | |
| October | | |
| November | | Winter stage unknown. |
| December | | |

**s** = summer or conidial stage

## 2. Scientific

### A. Conidial Stage

| | |
| --- | --- |
| APPEARANCE. | Reddish brown spot, coalescing. |
| LOCATION. | On leaves. |
| APPEARS IN | Spring. |

[1] Puddock Wendell. *New York Agr. Exp. Stn. Bull.* **163**, 331–369; **185**, 205–213.

| | |
|---|---|
| DURATION. | Throughout season. |
| MYCELIUM. | Embedded in leaf tissue. |
| CONIDIA. | Contained in a perithecium, $25 \times 10\mu$. |
| Appearance. | Oblong, coloured. |

**B. Ascigerous Stage**

Not yet known.

## APPLE MILDEW

Name *Podosphæra* Class *Erysiphaceæ* (*Powdery mildews*)
Order *Ascomycetes*

**Plants Attacked**

Apples.

**Related Diseases**

Other powdery mildews on gooseberry, hop, etc.

### 1. General

This disease resembles most of the other "powdery mildews." They have, in their summer (conidial) condition, a powdery, floury appearance, and the spawn or mycelium of the fungus is entirely on the outside of the plant, the fungus obtaining its nourishment by means of suckers or *haustoria* which bore into the leaves through the cell-wall.

On this account they are easier to deal with in their summer stage than those fungi which have their mycelium entirely embedded in the plant tissue.

**Occurrence and Symptoms**

The ends of the twigs show the first signs of the disease, and have the appearance of having been sprinkled with flour. This is the summer (conidial) stage of the fungus.

It is probable that a portion of the mycelium (spawn) of the mildew remains between the leaves of the bud throughout the winter and that this gives rise to the disease again in the following spring.

The winter (ascigerous) stage is comparatively rare, but has occasionally been found.

**Distinguishing Features**

Its white, floury appearance on the tips of the twigs.

**Effect on Plant**

The shoot is prevented from growing, owing to the exhaustion produced by the fungus it becomes stunted and in bad cases withered.

**How infection occurs**

1. From the summer (conidial) spores.
2. From portions of mycelium (spawn) wintering in the buds.
3. From the winter (ascigerous) spores—more rarely.

**Susceptible Varieties**

Lane's Prince Albert appears specially susceptible, it also frequently attacks Cox's Orange Pippin, Grenadier and Bismarck (Kitchener).

**Frequency of Disease**

Not uncommon—appears to be increasing in frequency.

**Distribution**

Widespread in Europe.

**Treatment (Control)**

1. Cut away all diseased tips as soon as these are seen, cutting well back beyond the limit of the attacked leaves.
2. Spray in the early spring with LIME-SULPHUR[1] or BORDEAUX MIXTURE[2] (winter strength) before buds burst, or later with a weaker solution of the same.
3. Spray when the fungus is seen with SOFT SOAP[3] and AMMONIUM POLYSULPHIDE[4].

**Calendar of Treatment**

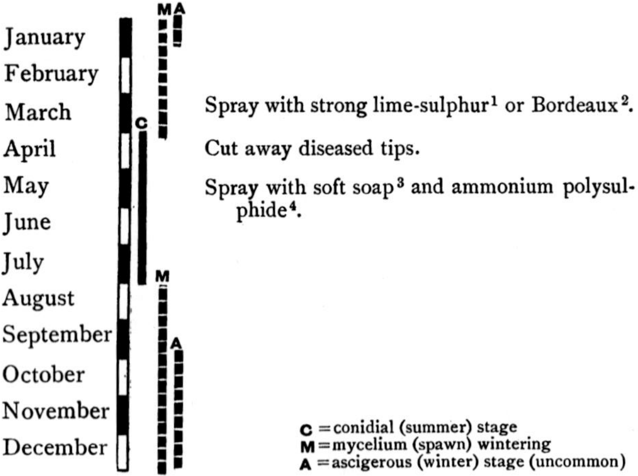


[1] See page 612. [2] See page 609. See page 452. [4] See page 602.

## 2. Scientific

| | |
|---|---|
| **A. Conidial Stage** | |
| APPEARANCE. | White patches resembling flour. |
| LOCATION. | On tips of shoots and terminal leaves. |
| APPEARS IN | Summer. |
| MYCELIUM. | White, external. |
| HAUSTORIA. | Numerous. |
| CONIDIA. | |
| Appearance. | White, oblong. |
| Arrangement. | Long chains. |
| **B. Ascigerous Stage** | |
| LOCATION. | On mycelium. |
| APPEARS IN | Winter. |
| PERITHECIA. | |
| Location. | On mycelium. |
| Appendages. | 1. From apex of perithecium, long, unbranched.<br>2. From base, wavy, slightly branched. |
| ASCI. | Oblong. |
| ASCOSPORES. | Elliptical, hyaline, $22—26 \times 12—14\mu$. |

## **APPLE SCAB** (Black Spot)

Name *Venturia inequalis* Order *Ascomycetes*

### **Plants Attacked**

Apples.

### **Related Diseases**

Pear Scab, Cherry leaf Scorch.

### **1. General**

This is the best known, and quite the most important of the fungi attacking apples. The losses annually occasioned by this disease must be very great. Even when the apples are marketable, they are very stunted in size, and realise correspondingly low prices.

Much has been discovered in recent years, both as to the life-history and as regards satisfactory treatment of this disease. It was at first thought to occur only on the fruit, but we now know that the shoots and leaves are affected, and that it is the diseased shoots that are mainly responsible for carrying the disease over from year to year.

The summer or conidial stage of the disease was the only form recognised until recently when Aderhold proved that this is the conidial form of the *Venturia* fungus.

### **Occurrence and Symptoms**

1. *The fruit.*

   This is where growers always notice this disease. It commences to show in the form of small, circular, greenish black spots on the skin of the apples. These increase in size and after a time the familiar scabby patches are produced, due to the dying of these parts of the skin.

2. *The leaves.*

   The fruit is usually *infected from the leaves* which often pass unnoticed. On the leaves the fungus appears as dark coloured spots, mainly on the upper surfaces. These increase in size and often run together, forming greenish black patches, which, when examined with a hand-glass are seen to have a fibrous appearance towards the margin. The skin of the leaf is ruptured, and spores are produced and set free which infest the fruit.

3. *The shoots.*

The leaves are usually infected from the *year old shoots.* If these are examined in the winter, or best in early spring, the skin or bark will be seen to be wrinkled and often torn and split, and even missing altogether, especially near the base of the previous year's shoot. The exposed portions are generally covered with black patches of the same stage (conidial) of the fungus as attacks the leaves and fruit. Spores are freely produced which are carried by the wind and other agencies on to the young leaves.

**Distinguishing Features**

This fungus is not readily mistaken for any other on the leaves and fruit. The wounds on the shoots are distinguished from *apple canker* (*Nectria ditissima*) by the appearance of *black* instead of *red* fructifications, and by the fact that the latter causes swelling and callus formation in addition to cracking and injury of the bark.

**Effect on Plant**

Apples affected with Scab are often distorted, and some varieties show deep cracks extending to the core. These are caused by the drying of the surface, which becomes stiff and unyielding, and the pressure set up by the growth of the fruit causes rupture.

**Degree of Damage**

Always considerable, varying with season.

Fig. 236. Apples badly attacked with apple scab fungus. Natural size.

**How infection occurs**

1. From conidia ("summer spores"), scattered by wind agency, insects, etc., from shoots to leaves and from leaves to fruit.
2. From ascospores (winter spores), from the "winter" (ascigerous) stage produced upon the old leaves.

**Conditions favourable to Fungus**

Cool and damp weather during spring and early summer.

**Susceptible Varieties**

Several varieties are so badly attacked that they are going out of cultivation, e.g. Wellingtons and Gladstones. Cox's Orange Pippin is usually very susceptible. Quick growing varieties, such as Bramleys, are less liable to attack.

**Frequency of Disease**

Very common.

**Distribution**

Occurs very widely.

**Treatment (Control)**

1. Cut away and burn all diseased shoots up to the point of the previous year's growth.

   This may be tedious and difficult work, but it is amply repaid, and the other treatment is of much less benefit if this be not first done.

2. Spray with CONCENTRATED LIME-SULPHUR, winter strength (see page 612), early in the year, but as late as it is safe. This is just before the opening of the leaf-buds (see page 616).

*Or*

Spray at same time with full strength BORDEAUX MIXTURE (see page 606).

Fig. 237. Fruit from two adjacent trees: unsprayed (above), winter sprayed with lime-sulphur (below).

Fig. 238. Effect of winter spray with lime-sulphur for apple scab—branch of sprayed tree.

3. If the fungus appears on the leaves in the spring (which should be carefully examined) spray twice, at intervals of 4 to 5 weeks, with summer strength lime-sulphur or Bordeaux (see pages 612, 606).

4. Cut well back and top graft with a quick growing variety, such as *Bramley Seedling*.

**Remarks on Remedies**

In the case of certain varieties of apples, particularly Cox's Orange Pippin, the leaves are injured by Bordeaux (Bordeaux injury, see page 605), and the summer strength lime-sulphur only must be employed.

**Calendar of Treatment**

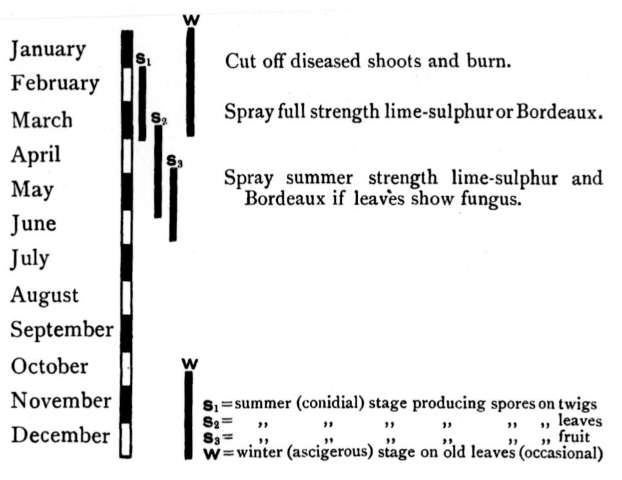


## 2. Scientific

### A. Conidial Stage

| | |
|---|---|
| APPEARANCE. | Greenish black spots and patches (fibrous on leaves). |
| LOCATION. | (1) on twigs, (2) on leaves, (3) on fruit. |
| APPEARS IN | (1) early spring, (2) spring and summer, (3) late spring and summer. |

| | |
|---|---|
| PROPAGATION. | Usually by conidia from this stage on twigs or by the ascospores from the winter (ascigerous) stage. |
| MYCELIUM. | Greenish black. |
| CONIDIOPHORES. | Brown, outline wavy; 50—60×4—6 μ. |
| CONIDIA. | Single, obclavate, yellowish olive; 30×7—9 μ. |

## B. Ascigerous Stage

| | |
|---|---|
| LOCATION. | On old leaves. |
| APPEARS IN | Spring (spores liberated). |
| PERITHECIA. | |
| APPEARANCE. | Spherical. |
| LOCATION. | Embedded in leaf. |
| ASCI. | Clavate to oblong; 55—75×6—12 μ. |
| ASCOSPORES. | 8, olive-brown, becoming 2-celled; 11—15×5—7 μ. |

# DISEASES OF THE CHERRY

*PLATE XV*

526 (*a*) Cherry Rot  526 (*b*) Cherry Scab
528 Cherry Witches' Brooms

chromo-lith. Cambridge University Press

PLATE XVI

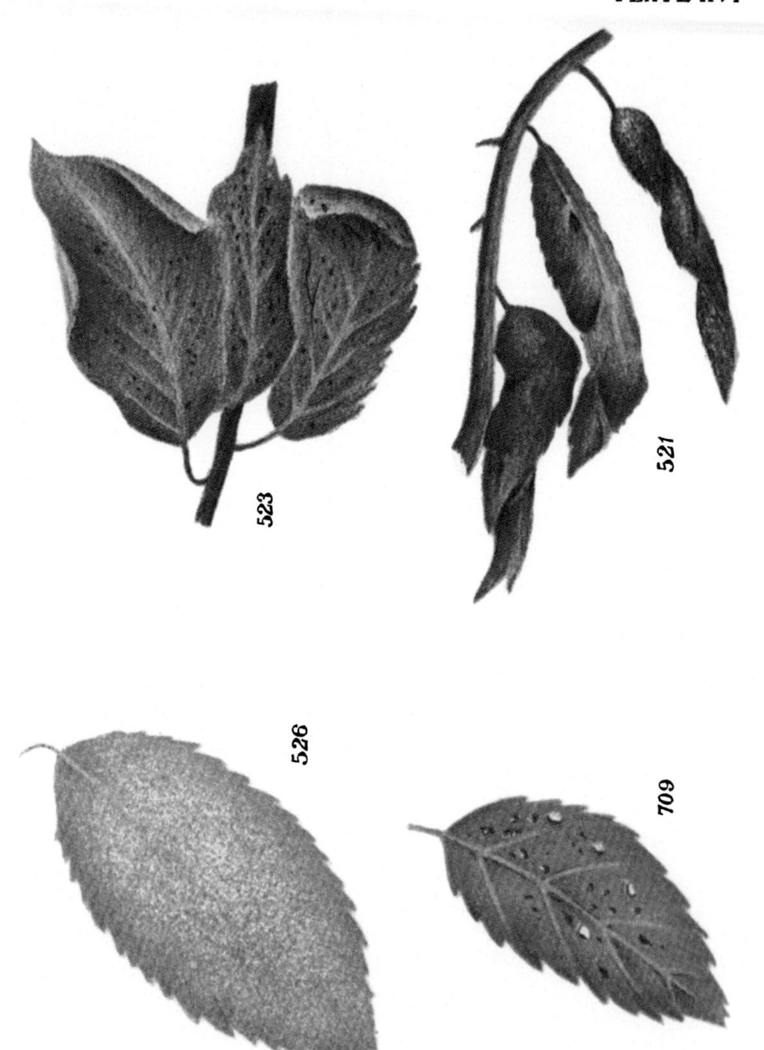


521 Cherry Leaf Curl 523 Cherry Leaf Scorch
526 Cherry Mildew 709 Cherry and Plum Leaf Blight

chromo-lith Cambridge University Press

# CHAPTER 31

## Diseases of the Cherry

## CHERRY BLOSSOM WILT AND CANKER

Name *Monilia cinerea*

This is the same fungus which attacks the Plum and the symptoms and treatment are as described for that disease on page 579. The attack has been investigated by *Wormald.*

## CHERRY LEAF CURL

Name *Exoascus minor* Class *Exoascaceæ* Order *Ascomycetes*

**Plants Attacked**

Cherries

**Related Diseases**

" Witches'-brooms " (*Exoascus cerasi*), page 528; Peach leaf curl (*Exoascus deformans*), page 557.

### General

This disease was first recorded in England by Professor Salmon[1] in the year 1907, and it has occurred in various localities since then from time to time.

**Occurrence and Symptoms**

When attacked by this disease the leaves of the cherry commence to curl up and become pinkish red in colour. Later, a delicate white bloom appears on the under-surface of the leaves, caused by the fungus fruits, in the form of minute "sacs" containing the spores. These are ejected from the leaves and infest neighbouring branches and trees. The affected leaves then turn brown, decay and drop off the tree. The disease is often confined to one or more leaves on a branch.

The mycelium (spawn) of the fungus passes the winter in the young buds of the cherry, ready to appear in the following spring. Fresh leaves are then affected, the spores appear and the fungus grows along to the new buds and so carries on the disease to a new season.

[1] *Wye Report* 1908, **17**, 320.

**Distinguishing Features**

The curled and red appearance of leaves in summer.

**Effect on Plant**

Not great on older trees, but very bad on nursery and young stock.

**How infection occurs**

By means of spores produced from fungus fruits on leaves in summer.

**Susceptible Varieties**

Sweet cherries principally attacked.

**Frequency of Disease**

Not uncommon.

**Distribution**

Widespread in Europe.

**Remarks**

The diseased leaves have a characteristic sweet-scented smell resembling new-mown hay.

**Treatment (control)**

1. Prune off each affected branch well below the affected leaves. This is necessary since the fungus passes the winter in the buds.
2. Spray with BORDEAUX MIXTURE[1] at the time of opening of the leaves.

**Calendar of Treatment**

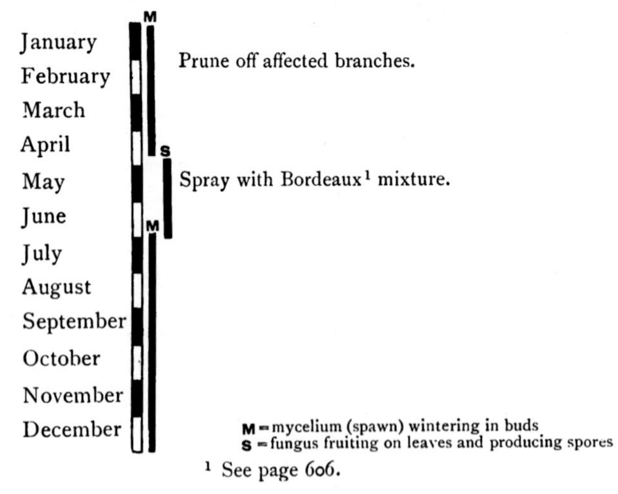


[1] See page 606.

## CHERRY LEAF SCORCH

Name *Gnomonia erythrostoma* Class *Hypomyces*
Order *Pyrenomycetes*

### Plants Attacked

Cherry.

### 1. General

This serious disease of cherries is now fairly widespread in England and in Europe. Attacked orchards are easily recognised in the winter and spring from the fact that the brown dead leaves remain hanging to the branches, and they are still to be seen even after the flowering of the tree.

### Occurrence and Symptoms

The leaves are first attacked in the early summer, and the disease appears as discoloured patches which increase in size. Yellowish at first, the patches soon turn brown and the whole leaf rapidly succumbs, presenting a scorched appearance. This is due to the spawn (*mycelium*) of the fungus spreading to the leafstalk, and cutting off the supply of food. Summer (conidial) spores are produced at this stage, and owing to the invasion of the leafstalk by the fungus, the healthy fall of the leaves in the autumn is prevented and these remain on the tree till they rot away in the following summer.

During the autumn and winter the ascigerous (winter) stage of the fungus is produced, giving rise to spores from the dead leaves in the following spring and early summer which infect the young leaves. The spores are ejected from their cases with considerable force.

The fruit is also attacked, causing it to fall or become distorted.

### Distinguishing Features

The scorched appearance of the leaves, and their persistence throughout the winter on the branches.

**Effect on Plant**

The affected trees gradually weaken, an increasing amount of dead wood is produced and the tree finally dies. The yield of fruit is progressively reduced.

**How infection occurs**

By means of the summer and winter spores. The latter carry the disease over from one season to the next.

**Susceptible Varieties**

The sweet cherries are apparently more susceptible to attack than the sour varieties.

**Frequency of Disease**

Fairly common.

**Distribution**

Widespread in Europe.

**Treatment (control)**

Since the new leaves in the spring and summer can become infected only by the spores from the winter stage of the disease on the dead leaves hanging on the branches, the most effective method is to REMOVE THE WITHERED LEAVES during the winter or early spring. The tree is then as healthy as before, because the mycelium (spawn) does not penetrate to the branches.

This method, however, still lays the former open to trouble by infection from the untreated orchards of neighbours; and emphasises the necessity for co-operative effort, or for compulsory powers[1].

Also it is often too big a job for the grower to undertake, and for both of these reasons it is preferable to spray the trees so as to cover the young leaves with a protective coating. This is successfully accomplished by spraying with BORDEAUX MIXTURE[2] once, just before the flowers open, and again just after they have fallen.

The disease was stamped out in Germany some years ago by the government compelling all growers to remove the dead leaves from their cherry trees in the affected districts. For this to be of service universal action is of course necessary.

[1] See page 7. [2] See page 604.

### Calendar of Treatment

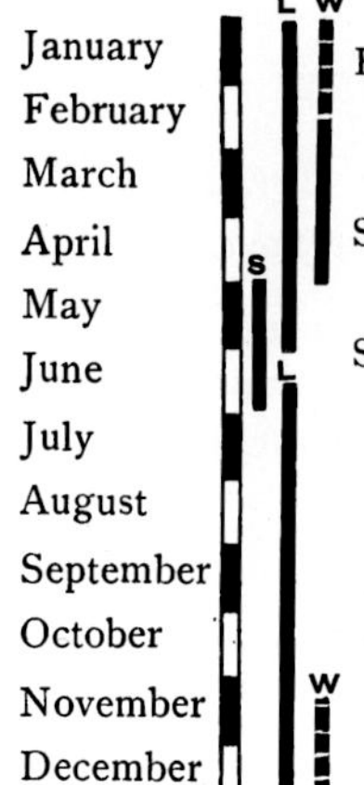


Remove all dead leaves. (January)

Spray Bordeaux[1] before flowers open. (April)

Spray Bordeaux[1] after bloom falls. (June)

S = summer (conidial) stage on young leaves
L = dead leaves hanging on trees
W = winter (ascigerous) stage on dead leaves: spores infecting new leaves

## 2. Scientific

### A. Conidial Stage

| | |
|---|---|
| APPEARANCE. | Yellow areas becoming brown: fruit in perithecia. |
| LOCATION. | On leaves. |
| APPEARS IN | Summer. |
| DURATION. | Few weeks—dead leaves persist on branches. |
| MYCELIUM. | In leaf and stalk tissues. |
| CONIDIA. | On branched sporophores. |
| Appearance. | Thread-like, hyaline, slightly curved. |
| Arrangement. | Terminal or nodal, 14—20 × 1—1·5 μ. |

### B. Ascigerous Stage

| | |
|---|---|
| APPEARANCE. | As minute black spots (beaks of fruit-conceptacles). |
| LOCATION. | On dead leaves. |
| APPEARS IN | Winter and spring. |
| PERITHECIA. | |
| Appearance. | Flask shaped with projecting beak. |
| Location. | In dead leaves. |
| ASCI. | 8 spores. |
| ASCOSPORES. | Hyaline, ovate, 1-septate, 16—18 × 5—6μ. |

[1] See page 604.

## CHERRY MILDEW

Name *Podosphæra oxyacanthæ* Class *Erysiphaceæ* (*Powdery mildews*) Order *Ascomycetes*

The fungus attacks cherries in America but is, so far as is known, confined to the Hawthorn in this country at present. Growers are advised to keep a watchful eye for its appearance in their plantations.

## CHERRY ROT

See page 497.

## CHERRY SCAB

Name *Fusicladium cerasi* Class *Sphæriaceæ*(?) Order *Ascomycetes* (?)

**Plants Attacked**

Cherries.

**Related Diseases**

Apple and pear scab (*Venturia*).

### 1. General

**Occurrence and Symptoms**

The disease appears in the form of blotches on the ripe cherries. These vary in size and there may be only one or several upon each fruit. They are greenish black and velvety in appearance and later on become scabby.

When the fruit is attacked in an early stage its development is stopped and it often remains on the tree in a partially withered condition for a considerable time.

The leaves and the young shoots are probably first attacked, but this is not at present definitely known.

This disease is probably very closely related to the apple and pear scab (*Venturia*), but, so far, the winter (ascigerous) stage has not been described.

### Effect on Plant

Renders the cherries unfit for eating, or prevents their development.

### How infection occurs

Probably from summer (conidial) spores scattered by wind, insects etc. from young shoots and leaves.

### Frequency of Disease

Not uncommon.

### Treatment (Control)

1. Spray with CONCENTRATED LIME-SULPHUR, winter strength (see page 614), early in the year, but as late as possible—i.e. just before the bursting of the buds.
2. When the first signs of the disease appear, spray with
   (*a*) LIME-SULPHUR (summer strength[1]) or
   (*b*) AMMONIUM POLYSULPHIDE and SOFT SOAP[2].

### Calendar of Treatment

As for Apple Scab (q.v.).

## 2. Scientific

### A. Conidial Stage

| | |
|---|---|
| APPEARANCE. | Olive black patches. |
| LOCATION. | On cherries. |
| APPEARS IN | May to July. |
| CONIDIA. | |
| Appearance. | Oblong with narrow ends, olive coloured. |
| Arrangement. | $20—25 \times 4—5\mu$. |

### B. Ascigerous Stage

Unknown.

[1] See page 612. [2] See page 601.

## CHERRY WITCHES'-BROOMS

Name *Exoascus cerasi* Class *Exoascaceæ* Order *Ascomycetes*

**Plants Attacked**

Cherry (wild and cultivated).

**Related Diseases**

Cherry leaf curl, peach leaf curl.

Fig. 239. Diagram showing cherry tree affected by the witches'-brooms disease.

### General

This disease shows itself in a curious tuft or broom of long branches growing out in portions of the tree. The branches are often much thicker than the remaining healthy ones. In the spring leaves appear but no blossom, and consequently this portion of the tree becomes quite barren.

### Occurrence and Symptoms

This abnormal growth of a portion of the tree is caused by the irritation of the mycelium (spawn) of the fungus in the tissue of the branch. The leaves, borne in the spring, soon turn a bright red colour, and afterwards a whitish "bloom" appears on their under surface. This is the fruit of the fungus, and spores are produced which infect the healthy leaves on the rest of the tree. On germinating, the fungus enters the leaf and then grows along to the branch, ultimately producing another "broom" on the tree.

### Distinguishing Feature

The "broom" formation.

### Effect on Plant

1. The part of the tree affected becomes barren.
2. The tree is weakened by the extra demand made upon its vitality.

### How infection occurs

By spores from the fungus-fruit on the leaves (v. s.).

### Frequency of Disease

Not uncommon.

### Distribution

Widely distributed in Europe.
Not common in America.

### Treatment (Control)

1. CUT OUT THE BROOMS as the fungus does not extend downwards into the tree, it will be healthy again.
2. Spray with BORDEAUX MIXTURE[1] in spring and early summer to prevent infection from neighbouring orchards.

[1] See page 606.

**Calendar of Treatment**

| Month | M | S | Treatment |
|---|---|---|---|
| January | M | | Cut out brooms. |
| February | M | | |
| March | M | S | Spray Bordeaux mixture[1]. |
| April | M | S | |
| May | M | S | Spray Bordeaux mixture[1]. |
| June | M | | |
| July | M | | |
| August | M | | |
| September | M | | |
| October | M | | |
| November | M | | |
| December | M | | |

M = mycelium (spawn) of fungus in branch of "broom"
S = spores of fungus produced on leaves of "broom" infecting healthy leaves

[1] See page 606.

# DISEASES OF THE CURRANT

533 Currant Leaf Spot 535 Currant Rust
583 Raspberry Rust 584 Raspberry Spot

chromo-lith. Cambridge University Press

# CHAPTER 32

## Diseases of the Currant

## CURRANT LEAF SPOT

Name *Pseudopeziza ribis* Class *Pezizineæ*
Order *Ascomycetes*

**Plants Attacked**

Black and Red Currants and Gooseberries.

### 1. General

This disease is frequently injurious and spreads with great rapidity in a plantation of currants and gooseberries.

**Occurrence and Symptoms**

The first signs of the fungus usually appear as small black spots mainly on the upper surfaces of the fully grown leaves. These represent groups of the fungus mycelium (spawn) on which the spores are later produced, when the skin of the leaf is ruptured. The spores, which are joined together, issue in the form of minute, viscid hairs, and the spores are liberated and washed off by means of rain.

The winter stage, which has not been found in this country, occurs on the dead leaves. Apparently the summer (conidial) stage is able to infect a plant continuously without the intervention of the winter fruit.

**Effect on Plant**

In bad cases, the leaves fall early in the season, and this not only checks the fruit crop of a given season, but influences the yield the following year.

**How infection occurs**

By means of the conidia (summer spores).

**Treatment (control)**

Spray as soon as the first signs of the disease appear with DILUTE

BORDEAUX[1] MIXTURE, summer strength LIME-SULHUR[2]. Growers are also advised to make trials with SOFT SOAP and AMMONIUM POLYSULPHIDE[3] ("A.P.S.").

## Calendar of Treatment

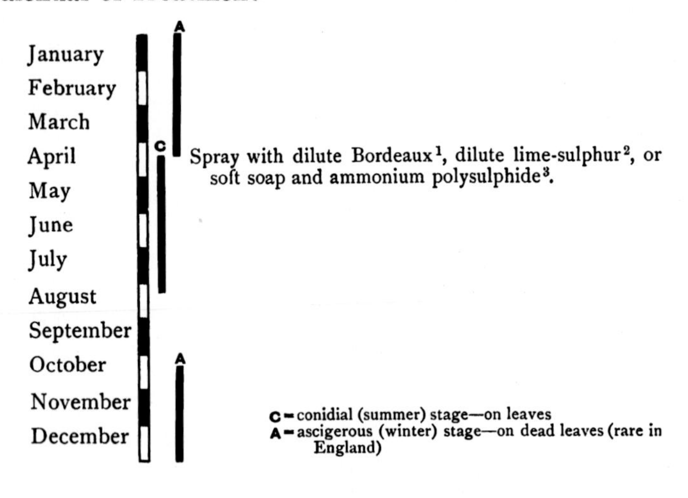


## 2. Scientific

### A. Conidial Stage

APPEARANCE. As black spots bursting through epidermis.
LOCATION. On upper surfaces of leaves.
APPEARS IN Summer.
MYCELIUM. Black: embedded in leaf-tissue.
CONIDIA.
Appearance. Minute, oblong, curved.
Arrangement. Agglutinated in viscid tendril, $10—12 \times 5—6\mu$.

### B. Ascigerous Stage (rare in England).

LOCATION. On dead leaves.
APPEARS IN Winter.
MYCELIUM. In tissue of leaf.
ASCI. With 8 spores.
ASCOSPORES. Elliptical, hyaline, $10 \times 6\mu$.

[1] See page 606. [2] See page 612. [3] See page 601.

## CURRANT RUST

Name *Cronartium ribicola* Class *Uredinaceæ*
Order *Hemi-basidiomycetes*

**Plants Attacked**

Currants of all kinds.

One stage of this disease occurs on the pine tree, and the disease on currants is therefore likely to occur in the vicinity of pine plantations. At present the disease has not assumed serious proportions in this country. Spraying has not been found of great success as a remedy.

## NETTLE-HEAD OR GOING WILD (REVERSION) OF CURRANTS

See page 704.

# DISEASES OF THE GOOSEBERRY

539 American Gooseberry Mildew
543 European Gooseberry Mildew
545 Gooseberry Black Knot
546 Gooseberry Cluster Cups

chromo-lith Cambridge University Press

# CHAPTER 33

## Diseases of the Gooseberry

## AMERICAN GOOSEBERRY MILDEW

Name *Sphærotheca mors-uvæ* Class *Erysiphaceæ*
Order *Ascomycetes*

**Plants Attacked**

Gooseberries and occasionally red and black currants.

**Related Diseases**

Powdery mildews of apple, cherry, hop, strawberry and vine: also European Gooseberry Mildew.

### 1. General

This serious disease of gooseberries has long been rampant in America, and within the last decade only too familiar to almost every gooseberry grower in England. Its establishment in this country offers one of the most flagrant examples of the danger of allowing untested stock to be imported. So serious did the outbreak become that official action was finally taken, and growers were compelled to notify the occurrence of the disease and were officially advised to grub their bushes when considered necessary. Probably as a result largely of this action the disease is now much less severe, but growers will do well to keep it severely in check by constant spraying.

A recent order dated September 23, 1919, forbids the moving of visibly affected bushes. Occurrence of the disease must be notified and powers are conferred to enforce spraying or grubbing. The two most important clauses are given below.

**Occurrence and Symptoms**

The disease usually appears first upon the plants about May. A delicate spidery film or web is seen upon the young leaves and buds which soon extends to the shoots and berries and grows more compact. At first pure white, it soon becomes grey owing to the formation of thousands of summer spores (conidia) which are scattered far and wide by the wind or by insect agency and so spread the disease.

Later on the fungus appears as a **brown felt** upon the berries and shoots. This is the winter or *ascigerous* stage of the disease. Spore cases are produced which liberate winter spores (out of asci) in the early spring and so re-infect the plants.

**Distinguishing Features**

It can be told from the "*European mildew*" (page 543) by its earlier appearance, and by the latter being usually confined to the leaves, and occurring as small scattered patches. Also by the brown felted appearance of the American mildew.

**Effect on Plant**

The disease seriously weakens and impoverishes the plant.

**Degree of Damage**

Not only is the plant affected and weakened, but the berries are rendered unsaleable by the brown felted covering. If marketed, they require special cleaning.

**How infection occurs**

Chiefly by the summer spores (conidia) which are produced in great abundance. Re-infection of the plant occurs by means of the winter spores (asci) produced the following spring. When a spore falls upon a healthy leaf it germinates and produces a "spawn" or *mycelium* on the leaf, giving off branches which penetrate into the tissue of the leaf and absorb the sap (called *haustoria*). Summer spores and then winter spores are produced in due course.

**Conditions Favourable to Fungus**

Dull, moist, "heavy" weather.

**Susceptible Varieties**

Varieties which were at first supposed to be immune have since been attacked.

**Frequency of Disease**

Very common, though rather less so than formerly.

**Distribution**

Widespread in England, the Continent, and in America.

**Treatment (Control)**

Winter spraying is ineffective owing to the very resistant spore cases (perithecia) of the fungus.

Spraying with *summer strength* LIME-SULPHUR[1] is effective, especially if several sprayings are given. It has the disadvantage of marking the berries with a fine deposit (sulphur) which is quite harmless, but renders them unsightly. These must therefore be cleaned before marketing. (Special machines are available for this purpose.)

The most effective wash, and one which does not mark the berries, is AMMONIUM POLYSULPHIDE[2] and SOFT SOAP[3] ("A.P.S." wash). For details of this spray see page 603. The grower is strongly advised to try this wash, especially as there is a prospect of its being produced at a cheaper rate[4] than has formerly been the case.

**Calendar of Treatment**

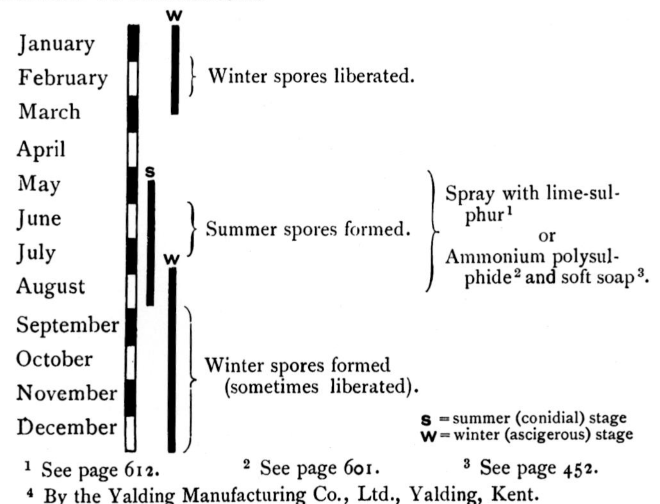


[1] See page 612. [2] See page 601. [3] See page 452.

[4] By the Yalding Manufacturing Co., Ltd., Yalding, Kent.

## American Gooseberry Mildew Order, 1919

The following are the main clauses affecting growers:

4. *Prohibition of Movement of Diseased Bushes.*—(1) No diseased bush shall be sold, offered for sale, or moved from any premises until it has been so pruned as to remove all visible traces of disease. An inspector, if he is not satisfied that the diseased bushes on any premises will be pruned before removal as required by this article, may, by notice served on the occupier of the premises, prohibit the removal of the diseased bushes from the premises until he has passed them as being free from visible disease.

(2) An inspector may, by notice served on the occupier of premises to which diseased bushes have been moved in contravention of this article, require him to move the bushes forthwith to the premises from which they were so moved, or alternatively to move the bushes to other suitable premises specified in the notice, and to remove by pruning and to destroy all visible traces of disease before the bushes are again moved.

5. *Powers of Dealing with Disease.*—(1) An inspector may, upon production if so required of his appointment, enter and examine any premises on which gooseberry or currant bushes are growing, and if he finds diseased bushes thereon, and he is not satisfied that the occupier is taking sufficient steps to control the disease, he may, by notice served on the occupier of the premises, require him to adopt one or more of the following measures: (*a*) To destroy all diseased fruit; (*b*) to spray the bushes with a suitable fungicide to the satisfaction of the inspector; (*c*) to prune out and burn forthwith all diseased shoots from each tree or bush.

(2) A notice under this Article may prescribe the time within which the adoption of any measures thereby prescribed shall be completed.

## 2. Scientific

### A. Conidial Stage

APPEARANCE. White, changing to grey.
LOCATION. On young leaves, shoots and berries.
APPEARS IN May and later.
DURATION. Throughout summer.

CONIDIA (summer spores).

Appearance. Oval with rod-shaped bodies in interior (differs from European mildew).
Arrangement. In chains on conidiophores.

**B. Ascigerous Stage**

APPEARANCE. As brown perithecia embedded in grey mycelium, turning brown.
LOCATION. On shoots and berries.
APPEARS IN Late summer and autumn.
DURATION. Throughout winter.
MYCELIUM. Grey, turning brown—felty.
PERITHECIA.
Appearance. Dark brown, with simple (unbranched) append ages (differs from European mildew).
Location. Embedded in mycelium.
Duration. Usually throughout winter, but not invariably.
ASCI. Broadly oblong.
ASCOSPORES. 8; $20-25 \times 12-15\mu$.

## EUROPEAN GOOSEBERRY MILDEW

Name *Microsphæra grossulariæ* Class *Erysiphaceæ*
Order *Ascomycetes*

**Plants Attacked**

Gooseberries and occasionally red currants.

**Related Diseases**

American Gooseberry Mildew, other powdery mildews.

### 1. General

This disease is fairly common and may be mistaken for the much more serious American mildew (page 539). It is easily however distinguished as follows:

1. European forms delicate threads (mycelium); American dense and woolly.
2. European on leaves chiefly; American chiefly shoots and berries.
3. European remains thin and scanty; American changes to brown felty patches.

The two species may occur on the same plant.

### Occurrence and Symptoms

It first occurs on the leaves as delicate greyish white patches of mould which later have a mealy appearance owing to the formation of spores (conidia). It may appear upon both surfaces of the leaf, and is somewhat uncertain in the time of its attack. Later, little black dots appear which are the "winter fruit."

### Effect on Plant

If the attack is severe, an early shedding of the leaves may occur.

### Degree of Damage

Not usually serious like the allied "American mildew."

### How infection occurs

By means of summer spores (conidia) and winter spores (asci).

### Frequency of Disease

Not uncommon.

### Treatment (Control)

Like all the "*powdery mildews*" the fungus, except for its "roots" (*haustoria*) which are embedded in the leaf tissue, is entirely upon the surface of the plant. It is therefore fairly open to attack.

1. Spray with *summer strength* LIME-SULPHUR (see page 612).
2. Spray with SOFT SOAP[1] and LIVER OF SULPHUR[2], PARAFFIN EMULSION[3] and LIVER OF SULPHUR[2], or, best of all, with SOFT SOAP[1] and AMMONIUM POLYSULPHIDE[4], especially if the attack is severe.
3. Dusting with SULPHUR[5] is effective if the fungus is treated in its earlier stages.

## 2. Scientific

### A. Conidial Stage

| | |
|---|---|
| APPEARANCE. | Greyish white, delicate. |
| LOCATION. | Both surfaces of leaves. |
| TIME OF APPEARANCE. | Erratic. |
| DURATION. | ,, |

### B. Ascigerous Stage

| | |
|---|---|
| APPEARANCE. | Small black scattered dots. |
| LOCATION. | In mycelium. |

[1] See page 452. [2] See page 624. [3] See page 417. [4] See page 601. [5] See page 626.

TIME OF APPEARANCE. Erratic.
DURATION. "
PERITHECIA.
Appearance. In scattered groups with characteristic, much branched appendages (different from American).
ASCI. 3—10 ovate.
ASCOSPORES. 3—6, variable in size, 20—30 × 12—15 μ.

## GOOSEBERRY BLACK-KNOT

Name *Plowrightia ribesia* Class *Sphæriaceæ*
Order *Ascomycetes*

### Plants Attacked

Gooseberries and Currants.

### General

This is one of the diseases which gains an entrance to the plant by means of minute wounds caused by aphis or scale attack. No sign of the fungus appears until the branch is dead. It occurs mainly in neglected gardens.

### Occurrence and Symptoms

The first sign of the disease is the wilting and yellowing of the leaves which soon afterwards fall off. The branch attacked as a rule lasts over the first year, but in the following spring the early buds only half open and the branch then dies. This is due to the fungus mycelium blocking the plant vessels and so cutting off the supply of sap from the roots.

When the branch is almost dead, large black warts appear on the bark. These are oval and placed with the long axis at right angles to the branch. These are the spore-producing bodies and are often so crowded as to give a blackened appearance to the branch.

### Prevention

Keep bushes clear of aphides and currant scale.

### Remedies

None at present known.

Wilting and yellowing branches should be at once removed.

**Calendar of Treatment**

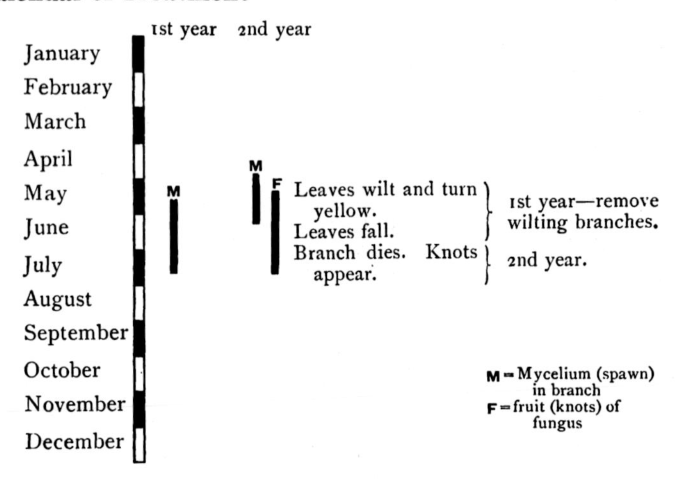


## GOOSEBERRY CLUSTER-CUPS

Name *Puccinia pringsheimiana* Class *Uredinaceæ*
Order *Hemi-basidiomycetes*

**Plants Attacked**

Gooseberries.

**Related Diseases**

Currant and other rusts.

### General

This disease occurs with some severity in certain seasons in various places. The reason for this is not at present explained, but weather conditions have probably a large influence. This fungus is one of those which have different forms of spores, and pass a stage of their development on another plant. It is not however essential for this to happen and the disease can propagate itself from year to year on gooseberries alone.

### Occurrence and Symptoms

The first sign of the trouble is the appearance of bright orange patches on the leaves and fruit. These patches become later dotted over with minute cup-like bodies, which have white, torn edges and are filled with orange spores.

Another stage of the disease occurs on the leaves of sedges, and appears as brown streaks similar to wheat rust. The winter spores of this stage of the disease infect gooseberries and develop into the characteristic cluster-cup forms. The gooseberry stage can however continue from season to season on the plants without the assistance of the sedges.

### Distinguishing Features

The orange patches and toothed edges of the "cups."

### Prevention

Cutting down neighbouring sedges in the spring and burning these has been advised.

### Remedies

No method has yet met with success. The disease does not often reach serious proportions and the safest plan is to collect and burn the infected leaves and fruit.

## DIE-BACK OF GOOSEBERRY

Name *Botrytis* Order *Hyphomycetes*

### Plants Attacked

Gooseberries.

### General

This disease is now fairly widespread in England and occurs on bushes almost wherever grown. It differs from many other diseases in being propagated from year to year by means of small hard masses of spawn (mycelium) called *sclerotia*. These are very resistant to frost and drought.

### Occurrence and Symptoms

The disease may first appear on single plants in a large plantation or in a small group of bushes amongst other healthy ones. The disease may also attack

1. The main stem.
2. The young wood.
3. The leaves.
4. The berry.

The stem is usually attacked near the ground, and the bush is finally killed at this spot, being "ringed" by the fungus. Several seasons usually elapse before this occurs. Often the spawn (mycelium) of the fungus travels up the stem and kills one or other of the branches. The appearance of dead branches on a bush is characteristic of the disease. At the end of a season's growth the bark peels off, sometimes in large pieces.

The leaves are often infected by the spores and turn yellow or grey at the edges. Sometimes a severe leaf fall occurs, more often the leaves fall normally.

The shoots are also attacked, and the berries are sometimes infected and some rot.

On the dead branches small grey tufts or cushions appear which change into powdery patches. These are the fungus spores. The *sclerotia* also appear as mentioned above.

**Treatment (Control)**

1. Remove all dead bushes or branches and burn them at once.
2. Any bushes with *main stem* diseased should be burnt.
3. In cases of severe infection, spray with COPPER SULPHATE solution (see page 611), 4 lbs. to the 100 gallons of water, just before buds burst. The main stems should be thoroughly wet.
4. Spray immediately the fruit is set with BORDEAUX MIXTURE (8 . 8 . 100), see page 606.

# DISEASES OF THE HOP

547 Die-back of Gooseberry 551 Hop Mildew
567 Plum (*a*) Leaf Blight, (*b*) Blister, (*c*) Rust
570 Plum Silver Leaf (*a*) leaf, (*b*) dead trunk, (*c*) dead branch

chromo-lith. Cambridge University Press

# CHAPTER 34

## Diseases of the Hop

## HOP MILDEW (mould)

Name *Sphærotheca humuli* Class *Erysiphaceæ* (*Powdery mildews*) Order *Ascomycetes*

### Plants Attacked

Hops: a kindred "strain" or "variety" of the same fungus occurs on strawberries and many wild plants, as geranium, meadow sweet, agrimony, etc., but Salmon has shown that this is not capable of infecting the hop.

### Related Diseases

Gooseberry, vine, apple, and other mildews.

### 1. General

The hop mildew or "mould" is a constant and serious menace to hop cultivation, second only in its harmful effects to the hop aphis ("fly").

It appears with varying degrees of severity practically every year, especially when the situation of the garden is much sheltered or the soil badly drained.

The great aim of the hop grower is to deal with the mildew before it has an opportunity to establish itself on the cone or "pin." Once this is infected the whole crop may be ruined as the flavour and keeping qualities of the hop are endangered.

It is safest to take prompt action whenever the least trace of mould is first seen and before it spreads over the garden and obtains a "hold."

**Occurrence and Symptoms**

The first sign of the mildew is the occurrence either on the upper or lower surface of the leaf of delicate white patches. These tend to increase in size until the whole leaf surface becomes covered.

Later the mildew becomes grey and "mealy," due to the formation of the summer spores (conidia) in enormous numbers. The winter spores (in perithecia) are formed later, or may not occur at all, the fungus disappearing from the leaf altogether under conditions which are unsuitable to its growth—such as dry, sunny weather. When found, the winter spores occur in minute black bodies.

**Distinguishing Features**

Not likely to be mistaken for any other fungus disease on hops.

**Effect on Plant**

Not usually serious as long as it is confined to the leaves.

**Degree of Damage**

Very serious if allowed to spread to the "burr" or "pin."

**How infection occurs**

Mainly by means of the summer spores (conidia), but the winter spores probably carry the disease over to the following season.

**Conditions Favourable to Fungus**

Dull "heavy" weather, and a too sheltered position of the garden, also lack of adequate drainage of the soil, producing an unhealthy "water-logged" condition.

**Frequency of Disease**

Common.

**Distribution**

Occurs wherever hops are grown.

## Treatment (Control)

1. PREVENTION

   Avoid low-lying situations.
   Ensure adequate drainage of soil.

2. REMEDIES

   Measures should be taken immediately any signs of the fungus are seen. In certain seasons the mould is very persistent and will require continuous treatment throughout the season. If the weather is bright and sunny, with light breezes, there will probably be little to fear, but as it is impossible to predict what the conditions will be, it is always the best plan to anticipate the worst. The most usual, and on the whole satisfactory, remedy is the dry spraying or dusting of SULPHUR, termed "sulphurating." The finest grades of sulphur obtainable should be used, these are "FLOWERS" and "PRECIPITATED SULPHUR" (see page 626).

   Certain weather conditions[1] are more favourable to the action of sulphur than others. Many growers can tell by the odour of their gardens after sulphurating if it is in an active condition. The exact manner in which it acts is not fully understood, there being rival theories on its action.

   Many growers like to use a small quantity of *liver of sulphur* in their final soap spray, when the burr is appearing. This has been shown to be of little value, unless sufficient is used to become dangerous to the hops.

   It is possible that ammonium polysulphide and soft soap ("A.P.S." wash, see page 601) used weak would be more efficient for this purpose, as has been proved in the case of American Gooseberry Mildew. This might be used with advantage in the early stages of growth. If tried when the hop is in burr, great caution must be employed in adjusting the strength of the spray, so that the "pin" is not injured. No recommendation as to strength can at present be made, the grower should himself experiment on a small scale before using any large quantities.

[1] Apparently bright sunny weather most favourable, but wind is detrimental.

**Calendar of Treatment**

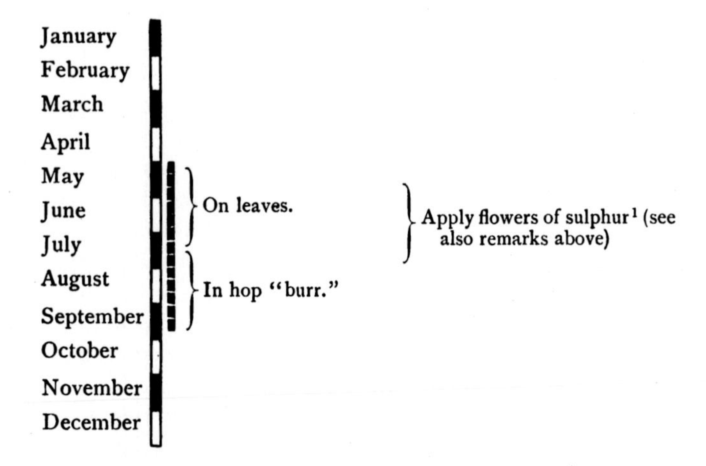


## 2. Scientific

### A. Conidial Stage

| | |
|---|---|
| APPEARANCE. | As delicate white patches becoming grey. |
| LOCATION. | On both surfaces of leaves. |
| TIME OF APPEARANCE. | Very uncertain. |
| DURATION. | Variable. |

### B. Ascigerous Stage

| | |
|---|---|
| MYCELIUM. | Greyish white. |
| PERITHECIA. | |
| APPEARANCE. | Small, black, with long brown appendages. |
| LOCATION. | In mycelium. |
| DURATION. | Variable. |
| ASCI. | Single, 8-spored. |
| ASCOSPORES. | Elliptical, $22 \times 15\,\mu$ (average). |

[1] See page 626.

# DISEASES OF THE PEACH

557 Peach Leaf Curl
588 Strawberry Mildew

chromo-lith. Cambridge University Press

# CHAPTER 35

## Diseases of the Peach

## PEACH LEAF CURL

Name *Exoascus deformans* Class *Exoascaceæ*
Order *Ascomycetes*

### Plants Attacked

Peaches, nectarines and almonds.

### Related Diseases

"Witches'-brooms" of Cherry, and Cherry Leaf Blister.

### General

This disease occurs fairly commonly on peaches, and is a serious menace to their cultivation in certain districts. It is more prevalent in the case of plants grown in the open. It occurs all over the world where peaches and allied plants are grown.

### Occurrence and Symptoms

The appearance of this fungus is very characteristic. The leaves remain green at first, but become distorted and crumpled, later they turn yellow with a reddish tinge and finally change to a rosy red. Later still a silvery bloom appears upon the surface, due to the fruiting stage of the fungus.

Unfortunately this disease penetrates into the interior of the shoots and remains there throughout the winter, passing into the new buds in the following spring. Spraying is therefore of little use, except as a preventive of attack.

### Distinguishing Features

The crumpled distorted leaves, turning red.

### How infection occurs

By means of spores carried by the wind or insect agency. The fungus threads apparently remain in the branches and occasionally grow up the expanding leaflets the following spring. More usually however plants are re-infected by spores which have lain dormant during the winter on the bud scales.

**Frequency of Disease**

Fairly common.

**Treatment (Control)**

Thorough spraying with BORDEAUX[1] or BURGUNDY[2] mixture is an effective treatment. The spray must be applied before the buds begin to swell in the spring. This period usually covers the time between middle February and the first or second week in March. Before spraying, all dead and diseased shoots should be cut away. One spray should suffice but a second application lessens any chance of failure.

[1] See page 606.

[2] See page 610. A double-strength Burgundy mixture of the following composition was found effective at Wisley (see *Jour. Board of Agric.* 1919, XXVI, No. 8): copper sulphate 2½ lbs., carbonate of soda (crystals) 2⅝ lbs., water 12 gallons.

# DISEASES OF THE PEAR

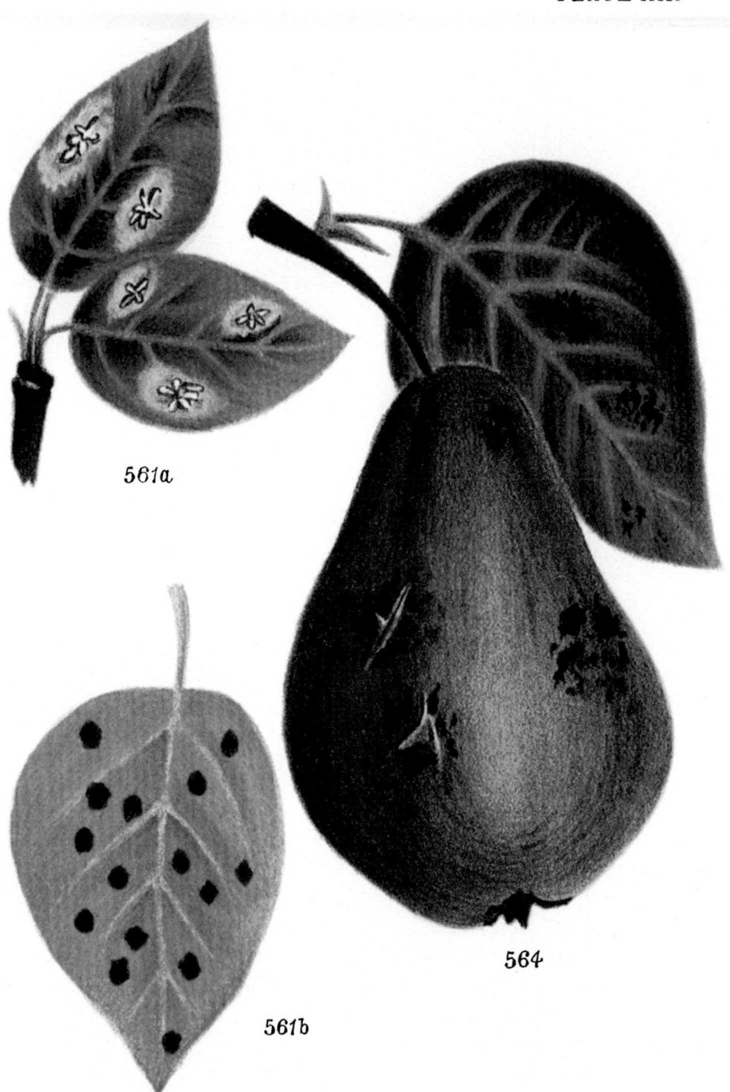


561 (*a*) Pear Cluster Cups   561 (*b*) Pear Leaf Spot
564 Pear Scab

chromo-lith. Cambridge University Press

# CHAPTER 36

## Diseases of the Pear

## PEAR LEAF CLUSTER-CUPS

Name *Gymnosporangium sabinæ* Class *Uredinaceæ*
Order *Basidiomycetes*

**Plants Attacked**

Pears.

**Related Diseases**

Gooseberry Cluster-cups (page 546).

**Occurrence and Symptoms**

The disease occurs as clusters of cup-like bodies on yellowish spots on the leaves. Another stage of the disease appears on juniper trees. When pears are badly attacked the leaves are shed early in the season.

**Treatment (Control)**

No treatment at present known appears to have much effect, although an early spraying with weak BORDEAUX MIXTURE[1] should be a preventive against attack, as the pears are probably infected from the junipers in the spring, or early summer.

## PEAR LEAF SPOT

Name *Mycosphærella sentina* Class *Sphæriaceæ*
Order *Ascomycetes*

**Plants Attacked**

Pears.

**Related Diseases**

Strawberry leaf blight.

[1] See page 606.

### 1. General

This is a common disease of pears, especially attacking nursery stock. It is one of those diseases in which the winter spores are developed on the dead leaves, and for this reason it is useless to expect to cure by spraying unless the fallen leaves are removed and destroyed.

**Occurrence and Symptoms**

Small round spots appear on the pear leaves, especially upon the upper surfaces. These spots become greyish as the leaf tissue dries up, and are surrounded by a brown border. If the under surfaces of the leaves are carefully examined, the fruiting stage of the fungus will be seen in the form of minute black points on the diseased areas (a hand-glass is sometimes necessary to distinguish them).

The winter fruit developes upon fallen leaves and the spores produced infect the young leaves in the following season.

**Effect on Plant.**

When the leaves are much spotted, they turn yellow and fall early in the season.

No direct damage is done to the fruit, but the crop naturally suffers through the weakening of the tree.

**How infection occurs**

A. By the "summer" spores on the living leaves.

B. In the following year, by the "winter" spores from the fruiting stage on the dead leaves.

**Frequency of Disease**

Fairly common.

**Distribution**

Very general wherever pears are grown

**Treatment (Control)**

Great benefit has been found from spraying with weak BORDEAUX[1] MIXTURE. LIME-SULPHUR[2] (summer strength) is however equally effective and less liable to injure the leaves. The first spraying should be given immediately after the petals fall, and two more at intervals of three weeks after.

Fallen leaves must be destroyed.

[1] See page 606. [2] See page 612.

**Calendar of Treatment**

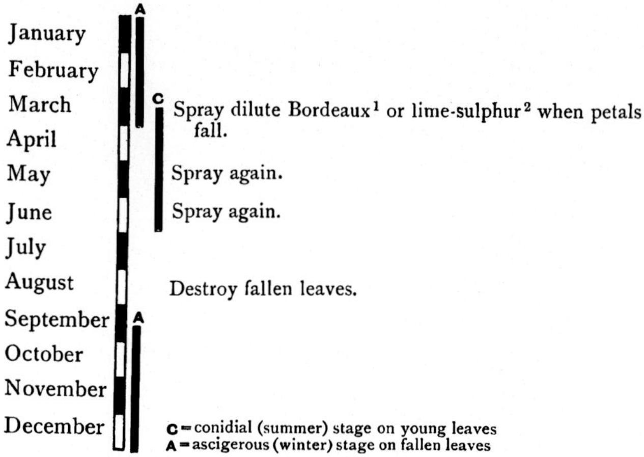

|  |  |
|---|---|
| January | |
| February | |
| March | Spray dilute Bordeaux[1] or lime-sulphur[2] when petals fall. |
| April | |
| May | Spray again. |
| June | Spray again. |
| July | |
| August | Destroy fallen leaves. |
| September | |
| October | |
| November | |
| December | |

C = conidial (summer) stage on young leaves
A = ascigerous (winter) stage on fallen leaves

## 2. Scientific

### A. Conidial Stage

| | |
|---|---|
| APPEARANCE. | As minute black perithecia on diseased areas. |
| LOCATION. | Under surfaces of leaves. |
| APPEARS IN | Summer. |
| CONIDIA. | |
| Appearance. | Long, slender, curved, 3-celled. |
| Arrangement. | As viscid tendril, $60 \times 3—4\mu$. |

### B. Ascigerous Stage

| | |
|---|---|
| APPEARANCE. | As globular perithecia with projecting mouth. |
| LOCATION. | On fallen leaves. |
| APPEARS IN | Winter and spring. |
| ASCI. | 8-spored, $60—75 \times 11—13\mu$. |
| ASCOSPORES. | Hyaline, slightly curved, $26—33 \times 4\mu$. |

[1] See page 606. [2] See page 612.

## PEAR SCAB

Name *Venturia pyrina* Class *Sphæriaceæ* Order *Ascomycetes*

### Plants Attacked

Pears.

### Related Diseases

Apple Scab (page 510).

### Occurrence and Symptoms

This disease very closely resembles Apple Scab (page 510), in all respects, except that the attacked fruit tends to *split* and *crack* much more often than in the case of apples. The winter fruit of the fungus occurs on the dead leaves.

### Treatment (Control)

As in the case of Apple Scab (page 513).

### Scientific

CONIDIAL STAGE

Velvety, olive-black, conidiophores short, outline wavy or knotted; conidia 28—30 × 7—9$\mu$.

ASCIGEROUS STAGE

Perithecia in colonies on dead leaves (under surface). Asci, 8-spored; spores 14—20 × 5—8$\mu$.

# DISEASES OF THE PLUM

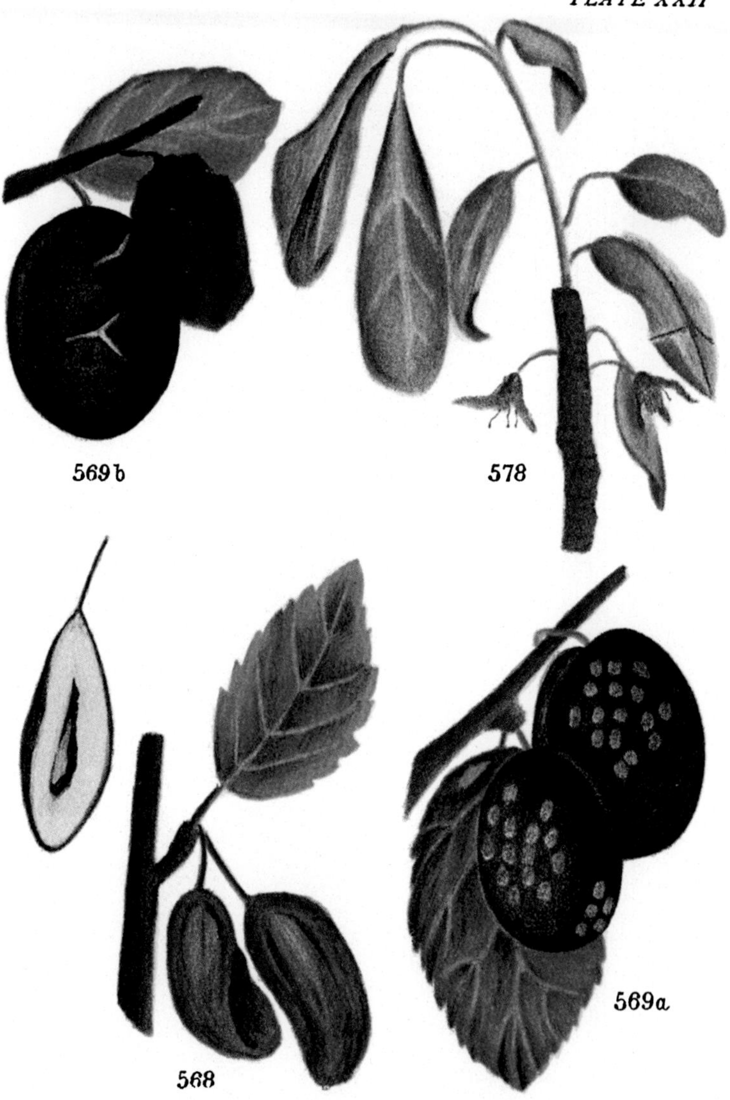


568 Plum Pockets
569 (*a*) Plum Rot
569 (*b*) Plum Scab
578 Plum Wither Tip

chromo-lith. Cambridge University Press

# CHAPTER 37

## Diseases of the Plum

### PLUM LEAF BLIGHT (*Cylindrosporium padi*)

See Cherry Leaf Blight, p. 709.

### PLUM LEAF BLISTER

Name *Polystigma rubrum*

**Plants Attacked**

Plums and damsons.

**Occurrence and Symptoms**

The fungus forms large orange-reddish patches of a dull appearance, best seen on the under surface of the leaf. With a hand-glass, very fine pits or holes can be seen on the patches which are the openings through which the spores are ejected. The winter form occurs on the fallen leaves.

**Control**

The disease is not often sufficiently severe to warrant the expense of spraying. All dead leaves should be cleared off the orchard and destroyed.

### PLUM LEAF RUST

Name *Puccinia pruni* Class *Uredinaceæ* Order *Ascomycetes*

**Plants Attacked**

Plums chiefly, also cherries and peaches.

**Related Diseases**

Raspberry Rust (see page 583).

**Occurrence and Symptoms**

The disease occurs as small brown spots on the under surfaces of the young leaves. In favourable weather the disease rapidly spreads, and frequently causes an early leaf fall, thus affecting the quality and yields of the fruit.

**Control**

Spraying with summer strength LIME-SULPHUR[1] or BORDEAUX[2] is the best way of dealing with an attack, if sufficiently serious. It is best to give two sprayings, one when the first signs of the disease are seen, and another about a month later.

## PLUM POCKETS

Name *Exoascus pruni* Class *Exoascaceæ* Order *Ascomycetes*

**Plants Attacked**

Plums.

**Related Diseases**

Peach Leaf Curl, Witches'-brooms of Cherry.

**Occurrence and Symptoms**

The disease is first seen about three weeks after the blossom has fallen. The young fruit then appears swollen owing to the excessive growth of the flesh of the fruit, the stone remaining quite small.

As it grows the fruit becomes much twisted and curved, and hollow in places. The colour, at first pale green, becomes afterwards reddish or purplish and much wrinkled.

The fruiting stage of the fungus is produced upon the diseased plums, appearing as a delicate bloom on the surface of the skin. Spores are produced which infect the shoots, and the fungus threads pass into the flowers and young fruitlets. The threads of the fungus remain in the tissue of the shoots all the year round and so infect the flowers each year.

**Treatment (Control)**

As the fungus remains in the shoot, spraying is useless as a remedy. Branches bearing diseased plums should be in all cases removed, if possible, before the fruiting of the fungus, i.e. as early as possible. Cut surfaces are best protected by a coating of Stockholm tar.

[1] See page 612. [2] See page 606.

## PLUM ROT

Name *Sclerotinia fructigena*

**Plants Attacked**

Apples, cherries, plums.

**General**

This is the same disease which attacks apples (see page 497). The disease on plums is not quite the same in appearance, but more like that on cherries (page 526). The fungus occurs in isolated patches, and not in circles as on apples, but the mummified fruit is formed in the same manner, and all other remarks apply as in the case of apples.

## PLUM SCAB

Name *Cladosporium carpophilum*

**Occurrence and Symptoms**

This is a disease very similar in all respects to Cherry Scab (see page 526) and is treated in the same manner. It is somewhat rare in this country.

## SILVER LEAF

Name *Stereum purpureum* Class *Thelephoreæ*
Order *Basidiomycetes*

### Plants Attacked

Chiefly plums, but apples are also not uncommonly attacked, and also cherries, peaches, gooseberries, currants, and apricots. Many other trees besides fruit trees liable to disease, the commonest being horse-chestnuts, sycamores and laburnums.

### General

The disease known as "Silver Leaf" is becoming, especially in certain districts, a very serious scourge to fruit-growers. In fact, it is not too much to say that unless energetic steps are taken to prevent its spread, it will seriously interfere with the cultivation of plums in this country. Many growers, as the author has found in his frequent visits to fruit-growing districts, are quite unaware of the risk they incur in allowing diseased trees to remain in their orchards year after year. Fortunately government action has now been taken and definite directions laid down to ensure that prompt measures are taken to control the disease. In the meantime, all growers should, in their own interests, lose no time in grubbing up, or at least felling[1], all badly diseased trees without waiting for the visit of the inspector.

The text of the order, referred to as Silver Leaf Order, 1919, is given below.

### Cause of the Disease

As far as our knowledge at present goes, it is not certain that all cases of "*silver leaf*" are caused by fungus attack. This

[1] If the roots are left, the main trunk must be covered over with at least six inches of soil (see page 576).

is however not of great importance to the fruit-grower. What has definitely been proved is that the fungus known as *Stereum*

Fig. 240. Two plum leaves. Healthy (left) and affected with silver leaf disease (right). Natural size.

can, and does, produce the silver leaf disease in plums, and that the spores of the fungus are able to carry and spread the disease far and wide.

It has also been proved that the disease on a tree is not infectious until the fruiting stage has appeared upon portions of dead wood on the branches or elsewhere.

If, therefore, a tree showing silver leaf is kept free from dead wood, there is no danger of infection to neighbouring trees. Since however it is easy to overlook small portions of dead branches, it is advisable, and in fact necessary in practice, to cut away all branches showing the silvering of the leaves before any portion is actually dead, and trees showing the disease all over should be ruthlessly grubbed up and *burnt at once.*

## Occurrence and Symptoms

The following is the normal course of the disease:

1. Generally speaking, the disease appears first on a single branch of the tree. The leaves develop an unmistakable silvery or leaden sheen (see figs. 240, 241).

   This appearance is caused by air spaces forming beneath the skin of the leaf, due to a splitting of the tissues in the interior of the leaf.

2. Other branches become affected in the same manner, some of the leaves often showing brown streaks and stains.

3. Affected branches commence to die back, or the entire branch dies suddenly. At this stage, the whole tree often becomes affected, all the leaves appearing "silvered."

4. The fruiting stage of the fungus appears upon the dead wood. It occurs usually as *purplish crusts* crowded together in irregular rows (fig. 242 and coloured plate). Sometimes it appears as a long strip on the under surface of the branch, or on the side of the trunk.

   In dry weather, the crusts shrivel up and darken in colour. They are not then so readily seen. Countless spores are liberated, especially during damp weather, and these are able to infect with the disease any tree on which they alight, provided that entrance can be obtained through a wound, which may be very slight.

Fig. 241. Two branches of plum; healthy (left), affected with silver leaf disease (right), reduced.

5. The whole tree finally dies.

   In addition to the fungus producing the silvering of the leaves, it also usually stains the wood of the branch brown, owing to the formation of a gummy substance. This can be seen on severing the branch and examining the cut surface.

**Effect on Plant.**

In the case of plums, the affected trees usually die sooner or later.

Affected apple trees sometimes recover of themselves and appear to be able to throw off the disease more readily. The yield and quality of the fruit is generally seriously affected almost at once, although exceptions occur to this.

**How infection occurs**

As stated above the disease is spread by a spore of the fungus gaining an entrance to a healthy tree through some wound. The wound may be so small as to be invisible to the eye, but there is of course more liability to infection if cut surfaces of the tree, etc. are left unprotected.

Spores may originate not only from diseased fruit trees but also from other trees and from dead wood.

**Susceptible Varieties**

Although *Victorias* and *Czars* appear most often attacked, this is probably because they occur more commonly.

No varieties have been found to be immune from the disease.

**Frequency of Disease**

Increasingly common.

**Distribution**

All districts where plums are grown in England are affected. It is less common on the continent and in America.

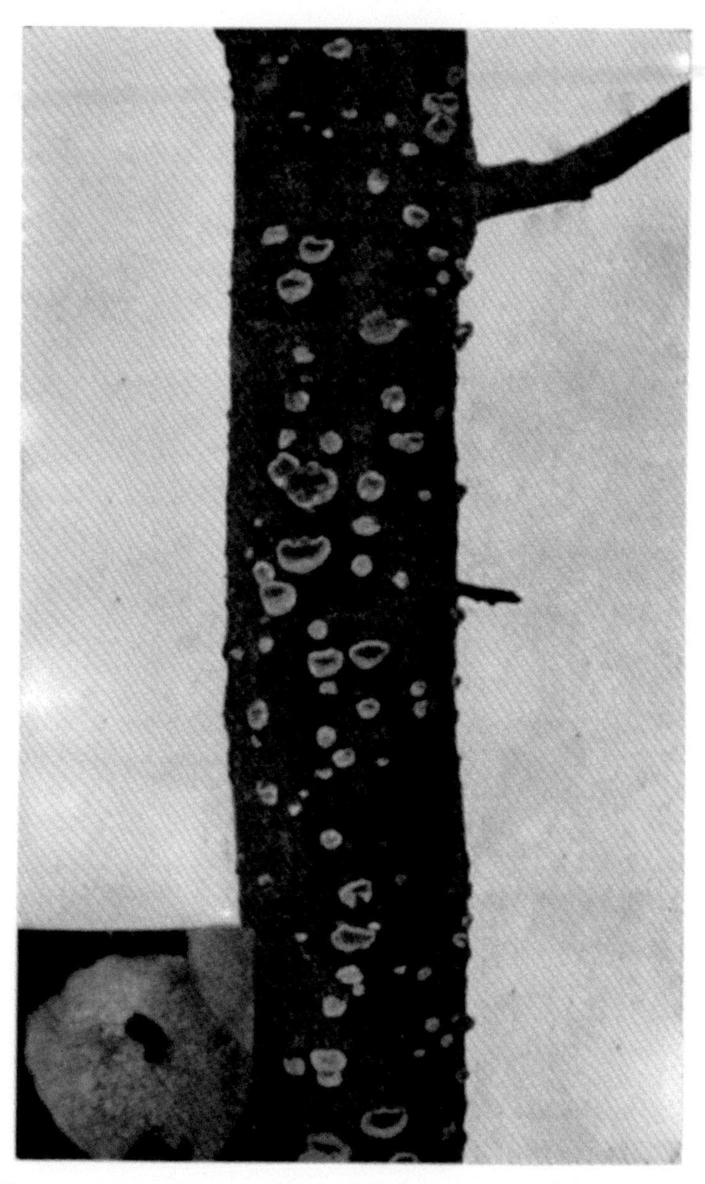

Fig. 242. Fruiting stage of *Stereum purpureum*, the fungus causing silver leaf disease, on dead branch of plum. Inset, a fungus growth magnified.

## Treatment (Control)

Although much research has been carried out, no reliable cure has been found for the disease.

Inoculation experiments appeared to give at one time hopeful results. Trees were plugged with sulphate of iron which was inserted in holes bored in the trunk. Success was not however achieved in practice and the method is not recommended.

It only therefore remains to take stringent precautions against the spread of the disease as follows:

A. SLIGHTLY AFFECTED TREES.

1. Cut back all branches showing silvered leaves to a point at which no brown stain appears in the wood.
2. Even up all cut surfaces, and coat them over with Stockholm tar.
3. Take care to *burn all branches at once* (otherwise the fungus may fruit on the dead branch).
4. If the work *cannot* be done when the leaves are on the trees, mark plainly those branches which are silvered so that they may be removed in the autumn. It is much safer to do the work at once, as the fungus usually fruits in the autumn.

B. TREES WITH DEAD BRANCHES.

The dead wood should be cut off and *burnt at once*, as grave risk is incurred by leaving any dead wood on affected trees.

C. DEAD OR DYING TREES (including trees with leaves entirely silvered).

These should be *completely grubbed up*. If it is not possible to take up the roots, it should be cut down to the ground as closely as possible and the stump well charred and covered over with at least six inches of earth. Otherwise the fungus is liable to fruit on the exposed wood and become a source of infection.

Large branches and trees should be immediately cut up for firewood. If this is not at once used, it must not be stored near the plantation, and should be placed under cover as far away as possible.

## SILVER LEAF ORDER, 1919

### THE OFFICIAL TEXT

The full text of the Order of the Board of Agriculture, dated Nov. 24, 1919, entitled the Silver Leaf Order, 1919, is as follows:—

The Board of Agriculture and Fisheries, by virtue and in exercise of the powers vested in them under the Destructive Insects and Pests Acts, 1877 and 1907, and for the purpose of preventing the spreading in England and Wales of the pest known as Silver Leaf, which is destructive to fruit trees and bushes, do order as follows:—

*Definitions.*—1. The expression "the Board" means the Board of Agriculture and Fisheries; the expression "plum trees" includes any stock, stool or cutting of a plum tree; "Inspector" means an Inspector of the Board or of the Local Authority; "The Local Authority" means as regards any District the Local Authority for the District under the Diseases of Animals Act, 1894.

*Destruction of Dead Wood.*—2. (1) The occupier of any premises on which plum trees are growing shall cut off and destroy by fire on the premises all the dead wood of each plum tree before April 1 of each year, and, where the dead wood in the trunk of any such tree extends to the ground, the occupier shall grub up and destroy by fire upon the premises the whole of any such tree, including the roots, before that date.

(2) An Inspector may at any time serve a notice on the occupier of any premises requiring him to cut off and destroy by fire within the time specified in the notice and on the premises any dead wood of any tree or bush of any kind whatsoever on the premises on which there are visible fruiting bodies of the fungus *Stereum purpureum*.

*Power of Entry.*—3. Any Inspector of the Board or of the Local Authority upon production if so required, of his appointment, may for the purpose of enforcing this Order enter any premises on which he has reason to suspect that a tree or bush to which this Order applies, is or recently has been, and examine any such tree or bush thereon and any wood cut from any such tree or bush.

*Service of Notices, &c.*—4. (1) For the purpose of this Order a notice shall be deemed to be served on any person if it is delivered to him personally or left for him at his last known place of abode or business, or sent through the post in a letter addressed to him there; and a notice purporting to be signed by an Inspector shall be *prima facie* evidence that it was signed by him as an Inspector.

(2) A copy of every notice served under this Order shall be sent to the Board by the Inspector by whom the notice is signed.

*Notification of Order.*—5. This Order shall be published by the Local Authority in accordance with any directions given by the Board.

*Offences.*—6. Every person shall be liable on conviction to a penalty not exceeding ten pounds who:—

(1) fails to comply with the requirements of this Order, or of any notice served under this Order, or

(2) wilfully obstructs or impedes any Inspector in the exercise of his powers or duties under this Order.

*Commencement.*—7. This Order shall come into operation on the first day of January, 1920.

*Application of this Order.*—8. The Order shall apply to England and Wales.

## PLUM WITHER-TIP AND BLOSSOM WILT

Name *Monilia cinerea*

### Plants Attacked

Plums; probably another "strain" of the same disease which attacks apples (as "Blossom Wilt" disease, page 494). Cherries are also attacked (page 521).

### Related Diseases

"Brown rot" of fruit (*Sclerotinia fructigena*).

### 1. General

This disease was first noticed and investigated by *Wormald*[1]. It is apparently a "strain" or variety of the same fungus which attacks apples, producing the "Blossom Wilt" of apples, a disease also investigated by Wormald.

[1] *Annals of Applied Biology*, 1918, **VI**. p. 28.

These are both allied to the Brown Rot disease of apples and other fruits.

The fact that it is not identical with the Blossom Wilt fungus is proved from the effect of innoculating apple blossoms with the Plum Wither-tip spores. If this is done, the apple blossoms are killed, but the disease extends no farther into the branch.

Although only recently investigated, the disease is probably quite an old one, but has passed unnoticed or been attributed to frost or other causes.

## Occurrence and Symptoms

The first sign likely to be noticed by the grower is wilting of the blossoms and the withering of the tips of the young twigs early in the season. These curve over in a characteristic manner. Infection is caused through the spores on the leaves, and the fungus then penetrates to the shoot, causing the death of the tip so affected. A discoloured area is formed on the shoot, which dries up at this point, and cuts off the sap to the young leaves. During the following winter and spring, grey pustules of the fungus fruit are borne on the dead twigs.

The fungus appears also as dark grey pustules on the fruit which becomes mummified on the trees.

## Distinguishing Features

The withering and wilting of the young tips of the twigs which remain on the tree throughout the winter. The dark grey fungus on the fruit.

## How infection occurs

Through the germination of the spores of the fungus, borne in the winter and spring on the withered tips, and infecting fresh blossoms and tips.

## Treatment (Control)

Spraying is of little use as far as our knowledge of the disease extends at present. All withered shoots and tips, and all fruit affected with "brown rot" should be removed from the trees before the pustules of the disease are produced, i.e. in autumn. Care should be taken to cut well below the diseased portions.

**Calendar of Treatment**

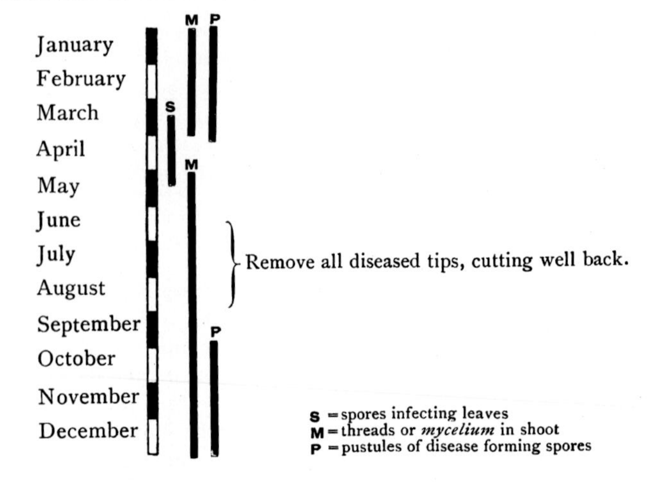


## 2. Scientific

### A. Conidial Stage

| | |
|---|---|
| APPEARANCE. | Occurring as grey pustules. |
| LOCATION. | On withered tips of shoots. |
| APPEARS IN | Late autumn or winter. |
| DURATION. | Till following spring and summer. |
| CONIDIA. | Size variable; from pustules on twigs, $5—10 \times 5{\cdot}5—17\mu$[1]. |

### B. Ascigerous Stage

Unknown in this country.

[1] Wormald, *Annals of Applied Biology*, 1918, **VI**, p. 34.

# DISEASES OF THE RASPBERRY

# CHAPTER 38

## Diseases of the Raspberry

## RASPBERRY RUST

Name *Phragmidium rubi-idei.* Class *Uredinaceæ*
Order *Basidiomycetes*

### Plants Attacked

Raspberries.

### Occurrence and Symptoms

This fungus produces three kinds of spores (as in all fungi of this class) and they all occur on the raspberry leaves.

The first appearance of the disease is in midsummer on the upper surfaces of the leaves, which become dotted with yellowish pustules, usually with a more or less circular arrangement.

Another fruiting stage occurs shortly after, similar in appearance to the first except that the pustules have an orange tint. The third form of the fungus fruit appears as small black clusters on the under surfaces of the leaves.

### Treatment (Control)

There is not much hope of a cure by means of spraying. As however spraying with liver of sulphur has been advised the author suggests that a spray with AMMONIUM POLYSULPHIDE and SOFT SOAP as soon as the disease appears (see page 601) would probably be better. Lime-sulphur cannot be used as it would disfigure the berries, unless these have been picked. The summer strength should then be employed[1]. Bordeaux mixture is liable to scorch the leaves severely.

### Scientific

| | |
|---|---|
| Ascidiospores. | Yellowish, 20—28μ diameter. |
| Uredospores. | Orange, 16—22μ. |
| Teleutospores. | Black, oblong, warted, 5—10 septate, 90—140 × 20—35μ. |

[1] See page 612.

## RASPBERRY SPOT

Name *Glœosporium venetum*

### Plants Attacked

Raspberries, also cloudberries.

### Occurrence and Symptoms

The canes become spotted with small red spots which increase in size and form blotches. Later they become paler with a dark red border. The leaves are occasionally attacked.

### Treatment (Control)

1. Cut out the infected canes as soon as these are seen, so as to prevent the spread of the disease, which is very rapid.
2. Spray with summer strength LIME-SULPHUR[1]; the author suggests a trial with AMMONIUM POLYSULPHIDE[2] and SOFT SOAP[3].

[1] See page 612. [2] See page 601. [3] See page 452.

# DISEASES OF THE STRAWBERRY

PLATE XXIII

587 Strawberry Leaf Spot

chromo-lith. Cambridge University Press

# CHAPTER 39

## Diseases of the Strawberry

## STRAWBERRY LEAF SPOT

Name *Sphærella fragariæ.* Class *Sphæriaceæ*
Order *Ascomycetes*

**Plants Attacked**

Strawberries, also wild varieties.

### General

This disease is fairly common and is known to most cultivators of strawberries. The fruit crop is not commonly injured seriously, unless the disease is severe.

**Occurrence and Symptoms**

The first sign of the disease is the appearance of small patches of red discoloration on the leaves. As time goes on these will increase in size and some overlap, forming large patches. The colour then changes in the centre to a grey tint due to the death of this part of the leaf. The patch is usually bordered by a ring of bright red. Later still, minute white tufts of the summer (conidial) fruit appear and these give rise to the summer spores which infect other leaves. On old and dead leaves the winter fruit of the fungus appears as minute black spots. These produce the winter spores, which are capable of infecting the young leaves in the following season.

**Effect on Plant**

In severe cases the fruit crop is badly affected, both in yield and quality, and the plant is severely weakened.

**How infection occurs**

By means of the winter and summer spores liberated from the two kinds of the fungus fruit. These are able to infect healthy leaves.

**Frequency of the Disease**

Common.

**Treatment (Control)**

Spraying should be commenced as early as possible and, if the disease has occurred the previous year, should be started before the spots actually appear.

Summer strength LIME-SULPHUR[1] may be used at this stage, and this has also the advantage of ridding the plants of *mildew*.

If it is late in the season there is danger of discolouring the fruit, the best spray to employ is AMMONIUM POLYSULPHIDE[2] and SOFT SOAP[3], or failing this LIVER OF SULPHUR[4] and SOFT SOAP.

It is preferable however wherever possible to spray before the opening of the flowers, and in bad cases two or three sprayings should be given at intervals.

## STRAWBERRY MILDEW

Name *Sphærotheca humuli* Class *Erysiphaceæ*
Order *Ascomycetes*

**Plants Attacked**

Strawberries, hops and many wild plants, e.g. meadow sweet, agrimony, etc. are attacked by the same species of fungus but, as Salmon has shown, it is a different "strain" and cannot infect the strawberry.

**Related Diseases**

Gooseberry mildew, vine mildew and other "powdery mildews."

### General

The fungus causing this serious and common disease of strawberries is a specialised form of that producing "mouldy hops." The disease may be controlled by similar methods to those employed in the case of the hop (page 553) and it is important to deal with it *before the berries are attacked.*

**Occurrence and Symptoms**

The first symptom of the disease is the gradual *turning upwards*

[1] See page 612. [2] See page 601. [3] See page 452.
[4] See page 624.

of the edges of the leaves. There are not, as in the case of the hop, conspicuous patches of white on the surfaces of the leaves. The curving of the leaves often continues until the whole of the under surface is exposed. If this be now examined with a hand-glass, a delicate white down is seen spreading over the leaf.

The attack may not happen until late in the season, when the crop has been gathered. It is when it occurs on the young leaves that it is most serious, since the fruit is likely to be attacked and spoilt. The summer spores are produced in great profusion on the down (or mycelium) and the winter stage probably occurs on the old leaves, though it has not often been detected. As stated above, the plants may become infected each year from neighbouring weeds.

**Effect on Plant**

The plant is weakened, and in serious cases killed by the fungus. If the fruit is attacked it is rendered quite unsaleable.

**Frequency of Disease**

Common.

**Distribution**

Widespread.

**Treatment (Control)**

It is especially important in the case of this disease to keep a careful watch for its first appearance, so that it may be controlled before the fruit is attacked.

The plants are at once sprayed with either :

1. LIVER OF SULPHUR and SOFT SOAP.
2. LIME-SULPHUR (half summer strength), or, best of all, if obtainable,
3. The author also suggests the experimental use of AMMONIUM POLYSULPHIDE and SOFT SOAP. Care must be taken to adjust the quantity so that the leaves are not scorched.

Blowing SULPHUR on the plants with a hand-bellows or sulphurator is also of proved benefit, but cannot be done for some weeks before the fruit is ripe. If the disease occurs after the fruit is picked, advantage has been derived from placing straw and other inflammable materials around the plants, and burning the leaves. The new leaves in the following spring are said to be strong and healthy.

All weeds should be kept off the garden and, as far as possible, the neighbouring ground should also be kept clear.

For scientific data, see page 554.

# DISEASES OF THE VINE

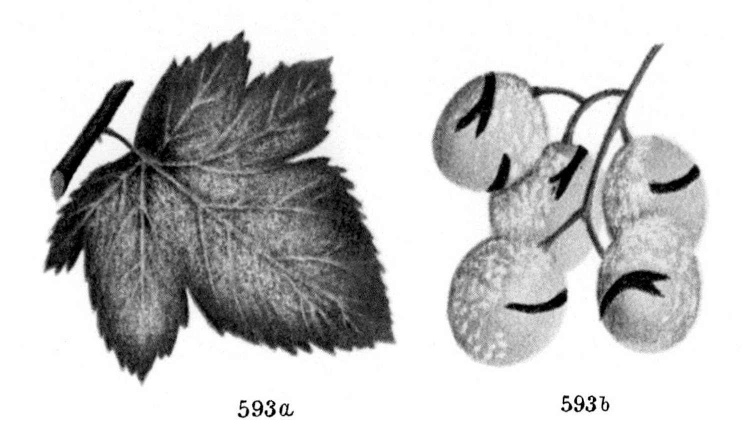

593 Vine Mildew (*a*) on leaf, (*b*) on fruit
594 Vine Sclerotinia

chromo-lith. Cambridge University Press

# CHAPTER 40

## Diseases of the Vine

## VINE MILDEW

Name *Uncinula spiralis* Class *Erysiphaceæ*
Order *Ascomycetes*

### Plants Attacked

Vines.

### Related Diseases

Other powdery mildews (hop, gooseberry, apple, etc.).

### General

This is a common and very destructive disease of grape vines, and was first noticed at Margate in the year 1845, since when it has spread over the entire globe. Many countries gave up the cultivation of the grape in consequence of the immense amount of damage to the fruit caused by the fungus. This was before the days of spraying. By prompt action it is not difficult to keep the disease in check, as the fungus, in common with the other powdery mildews, is very open to treatment, the fungus threads (or mycelium) being on the surface of the leaf.

The damage is done by the "roots" of the fungus (haustoria) which penetrate the leaf and withdraw the sap from the leaf cells, in this manner finally causing their death.

### Occurrence and Symptoms

The fungus appears as whitish patches on the upper surfaces of the leaves, or on the shoots, flowers or fruit. These rapidly increase in size until the whole surface is covered. The fungus later on becomes grey and powdery owing to the formation of millions of "summer" fruit (conidia) which disperse, and infect other plants. Later on, the winter (ascigerous) stage of the fungus is produced among the same threads (mycelium) as minute yellowish bodies, turning brown. These are winter-spore cases, and are able to stand severe frosts and liberate spores which re-infect the plants the following year.

**Effect on Plant**

1. The young *leaves* are often withered, old leaves become dry and develop brown stains.
2. Attacked *shoots* of young blacken and die, older shoots become discoloured and weakened.
3. *Fruit* usually cracks and dries up, becoming quite useless.

**Treatment (Control)**

The safest and most successful treatment is the application of FLOWERS OF SULPHUR[1], see page 626, with a hand bellows.

The time of application and the temperature are of importance. It should be applied

1. Before the flowers open (about a week previous).
2. When in full bloom.
3. About a month later if the fungus is still seen.

During the application (dusting or blowing) the thermometer should stand between 80 and 100° F.

Good results have also been obtained by the use of LIVER OF SULPHUR[2] ($\frac{1}{2}$ oz. to one gallon of water) sprayed on the plants. The author does not recommend this owing to the liability of variation in its composition (see page 624).

## VINE SCLEROTINIA

Name *Sclerotinia fuckeliana* Class *Pezizineæ*
Order *Ascomycetes*

**Plants Attacked**

Vines.

### 1. General

**Occurrence and Symptoms**

This disease occurs as a dense, dark greyish, fluffy covering over the trees and fruit, and other portions of the plant.

Countless spores are produced which readily float in the air and infect other portions of the vine. Meanwhile, black portions of the fungus, termed *sclerotia*, are formed in the interior of the plant underneath the part attacked by the mouse-coloured mould.

[1] Some growers have placed sulphur on the hot water pipes but this is too risky to recommend.

[2] See page 624.

These remain in position throughout the winter, and in the following spring produce either "summer" or "winter" fruit, according to weather conditions, etc.

In either case, spores are produced which infect other portions of the plant.

### Effect on Plant

Very serious injury is caused by this disease. The fruit, if attacked, is quite destroyed, and the plants do not long survive if prompt measures are not taken to remove infected portions.

### Frequency of Disease

The disease is widespread. In this country it frequently attacks grapes grown under glass, the disease requiring a damp atmosphere for its growth.

### Treatment (Control)

*Adequate ventilation* is of great importance as a preventive of attack.

When the disease has appeared, it is recommended to spray with LIVER OF SULPHUR[1], but the author advises an experimental trial of AMMONIUM POLYSULPHIDE[2] and SOFT SOAP[3], as this has proved of great efficiency in other cases.

In any case, infected leaves and fruit *should be at once* removed, as the fungus harbours in the interior of the tissues if these are allowed to remain, and grows out again the following year.

## 2. Scientific

### Conidia

Borne in clusters upon simple or branched hyphæ, forming dense tufts, spherical, 10—12$\mu$.

### Ascospores

2 to 3 on black sclerotium, brownish, spores 10—11 × 6—7$\mu$.

[1] See page 624. [2] See page 601. [3] See page 452.

# SECTION VIII

## FUNGICIDES

# SECTION VIII

## FUNGICIDES

## CHAPTER 41

### Introduction

The fungicides at present of practical interest and value are all either

1. COPPER COMPOUNDS, or
2. SULPHUR AND ITS COMPOUNDS.

1. **Copper compounds**

A good many of these have been proposed from time to time, but two only are of importance as far as the fruit grower is concerned. The first of these is

COPPER SULPHATE (BLUESTONE),

which is an excellent and powerful *winter spray* for all fungus diseases which are amenable to treatment. It is much too powerful to use when trees are in leaf. At such times, an insoluble compound of copper, known as

BORDEAUX MIXTURE

is used. It is chiefly valuable as a preventive against infection by fungus spores, the deposit remaining on the leaf for many weeks. Another similar compound is

BURGUNDY MIXTURE,

which is largely used in spraying potatoes.

Copper compounds *cannot be used with soft soap*. This is because the copper is thrown down as a curd by the soap, with which it forms an insoluble compound.

2. **Sulphur and its compounds**

It is advisable to distinguish between those compounds of sulphur *which can, and should be used with soft soap*, and those which cannot, or are not so used. To the first class belong

AMMONIUM POLYSULPHIDE,
LIVER OF SULPHUR,
SODA-SULPHUR.

One great advantage of these fungicides is that they can all be sprayed with an insecticide, such as nicotine, and so be used to "kill two birds with one stone."

The first-named is a recent introduction, but will undoubtedly prove of great value, and will replace Liver of Sulphur, the great disadvantage of which is its liability to variation. By "Soda-sulphur" is meant the substitute for Liver of Sulphur, manufactured with soda instead of potash. It is even less satisfactory, having a greater tendency to scorch. The two remaining fungicides which are used without soap are

SULPHUR itself, and

LIME-SULPHUR.

Sulphur is used by *blowing* or *dusting* it on to the plants, as a fine powder. It is useful for those fungus diseases of which the threads (or *mycelium*) live on the surface of the leaves, etc. These are known as the POWDERY MILDEWS, and include the hop mould, vine mildew, strawberry mildew, etc.

Lime-sulphur is at the present time the most popular all-round fungicide. It is largely used during the dormant season as a cleansing wash, and is thought by some to have valuable insecticidal properties, especially against aphis[1]. It is also employed in place of Bordeaux for spring and summer spraying in the case of varieties of fruit trees which are injured by copper sprays. It is the standard wash to employ against American Gooseberry Mildew. Although it cannot be used with soft soap it has recently been found that the use of *saponin* increases its efficiency.

[1] See for instance Seabrook's *Modern Fruit Growing*, pages 91, 94.

# CHAPTER 42

## Fungicide Materials

## AMMONIUM POLYSULPHIDE

### 1. General

#### Employment

This is a comparatively new spraying material and will probably be found to have a wide application. In particular, it should be useful for the spraying of trees or bushes in fruit in place of lime-sulphur, as, unlike the latter, it leaves no deposit on the berries, etc. It has been very successfully used in the case of AMERICAN GOOSEBERRY MILDEW (page 539), and can be employed in any stage without the fruit being marked.

It has the great advantage that it can be used with soft soap which enables it to **penetrate** the fungus. Insecticides may also be used at the same time (e.g. nicotine, quassia, etc.).

#### Description

The use of ammonium polysulphide as a remedy for fungus diseases, and especially against American gooseberry mildew, originated from researches conducted at Wye college, by Professor Salmon and Dr Eyre. They showed that, used with soft soap, it is a fungicide of great power, and very much superior to liver of sulphur, the only other material for the purpose which could be used in a soap wash.

Growers are advised to experiment with this wash as a substitute for liver of sulphur. Soap and insecticides can be used with it. It must be cautiously applied, as an excess may easily scorch the plants.

#### Properties

The concentrated material is a dark orange-red liquid, very pungent, and smelling strongly of ammonia. Care should be taken, especially in hot weather, to stand clear of the vapour when handling the strong liquid. It is also very necessary to store in a cool place and in tight packages, and if a can or barrel is opened, it is best to use the whole of it at once if possible. When diluted, it is not unpleasant in use, is harmless, and non-poisonous.

## Preparation

As this involves the use of air-tight vessels and the employment of an extremely poisonous gas[1], necessitating skilled scientific control, it is out of the question for the grower to prepare his own material.

## Tests

It is not possible for the grower to apply any test for strength or quality—the specific gravity test is worthless in this case. It is therefore very important to purchase from a firm of high standing which is prepared to guarantee their product being prepared according to the standard formula, and of the requisite strength.

The active ingredient is the ammonium polysulphide, and the strength is therefore reckoned on the percentage of this constituent.

According to tests made by the author, the "1918" formula solution should contain not less than the following percentages of ingredients:

Ammonia 18·5 per cent.
Sulphur 22·0 " "
as sulphide 9·0 per cent.
as polysulphide 13·0 per cent.

Stronger concentrates than this can however be prepared.

## Action on Fungi

Wherever applied under the proper conditions, and with a sufficient amount of soft soap, it appears not only to stop the growth and spread of the disease, but in many cases to actually destroy the fungus itself (the "threads"—*mycelium*—becoming brown and withered). This particularly applies to the "powdery mildews"—American gooseberry mildew, hop mould, apple mildew, strawberry mildew, etc.

## Action on Plants

Damage is liable to occur to the leaves of the plant sprayed, and it is well to give a trial application in the case of gooseberries, and especially in regard to other mildews.

Some varieties of gooseberries stand the wash very well, but others are notably susceptible to all sulphur washes.

[1] Even under skilled control, slight leaks of the gas have resulted in serious effects upon the operators.

## 2. Method of Use

### Strength of Spray

There are two formulæ suggested by Wye college conveniently referred to as "A.P.S. 1918" and "A.P.S. 1919." The latter is approximately double as strong as the former.

Two strengths are advised according to whether two applications are given or a single spraying only. The weaker strength applied twice is advised in all cases.

A. FOR DOUBLE SPRAYING (and in all cases of doubt or first trial) per **100 gallons** of spray—**1 gallon** of concentrated ammonium polysulphide ("A.P.S. 1918") or ½ **gallon** ("A.P.S. 1919").

B. FOR A SINGLE SPRAYING (only if previously tried and successful) per **100 gallons** of spray—**2 gallons** of concentrated ammonium polysulphide ("A.P.S. 1918") or **1 gallon** ("A.P.S. 1919").

### Soft Soap

In all cases, soft soap *must* be employed with the spray to ensure that the fungus is reached (or penetrated). There should be not less than 4 lbs. of *free soap* (see page 458).

The following will be found a fair guide:

| | | |
|---|---|---|
| SOFT WATER | **5— 7** | **lbs. soft soap.** |
| MEDIUM WATER | **8— 9** | ,, ,, ,, |
| HARD WATER | **10—12** | ,, ,, ,, |

A much better method is to get an idea of the hardness of the water, and make use of the following table:

| Hardness of water (see page 460) | Amount of soft soap[1] to use (lbs. per 100 gallons water) |
|---|---|
| 5— 8 | 5 |
| 8—11 | 6 |
| 11—14 | 7 |
| 14—17 | 7¾ |
| 17—20 | 8½ |
| 20—23 | 9¼ |
| 23—26 | 10 |
| 26—30 | 11 |

### Preparation of Spray

Boil the soft soap with a small bulk of the water in the usual way; dilute to 100 gallons; then add the ammonium polysulphide and stir well.

[1] Of at least 40 per cent. fatty acids (page 452).

The wash should be used at once and not allowed to stand for any length of time.

NOTE. Wooden and iron vessels are suitable, but copper or brass should not be used. The spraying machine and nozzles should be well washed through after use with water (as for lime-sulphur).

## When to Spray

In the case of "American Gooseberry Mildew," the first spraying with the "A" formula (1—100) should be given as soon as the mildew is seen and another application made 10—14 days after, *or* the mildew is sprayed once with "B" solution (2—100) as soon as it shows signs of becoming "powdery," as it is more readily killed at this stage than previously.

For other fungus diseases, refer to the detailed descriptions in Section VII.

## How to Spray

As for soap (see page 452).

# 3. Scientific

## Chemical Composition

It consists of the sulphide and polysulphides of ammonium, together with some free ammonia, and probably some ammonium thiosulphate.

Material made on the Wye formula ("A.P.S. 1918") contains on an average the following parts per 100:

Ammonia 19·0
Sulphur 24·0
as sulphide 10·5
as polysulphide 13·5.

The "A.P.S. 1919" formula contains about double the percentage of polysulphides.

## Properties

Pungent reddish-orange liquid. Readily soluble in water.

In dilute solutions becomes cloudy on standing owing to deposition of sulphur.

As made by the Wye formula ("A.P.S. 1918") the specific gravity is 1·036 at 17° C.

## Analysis

The ammonia, sulphide, sulphur and polysulphide sulphur should be determined, and should conform roughly to the above.

## BORDEAUX MIXTURE

### 1. General

**Employment**

As a spring and summer spray for the treatment of most fungus diseases.

**Description**

Bordeaux mixture results from treating a solution of copper sulphate with lime. By this means the copper is thrown out of solution in the form of what are known as "*basic sulphates.*"

The exact action of the mixture on the fungus is still a matter of controversy. Either the copper becomes gradually soluble, and so active, by the action of the air[1] (actually by the "carbon dioxide" existing in the atmosphere in small quantities) or the action of the fungus itself upon the Bordeaux renders it soluble and capable of being absorbed by the former[2].

As the name indicates, the preparation was originally used on vines in France. It finds its chief use against the potato blight in this country.

As regards fruit trees, the use of Bordeaux, which was pretty general a few years ago, has rapidly declined in favour of lime-sulphur, owing to the liability of scorching action on the leaves and fruit ("Bordeaux injury"). Certain varieties of trees will not stand even a weak application of copper compounds; not only are the leaves and fruit affected, but the leaves are shed.

A great advantage of Bordeaux mixture properly applied is that a thin continuous film of the material remains on the leaf for long periods after the application, thus ensuring protection from fungus spores.

**Bordeaux injury**

This is shown:

1. By certain varieties of apples under all conditions.
2. In moist or damp weather.
3. When the wrong kind of spray is used.

[1] The "chemical theory" advocated by Pickering (see *Woburn Reports*).

[2] The "biologic theory" supported amongst others by Professors Barker and Gimmingham (see *Reports of the Fruit and Cider Institute*, Long Ashton).

Amongst the most susceptible of injury are Cox's Orange, Beauty of Bath, Gladstone, Lady Sudeley.

The injury to the leaves shows itself by (1) holes ("shot holes") produced by dead portions of the leaf which drop out. This kind of injury is usually the result of a too heavy or too coarse spraying. (2) The leaves may become yellowed all over, and this is the case when certain susceptible varieties are sprayed. Curiously enough, varieties which in one district are injured appeared to stand the treatment quite well in a different part of the country.

The injury to the fruit takes the form of a corky roughening of the skin (known as "russetting") due to the death of these cells by the action of the copper. In severe cases the fruit cracks open.

## Preparation

A. FOR A SMALL ACREAGE.

| *Formula.* | |
|---|---|
| COPPER SULPHATE ("Bluestone") | **8 lbs.** |
| QUICKLIME (lumps) | **8 lbs.** |
| WATER | **100 gallons.** |

*Utensils required.*

Two wooden vats to hold 40 gallons and 100 gallons respectively (a barrel with the head out is suitable for the former).

A good strainer of copper gauze.

*Method.*

1. Place about 20 gallons of water in the barrel and dissolve the copper sulphate therein by wrapping it up in a piece of sacking and suspending it just under the surface of the water. It is best left overnight to dissolve.
2. Place the quicklime—rejecting all but good lumps—in the larger vat, and slake it by adding water in small quantities at a time. When slaked, make up to 80 gallons with water and stir well.
   This is known as milk of lime.
3. Pour the copper sulphate solution slowly into the milk of lime, stirring well till all is added.
4. Use immediately, pouring the well-stirred mixture through a strainer into the spraying machine, or knapsack, which should also have a strainer of its own.

*Cautions*

*a.* Use the purest copper sulphate obtainable (98 per cent. purity), not so-called "agricultural bluestone."

*b.* Freshly burnt lime of high quality is necessary, rejecting any powder (which is "air-slaked" and useless).

*c.* Stir the slaked lime thoroughly so as to produce a fine "milk."

*d.* Stir well when adding the copper solution.

*e.* Never use iron or zinc vessels but only those of wood or copper.

B. FOR LARGE SCALE OPERATIONS.

The same procedure is followed, except that a double platform is advisable, and three vats are then used. On the top stage are placed, side by side, two 50 gallon tubs, each with a tap, so that the liquid can be run into a 100 gallon vat placed on the lower stage. This latter is just high enough to feed with a tap direct into the spraying machine.

The copper sulphate is made up into a "*stock*" *solution* of 100 lbs. to 100 gallons in another vat, and for each mixing 8 gallons are measured out and placed into one of the tubs on the top staging.

Water is now added up to the 50 gallons mark (it is a great convenience if water can be laid on to this stage) and the contents stirred well.

In the other tub the lime is made up as in § 2 above, but 50 gallons of water only are added.

The mixture is now prepared by opening both taps and running the two liquids into the vat beneath (under the two taps it is best to place the strainer). When all is run in and well mixed in the lower vat, it will contain 100 gallons of prepared mixture. This is now fed out through the tap into the spraying machine which should have a fine copper gauze strainer.

**Commercial Brands**

There are several ready-prepared Bordeaux powders and pastes on the market, but it has been proved beyond doubt that the freshly prepared mixture, made as before described, is superior for spraying fruit to either the powder or the paste.

Of the two, the paste is preferable to use, as it is in a much finer condition, is less likely to cause damage and mixes more easily with the water.

The disadvantage of even the best pastes however is that they have not the adhesive properties of the freshly prepared mixture. On careful trials, it was proved that over half of the copper was soon washed off the trees. Whereas, with the fresh mixture, only about 5 per cent. was lost in the same time.

If however it is not possible for the grower to prepare his own mixture, a good make of paste from a manufacturer of repute should be used.

## Tests

If insufficient lime is present in Bordeaux mixture, there will be some free or soluble copper in the wash which is liable to cause grave injury to the plant. The same remark applies of course to the commercial pastes and powders.

In order to test for free copper in the liquid, proceed as follows:

1. Purchase from a druggist an ounce or two of POTASSIUM FERROCYANIDE.
2. Dissolve in water in a medicine bottle.
3. Take a little of the well-settled Bordeaux mixture and place in a white saucer. Add a few drops of the solution from the bottle. If a *chocolate brown colour* is produced there is *free copper* present.

The way to cure a mixture with free copper in solution is to add a little milk of lime to it and stir well.

The test as above may then be again applied.

The test using a piece of bright steel is not sufficiently delicate and should not be relied upon.

## Remarks

Mr Spencer Pickering has made a very careful investigation of the chemistry of Bordeaux mixture. As a result he recommended the use of much less lime in the mixture, which would produce more "available copper" and therefore be stronger and more efficient. Unfortunately, the Bordeaux so prepared has not the adhesive properties of the Bordeaux normal mixture and so is liable to be readily washed off the trees. On the whole therefore it is best for the grower to stick to the old and well-tested formula as above.

## 2. Method of Use

### Strength of Spray

See under *preparation* above.

This is the standard strength. For summer spraying and for some of the more susceptible varieties, the spray may be diluted with an equal amount of water.

In the case of the various pastes and powders, the manufacturer's directions should be followed.

For winter spraying, it is best to use copper sulphate, or lime-sulphur at winter strength.

### When to Spray

For almost all diseases of fruit, it is best to give two sprayings:

1. When the leaves are just fully expanded—which is generally soon after the fruit has set.
2. A month or five weeks later.

It should be borne in mind that Bordeaux is a *preventive* spray and growers should not wait for signs of the disease on the leaves or other parts of the plant. It is a good plan in the case of some of the more susceptible varieties of apples to give a spraying of Bordeaux first, and to use lime-sulphur at summer strength for the second spray a month or so later.

### How to Spray

A type of nozzle must be employed which gives a very fine spray —practically a "fog." This is even more necessary for Bordeaux spraying than for arsenate of lead.

A fine hole should be used, and a high pressure at the pump is essential. There should be no large drops or jets of liquid.

The spray must be sparingly given and should not run off the leaves. Both surfaces of the leaf should be coated equally all over.

In order to reach the high branches, a long lance (preferably of bamboo) is necessary. Most spraying outfits have fittings for extension rods for this purpose.

Knapsack sprayers are useful for the smaller trees and bushes. Suitable hand sprayers are shown on pages 640—645 and petrol driven machines on pages 649—657; while the correct adjustment for spraying nozzles is as figs. 252, 253 on page 635. A sprayer for ground crops is shown on page 648.

### 3. Scientific

#### Chemical Composition

Pickering[1] has investigated the composition of Bordeaux mixtures very closely. He finds that, according to the proportion of lime employed, various insoluble basic sulphates are produced. Thus, using lime water, and employing varying proportions of lime, the following are formed:

| Lime to copper sulphate | Basic sulphate formed |
|---|---|
| 1 : 6 | $4CuOSO_3$ |
| 1 : 5½ | $5CuOSO_3$ |
| 1 : 5 | $10CuOSO_3$ |

Using the proportions of ordinary Bordeaux mixture (equal parts of lime and crystallised copper sulphate) the $10CuOSO_3$ sulphate is chiefly produced, together with calcium sulphate. Other secondary reactions also occur.

#### Action of Bordeaux

As previously stated, this is still a matter of controversy, and must at present be left for the experts to decide.

#### Analysis

As the action of Bordeaux depends upon the gradual setting free of copper in a soluble form, and as there is always excess of insoluble copper present, it is not possible to value a sample of paste or powder by its *copper content*, other properties, such as fineness, absence of free copper, or adhering powers being much more important.

Valuation is therefore only possible on an actual experimental basis, under the conditions obtaining in practice.

## BURGUNDY MIXTURE

This has been largely used as an alternative to Bordeaux mixture for potato spraying. As far as is at present known, it offers no advantage over Bordeaux as a fruit spray. It is therefore not described in detail here.

#### Description

Burgundy mixture is similar in most respects to Bordeaux, with the exception that the place of lime is taken by carbonate of soda (washing soda), a usual formula being:

| | |
|---|---|
| Copper sulphate (bluestone) crystals | 10 lbs. |
| Washing soda crystals (carbonate of soda) | 12½ lbs. |
| Water to | 100 gallons. |

[1] 11th Report of the Woburn Experimental Fruit Farm.

The copper sulphate is dissolved in say 70 gallons of water, the washing soda in the remainder, and this is then stirred into the copper solution. Practically all the other directions given under Bordeaux are applicable to Burgundy mixture also.

## COPPER SULPHATE

(bluestone)

### 1. General

**Employment**

As a winter wash for fruit trees and in the preparation of Bordeaux and Burgundy mixtures.

**Description**

Copper sulphate is a powerful poison to all fungi. It cannot however be used on trees in leaf, owing to its violent scorching action. Scorching has been found to occur at a strength of 12 ozs. per 100 gallons of the crystallised copper sulphate. For this reason, insoluble compounds of copper are used upon trees in leaf (see pages 605, 610).

As a winter wash it has given very successful results. Fungus diseases, such as Apple Canker, Black Spot (Scab), and Brown Rot have been eradicated by winter spraying with 10 lb. per 100 gallons. In the case of old and neglected or cankered trees, it is best used some weeks after treatment with caustic soda (see page 413). Some successful growers make a practice of spraying their trees with caustic soda and copper sulphate in this way, every three years.

**Properties**

It occurs as blue crystals containing about 64 per cent. of true or dry copper sulphate, the rest being water.

It should be free from iron, but usually contains traces of nickel. It is a poisonous substance and should be handled with care. As it is rather troublesome to dissolve, it is best to buy the small crystals or powder.

**Remarks**

The grower should purchase the pure article only (98 per cent. guaranteed). The so-called "agricultural bluestone" should be avoided.

### 2. Method of Use

#### Strength of Spray

*For spraying dormant trees only,*

**4—10 lbs. per 100 gallons**[1].

Dissolve in a pailful of water (hot is quicker) and stir into the bulk of the water. Another way is to tie up the crystals in a piece of sacking and suspend just beneath the surface of the liquid. It will *not* mix with soap.

## LIME-SULPHUR

### 1. General

#### Employment

As a winter and summer wash for all fruit trees, for American Gooseberry Mildew and for fungus diseases generally, especially *powdery mildews*, Apple Scab (Black Spot), etc. Also for Big Bud mite in spring[2]. With regard to its use as an insecticide, see page 430.

#### Description

Lime-sulphur, sometimes wrongly described as "lime and sulphur" is the name given to the liquid produced by boiling lime and sulphur with water. Both the strength and the composition vary a great deal according to the amount of each ingredient used, the manner of boiling, and the length of time taken in its manufacture. On the exact chemical nature of the constituents of lime-sulphur, scientists are not yet agreed, but it may be taken as a general rule, THAT THE MORE SULPHUR CONTAINED IN SOLUTION IN THE WASH THE MORE VALUABLE IT IS FOR SPRAYING PURPOSES against fungus diseases (see however below).

Formerly many growers boiled their own lime-sulphur, but this practice is rightly dying out, as the home-boiled substance is much less efficient, and much more wasteful of sulphur than the

[1] Copper sulphate has been used at a strength of 100 lbs. per 100 gallons without causing injury (see *Wye Journal*, No. 20, p. 405).

[2] Professor Lees. See also note on page 371 and fig. 246.

high grade factory produced article. This is because it is not possible, without producing large amounts of wasteful residues, to boil high-strength lime-sulphur in the ordinary manner. When weaker solutions are made, much of the sulphur is used up in the formation of a useless substance called *thiosulphate*, which is only produced in small quantities in the stronger liquids.

The active substance in lime-sulphur is sulphur, which is combined with the lime, forming substances known as *sulphides*.

A good deal of the sulphur may be in solution and not combined with the lime, but it is usually taken for granted that substances termed POLYSULPHIDES are produced. It can now be taken as definitely established[1] that these polysulphides form the active ingredients in the wash. Besides destroying any germinating spores, these substances, when exposed to the air, deposit sulphur in a fine form, and in this way probably also protect the leaf against further fungus attack.

**Preparation**

For the boiling of lime-sulphur, an iron pot, mounted in brickwork with an iron grate underneath, is suitable. It is better to have a mechanical stirring gear, but for small quantities this is not essential.

The lime, which should be in lumps and not powder, is placed in the pot, and slaked with a little water first, then the bulk of the water added. The fire is now made and the contents brought to a boil, and kept gently boiling while the sulphur is added. This should be in small quantities at a time. The boiling should continue about an hour after the addition of the last of the sulphur. Water should be added from time to time to make up the loss through evaporation. The finished liquor is tested and used as described on page 619.

**Quantities**

A suitable quantity to boil is 100 gallons, the charge being

| | |
|---|---|
| Lime | 1½ cwt. |
| Sulphur | 2 ,, |
| Water to | 100 gallons. |

If the lime is extra good, less may be used.

[1] See Eyre, Salmon and Wormald, *Journal of Agricultural Science*, Vol. IX., Sep. 1919, p. 283.

### Commercial Brands

High grade lime-sulphur is at present sold by various makers on its *specific gravity*, i.e. its relative weight compared with water. The agreed standard among manufacturers is 1·3 specific gravity. This means that each gallon of the concentrated liquor weighs 13 lbs. This may also be described as 33 degrees Beaumé or 60 degrees Twaddell.

Lower grades are also marketed, but it is not economical or advisable to use them.

It may here be pointed out that the specific gravity, although a good indication of strength, is not, in itself, a final guarantee of the efficiency of a brand. It is quite possible to obtain a S.G. of 1·3 with the use of only a relatively small proportion of sulphur. Growers are therefore advised to enquire the percentage of *poly-sulphide-sulphur* present in a brand as well as the specific gravity, before placing their orders.

### Action

Lime-sulphur appears to have a three-fold action when used for spraying trees. Its *caustic* properties (due probably to the *sulphides*) are responsible for its cleansing action on the bark of the tree. In the second place, fungi and possibly fungus spores are killed at once by these ingredients, while its third action is a preventive one, due to the deposition of the sulphur in fine particles[1] on the leaf. This probably prevents the fungus spores germinating.

## 2. Method of Use

### Strength of Spray

For **winter spraying** of trees in a dormant state, the usual strength is **5 gallons concentrated lime-sulphur (1·3 specific gravity) per 100 gallons of wash** (or 1 in 20).

In the case of very neglected trees, and to remove moss and lichen, a quarter to half as much again has been employed with benefit.

For **summer spraying, 1½ gallons of concentrate (1·3 specific gravity) per 100 gallons of wash** (or 1 in 60), may be regarded as FULL SUMMER STRENGTH. Whether this can be safely used depends upon the VARIETY OF TREE and also, to some degree, on weather conditions. One half to one quarter this strength is sometimes advised (see under description of diseases, Section VII).

[1] Probably by the splitting up of the "polysulphides."

## Preparation of Spray

There are two ways in which the spray may be prepared. The correct amount of the strong lime-sulphur can be measured out and stirred into the bulk of the water. Another way which may

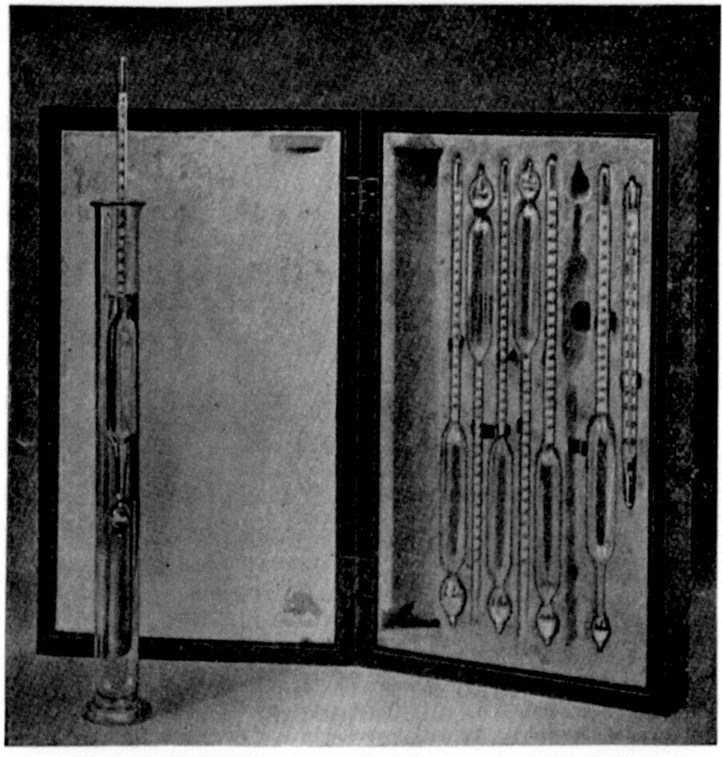

Fig. 243. Set of hydrometers for liquids of various specific gravities, with thermometer.

be used (and which must be employed in the case of "home-boiled" lime-sulphur) is to use a simple instrument called a *hydrometer*. (The accompanying figure (243) shows a case of hydrometers and the method of use). For careful work it is best to use the hydrometer for checking the strong liquid (as bought or made) and then to measure out the correct quantity. This is because it is not an easy matter to use the ordinary kind of hydrometer accurately with very weak solutions.

**How to use the hydrometer**

The hydrometer is simply a float made of a glass bulb and weighted with mercury or lead, and having a stem with marks

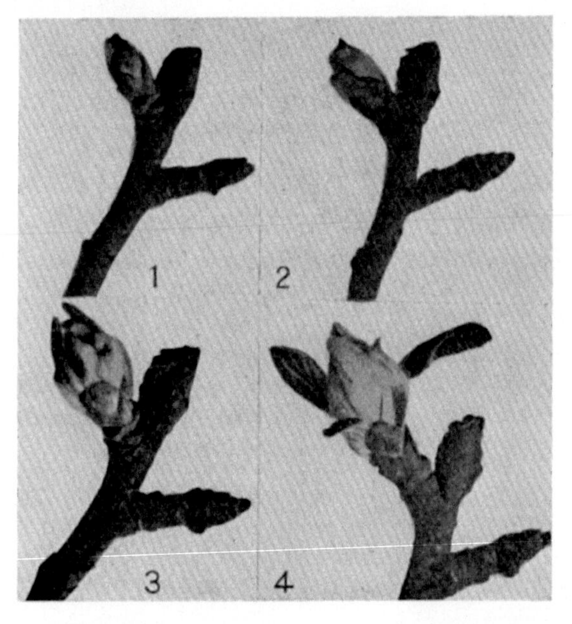


Fig. 244. Apple—Cox's Orange Pippin. Various stages of development of same bud. Full winter strength lime-sulphur is perfectly safe up to stage 3 and did no ultimate harm after careful trials of spraying at stage 4.

upon it at equal distances. According to the relative weight or *density* of the liquid (i.e. its specific gravity) the hydrometer sinks either higher or lower, and more or less of the stem is immersed. The reading of the specific gravity is taken at a point on the stem level with the surface of the water.

This should be read with the eye on a level with the surface of the liquid, which is best placed in a glass cylinder on a table.

There are three scales of measurement in common use on hydrometers:

| | |
|---|---|
| Actual Specific Gravity | (water = 1·000), |
| Beaumé | (water = 0), |
| Twaddell | (water = 0). |

The following table shows how the scales correspond with each other:

| Twaddell | Beaumé | Specific gravity | Twaddell | Beaumé | Specific gravity | Twaddell | Beaumé | Specific gravity |
|---|---|---|---|---|---|---|---|---|
| 0 | 0 | 1·000 | 21 | 13·6 | 1·105 | 44 | 26·0 | 1·220 |
| 1 | 0·7 | 1·005 | 21·6 | 14·0 | 1·108 | 45 | 26·4 | 1·225 |
| 1·4 | 1·0 | 1·007 | 22 | 14·2 | 1·110 | 46 | 26·9 | 1·230 |
| 2 | 1·4 | 1·010 | 23 | 14·9 | 1·115 | 46·2 | 27·0 | 1·231 |
| 2·8 | 2·0 | 1·014 | 23·2 | 15·0 | 1·116 | 47 | 27·4 | 1·235 |
| 3 | 2·1 | 1·015 | 24 | 15·4 | 1·120 | 48 | 27·9 | 1·240 |
| 4 | 2·7 | 1·020 | 25 | 16·0 | 1·125 | 48·2 | 28·0 | 1·241 |
| 4·4 | 3·0 | 1·022 | 26 | 16·5 | 1·130 | 49 | 28·4 | 1·245 |
| 5 | 3·4 | 1·025 | 26·8 | 17·0 | 1·134 | 50 | 28·8 | 1·250 |
| 5·8 | 4·0 | 1·029 | 27 | 17·1 | 1·135 | 50·4 | 29·0 | 1·252 |
| 6 | 4·1 | 1·030 | 28 | 17·7 | 1·140 | 51 | 29·3 | 1·255 |
| 7 | 4·7 | 1·035 | 28·4 | 18·0 | 1·142 | 52 | 29·7 | 1·260 |
| 7·4 | 5·0 | 1·037 | 29 | 18·3 | 1·145 | 52·6 | 30·0 | 1·263 |
| 8 | 5·4 | 1·040 | 30 | 18·8 | 1·150 | 53 | 30·2 | 1·265 |
| 9 | 6·0 | 1·045 | 30·4 | 19·0 | 1·152 | 54 | 30·6 | 1·270 |
| 10 | 6·7 | 1·050 | 31 | 19·3 | 1·155 | 54·8 | 31·0 | 1·274 |
| 10·2 | 7·0 | 1·052 | 32 | 19·8 | 1·160 | 55 | 31·1 | 1·275 |
| 11 | 7·4 | 1·055 | 32·4 | 20·0 | 1·162 | 56 | 31·5 | 1·280 |
| 12 | 8·0 | 1·060 | 33 | 20·3 | 1·165 | 57 | 32·0 | 1·285 |
| 13 | 8·7 | 1·065 | 34 | 20·9 | 1·170 | 58 | 32·4 | 1·290 |
| 13·4 | 9·0 | 1·067 | 34·2 | 21·0 | 1·171 | 59 | 32·8 | 1·295 |
| 14 | 9·4 | 1·070 | 35 | 21·4 | 1·175 | 59·4 | 33·0 | 1·297 |
| 15 | 10·0 | 1·075 | 36 | 22·0 | 1·180 | 60 | 33·3 | 1·300 |
| 16 | 10·6 | 1·080 | 37 | 22·5 | 1·185 | 61 | 33·7 | 1·305 |
| 16·6 | 11·0 | 1·083 | 38 | 23·0 | 1·190 | 61·6 | 34·0 | 1·308 |
| 17 | 11·2 | 1·085 | 39 | 23·5 | 1·195 | 62 | 34·2 | 1·310 |
| 18 | 11·9 | 1·090 | 40 | 24·0 | 1·200 | 63 | 34·6 | 1·315 |
| 18·2 | 12·0 | 1·091 | 41 | 24·5 | 1·205 | 64 | 35·0 | 1·320 |
| 19 | 12·4 | 1·095 | 42 | 25·0 | 1·210 | 65 | 35·4 | 1·325 |
| 20 | 13·0 | 1·100 | 43 | 25·5 | 1·215 | 66 | 35·8 | 1·330 |

## Strength required using the hydrometer

The full WINTER STRENGTH of 1 in 20 (5 gallons to 100) using high grade lime-sulphur of 1·3 specific gravity is equal to the following on the hydrometer:

Specific gravity 1·015
Beaumé 2·1 degrees
Twaddell 3 „

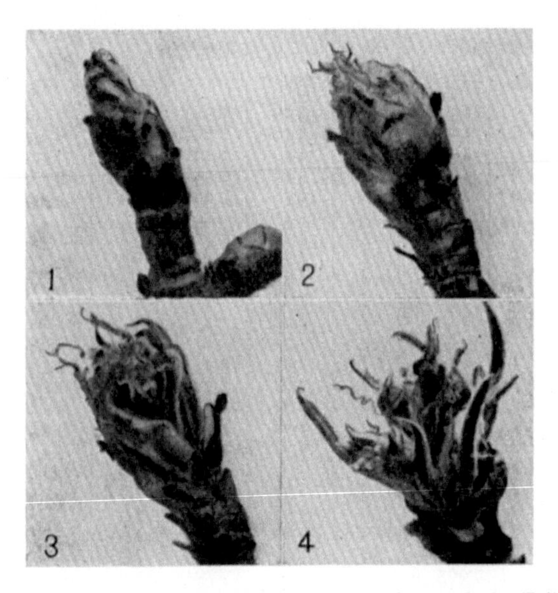


Fig. 245. Pear—various stages of development of same bud. Full winter strength lime-sulphur was tried in all stages shown and though it caused scorching in stages 3 and 4, there was no ultimate injury.

The full SUMMER STRENGTH of 1 in 60 ($1\frac{1}{2}$ gallons to 100) under the same conditions is:

Specific gravity 1·005
Beaumé 0·7 degrees
Twaddell 1 „

## How to dilute home-boiled lime-sulphur

1. Cóol[1] a sample of the settled finished liquid (it should be quite clear). Place in a glass vessel and dip the hydrometer into it.
2. Take the reading in the usual manner, previously described, and according to this figure make the correct quantity up to 100 gallons with water as under:

## Amount of lime-sulphur (home-boiled) of various specific gravities required for 100 gallons of wash

| Specific gravity | Degrees Twaddell | Degrees Beaumé | For full winter strength | For full summer strength |
|---|---|---|---|---|
| 1·25 | 50 | 29 | 6 galls | 16 pints |
| 1·24 | 48 | 28 | 6¼ | 17 |
| 1·23 | 46 | 27 | 6½ | 17½ |
| 1·22 | 44 | 26 | 6¾ | 18 |
| 1·21 | 42 | 25 | 7¼ | 19 |
| 1·20 | 40 | 24 | 7½ | 20 |
| 1·19 | 38 | 23 | 8 | 21 |
| 1·18 | 36 | 22 | 8½ | 22 |
| 1·17 | 34 | 20 | 8¾ | 23½ |
| 1·16 | 32 | 20 | 9½ | 25 |
| 1·15 | 30 | 19 | 10 | 27 |
| 1·14 | 28 | 17¾ | 10¾ | 29 |
| 1·13 | 26 | 16½ | 11½ | 31 |
| 1·12 | 24 | 15½ | 12½ | 33 |
| 1·11 | 22 | 14 | 13¾ | 36½ |
| 1·10 | 20 | 13 | 15 | 40 |

It should be pointed out that different boils will vary in efficiency according to the percentage of the different ingredients. As previously mentioned the high strength factory product when diluted to the same specific gravity as a home-boiled lime-sulphur will have much greater working strength.

## Use of Saponin

In order to increase the wetting action or *penetration* of lime-sulphur, especially for summer use, it is advised to use SAPONIN

[1] It should be cooled to 60° F. (15·5° C.), i.e. about the temperature of an ordinary living room.

with it. The commercial preparation[1] should preferably be employed, using 1—2 gallons to 100 gallons of the made-up wash (equivalent to 0·05 to 0·1 per cent. of pure saponin).

### When to Spray

A good deal of difference of opinion exists as to the best time to apply lime-sulphur in the dormant state of the trees. The balance of evidence is in favour of as late a use as possible in the early

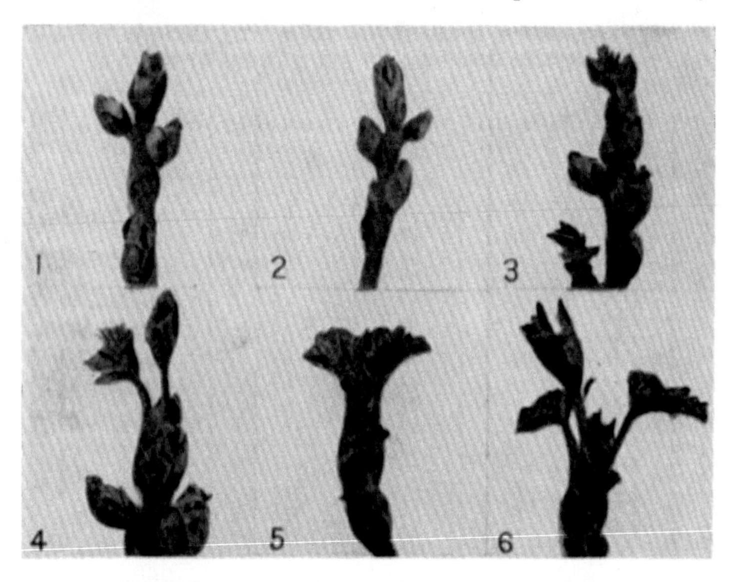


Fig. 246. Black currant, showing development of buds. The best time to spray with lime-sulphur for big bud mite is when shoot is at stage 6. Full winter strength should be used, or even stronger (6 or 8 gallons to 100 gallons). See note on page 371.

spring. The author has made many experiments as to the effect of applying the winter strength lime-sulphur during the opening of the buds, and finds that, although severe scorching of the petals of the opening blossoms may occur, no damage is done to the essential parts of the flower (see Chapter 4), and the fruit develops in the normal way.

[1] Sold by the Yalding Manufacturing Co., Ltd., Yalding, Kent. See page 449.

This may seem drastic treatment to many, and the author does not advise such late spraying in all cases, as it is easy to overstep the safe period and so cause damage. The following photographs

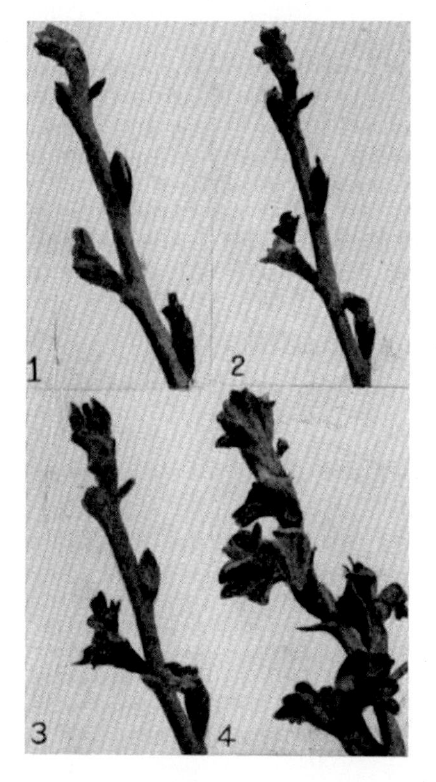


Fig. 247. Gooseberry—stages of shoot development. Spray for American gooseberry mildew immediately first signs of fungus are seen.

show what may be taken as quite safe stages of the opening buds in which to spray at the winter strength in the case of various fruits. In the case of big bud disease on black currants, the plants should only be sprayed at the stage specified (fig. 246, stage 6).

## How to Spray

For winter spraying of trees without foliage a high pressure and a small cone of spray are best. If the spray is too much spread out and too misty, a good deal of waste is unavoidable.

A windy day is very unsuitable for spraying as only one side of the trees will be wet.

The spray has a smarting action on sensitive skins and especially on the eyes. The men should therefore be PROTECTED by the use of goggles for the eyes and rubber gloves, and only old clothes should be worn, or overalls used, while oil or fat rubbed well into the skin will prevent the face smarting.

Copper vessels are unsuitable for handling the spray—iron and wood are unaffected. The spray machines after use should be well rinsed through with water, otherwise the brass nozzles and connections are liable to become corroded.

## Quantity of Spray Used

1. WINTER AND EARLY SPRING WASHING.

The following are, in the author's experience, average figures which may be taken as a fair guide:

| Spread of trees, feet | Amount of diluted spray per 100 trees | Amount of concentrated lime-sulphur (1·3 S.G.)[1] |
|---|---|---|
| 7 | 150 gallons | 7½ gallons |
| 10 | 200 ,, | 10 ,, |
| 15 | 350 ,, | 17½ ,, |
| 20 | 530 ,, | 26½ ,, |
| 25 | 800 ,, | 42½ ,, |
| 30 | 1150 ,, | 57½ ,, |
| 35 | 1600 ,, | 80 ,, |
| 40 | 2100 ,, | 105 ,, |

To find the quantities per acre, multiply the above figures by the following factors; according to the distance apart:

6 ft, 12·1; 10 ft, 4·35; 12 ft, 3·02; 15 ft, 1·93; 18 ft, 1·34; 20 ft, 1·08; 24 ft, 0·75; 30 ft, 0·48; 40 ft, 0·28.

[1] Using full winter strength (1 in 20).

2. SUMMER SPRAYING.

The amount is variable, according to the leafage. The following is therefore an indication only and represents average results:

| Spread of trees, feet | Amount of diluted spray per 100 trees | Amount of concentrated lime-sulphur (1·3 S.G.)[1] |
|---|---|---|
| 7 | 225 gallons | 3¾ gallons |
| 10 | 300 ,, | 5 ,, |
| 15 | 520 ,, | 8½ ,, |
| 20 | 800 ,, | 13¼ ,, |
| 25 | 1250 ,, | 21 ,, |
| 30 | 1700 ,, | 28¼ ,, |
| 35 | 2450 ,, | 41 ,, |
| 40 | 3000 ,, | 50 ,, |

To find amount required per acre, use figures as above.

## 3. Scientific

### Chemical Composition

This is a matter of some dispute at the present time. While the general opinion is that polysulphides are produced there is evidence to show that the sulphur may be in *physical solution* only, being dissolved in a solution of di-sulphide and thiosulphate with a little hydroxyhydrosulphide, sulphates and sulphites[2]. It is certain that the sulphur is very loosely held and is easily deposited by the action of the air. In any case, it may be taken as proved that it is sulphur in this condition which is the active ingredient. Assuming the higher sulphides are produced, the following are the chief variations in commercial products:

The POLYSULPHIDE SULPHUR varies from about 15 to 25 per cent. in high grade products of specific gravity 1·3, while it is very much lower in the case of home-boils.

The SULPHIDE SULPHUR does not vary much, being about 3 to 4 per cent. in factory and home-boiled products.

The THIOSULPHATE SULPHUR, which is practically useless as a spraying ingredient, is small in the factory-boiled product, being as a rule not over 2 to 3 per cent. In the case of home-boils, it frequently reaches high proportions, thus in many cases over a quarter of the total sulphur present exists as thiosulphate.

[1] Using full summer strength, e.g. 1 in 60.

[2] A. A. Ramsay, Dept of Agriculture, *N. S. Wales Bulletin*, 15, 1915.

### Analysis

There are several methods advised, and the analyst is referred to the original paper:

"Tartar—Chemical Investigations in the Lime-Sulphur Spray": *Oregon Agr. Coll. Exp. Station Bulletin*, March 1914.

## LIVER OF SULPHUR

### 1. General

### Employment

For general spraying against fungus diseases, usually with soft soap. It is also used as an insecticide for red-spider (see page 376).

### Description

Liver of sulphur, being manufactured in a somewhat crude manner, and being also very liable to change on keeping and on exposure to air, is a somewhat indefinite substance. In many ways it is at present *very unsatisfactory* for spraying purposes. As ordinarily prepared it has been shown by Professor Salmon to be ineffective against mildews, except when used in such quantities as to cause injury to the foliage. There is no doubt that it has, in the past, caused severe scorching to foliage, especially in the case of hops, when used in the proportions commonly recommended. It is therefore necessary to be VERY CAUTIOUS in using it, and it is best, in all cases, to make a trial at the strength at which it is intended to employ it.

### Properties

Liver of sulphur is a mixture of sulphides (mostly polysulphides) of potash together with other substances (see page 431) and is liable to contain qualities of potash as carbonate or caustic. It is not uniform in composition but varies according to the manner of manufacture. It is a dark brown or greenish substance, with an unpleasant sulphurous odour, and should be entirely dissolved by water giving a fairly clear solution. Recently large quantities of so-called liver of sulphur have been prepared with soda[1] in place of potash. Growers should *insist upon a guarantee* of its preparation from pure potash.

[1] See *soda-sulphur*, page 626.

### Preparation

It is manufactured by gently heating together sulphur and carbonate of potash in a closed vessel. The temperature and the proportions used of each ingredient have a great influence on its final composition.

## 2. Method of Use

### Strength of Spray

The quantity generally advised is from

**2—3 lbs.[1] per 100 gallons of wash.**

It is best used with SOFT SOAP sufficient to produce a good lather in the spray.

### Preparation of Spray

The correct amount of liver of sulphur is weighed out, dissolved in half a pailful of water, and well stirred into the bulk of the soapy liquid.

### When to Spray

See under description of diseases (Section VII).

### How to Spray

As for soap (page 455).

### Remarks

The author advises the trial of ammonium polysulphide (with soap) or lime-sulphur (without soap) in place of liver of sulphur, as at present manufactured.

## 3. Scientific

### Chemical Composition

Liver of sulphur is often *misnamed potassium sulphide*[2]. It is a complex, and by no means uniform, mixture of various polysulphides of potash, together with varying, but often large, amounts of sulphate and thiosulphate.

### Analysis

It should be valued upon its content o *sulphide-sulphur*. Sulphates and thiosulphates are harmless to the plant, but caustic potash or carbonate is a source of danger. *Soda* should be absent.

[1] See caution on page 624.

[2] See Board of Agriculture Leaflets 133, 185, 195, etc.

## SODA-SULPHUR

This is a newly coined name which may conveniently be used for *liver of sulphur* prepared with carbonate of soda in place of potash.

It has much the same composition and action as the other compound (see page 624) but is more liable, if anything, to scorch the leaves. In the author's opinion, neither of these substances is, as at present manufactured, sufficiently safe to use, if efficient proportions are to be employed against fungus diseases. For use with soap, a trial with ammonium polysulphide[1] is advised, and where soap is not required, lime-sulphur[2] is of proved efficiency.

## SULPHUR

### 1. General

#### Employment

For dusting or blowing on to plants affected with fungus diseases, especially the *powdery mildews* (hop, vine, etc.). Also in the manufacture of lime-sulphur.

#### Description

Sulphur exists in commerce in several forms. There is the crude or *rock* sulphur, most of which comes from the mines of Sicily. This is usually classed in three grades. *Roll* sulphur or brimstone represents the rock sulphur, refined by melting and straining and cast into sticks, while the highest grade is known as *flowers of sulphur*. This is produced by vaporising or "subliming" the sulphur which solidifies in a very fine state.

In addition to the natural product, sulphur is now recovered in various industries, and is known in this form as "recovered sulphur."

One of the forms in which recovered sulphur may be obtained is as "precipitated sulphur." Although in an exceedingly fine state it usually contains only a relatively low percentage of sulphur and this should be taken into account when purchasing.

[1] See page 601. [2] See page 612.

### Commercial Brands

The only suitable brand for this purpose is the finest flowers of sulphur. The "precipitated sulphur" has been successfully employed but it should be valued upon the percentage of free sulphur contained in it.

### Action

It has been suggested that the action of sulphur is due either to sulphur dioxide (the suffocating gas formed on burning sulphur) or sulphuretted hydrogen (the gas emitted from rotten eggs). It is however almost certainly neither of these, but the sulphur vapour itself which is responsible for its action on fungi.

On plants, sulphur has a distinct stimulating action. In the case of hops, the growth of the bine is increased and it is for this reason that a final dusting with sulphur, just as the burr forms, is valuable in hop cultivation.

## 2. Method of Use

### How to Apply

On a large scale, the sulphur is blown on to the hops by a special machine known as a sulphurator (see page 659). Care should be taken to avoid getting the dust into the eyes; goggles may be worn as a preventive. Other mildews, on strawberries, vines, etc., should be treated in a similar manner using hand sulphurators along strawberry rows and an instrument of the syringe type for grapes under glass.

### When to Apply

Sulphur should be applied as soon as any sign of the "mould" is seen on the leaves, or if the disease is known to be in the district, it is well not to wait for its appearance.

Many growers "sulphur" their hops as a standard practice, near the end of the season. This is for its stimulating effect on the hop and to prevent any possibility of mould attacking the cone (or burr).

## 3. Scientific

### Chemical Composition

Sulphur is a chemical element, represented by the symbol S, atomic weight = 32.

### Properties

Sulphur as usually found is a light yellow solid, nearly insoluble in water, capable of being evaporated or "sublimed" with heat.

It dissolves readily in *carbon bisulphide.*

### Analysis

The physical condition, especially the degree of fineness of the particles, should be noted as well as its chemical purity. The percentage of *free sulphur* should be determined.

# PART III

## SPRAYING IN THEORY AND PRACTICE

## SECTION IX

### SPRAYING APPLIANCES AND METHODS

# SECTION IX

## SPRAYING APPLIANCES AND METHODS

## CHAPTER 43

### Introduction

There are several important requirements to be taken into account in the selection of spraying appliances and machinery. Of these the chief are:

A. A nozzle of suitable and efficient construction.

B. The provision of adequate pressure to produce the required type of spray, and to reach as high as necessary.

C. A type of machine suited to the size and conditions of the plantation to be sprayed.

### A. The Nozzle

Since most machines will be called upon to spray several different materials, it is essential to have a nozzle supplied which can be adapted for each spray and for all conditions of work. Such requirements are fulfilled in the modern ADJUSTABLE NOZZLE. The nozzle shown in figs. 257, 258 is not adjustable and is only suitable when very fine misty sprays are required. A nozzle of the adjustable type consists of the following parts (see figs. 248, 249):

(*a*) The stem or body.

(*b*) A cap bored with two oblique holes to give a circular motion to the liquid.

(*c*) A collar of conical shape which is adjustable by screwing up or down on the stem, and so controlling the amount of liquid passing through.

(*d*) A locknut, with rubber packing ring for clamping the collar (*c*) in any desired position.

(*e*) An outside cap into which fit various removable steel discs, bored with different sized holes.

USE.

1. To obtain *powerful but relatively coarse sprays*, the collar (*c*) is screwed almost to the top of the stem, and there locked. The amount of liquid issuing in the spray is controlled by the size of disc used.

Fig. 248. "Mistifier junior[1]" adjustable nozzle. *a—e*, see in text. *f'*, *f''*, *f'''*, varying size disc jets. *g*, washers. *h*, complete nozzle (screwed down for fine spraying).

Such a type of spray is generally regarded as suitable for spraying aphides or caterpillars with a CONTACT wash. Being relatively powerful, many of the aphides are "hit" and some of the larger ones dislodged by the force of the spray. It is also required when reaching up to the top branches of high trees. Since this adjustment of the nozzle produces a narrow cone of spray, it is more suitable when it is wished to concentrate the spray upon a small portion—such as the bare branch of a tree, etc. For this reason, it is probably the best adjustment for all winter spraying, such as LIME-SULPHUR, CAUSTIC ALKALI, etc.

Although a special form of nozzle ("seneca" nozzle) has been introduced for LIME spraying (see fig. 254), this adjustment of the standard nozzle is very suitable, provided that the lime liquid has been properly STRAINED. The advantage of the seneca nozzle is that it can be rapidly reversed in case of blocking.

2. To adjust the nozzle for *fine misty sprays*, the collar (*c*) is screwed down as far as it will go and locked there in position by means of the nut.

This has the effect of spreading out the spray and of rendering it very fine, especially when a small disc is used.

[1] Manufactured by Messrs Drake and Fletcher, Maidstone.

This class of spray is essential for ARSENATE OF LEAD and other poison washes for caterpillars, for BORDEAUX MIXTURE, for summer spraying of LIME-SULPHUR, and for hops early in the season.

For other purposes an intermediate position of the collar may be used to produce a spray of medium fineness.

Fig. 249. "Multi-spray[1]" adjustable nozzle. *a—e*, see in text. *e'*, large cap-form jets. *f'*, *f''*, *f'''*, various size discs. *g*, washers. *h*, complete nozzle (screwed down for fine spraying).

## B. Adequate Pressure

This is essential in successful spraying. It is particularly necessary when the nozzle is adjusted for producing a *fine misty spray*. Without adequate pressure the liquid cannot be properly broken up or "atomised." This is the great advantage of using *power spraying machines* wherever possible. Such machines produce uniform high pressures, and are

[1] Manufactured by Messrs Weeks, Maidstone.

Fig. 250. Collar of nozzle screwed up, using a large disc opening. This adjustment gives a powerful spray reaching to a considerable height, but relatively coarse.

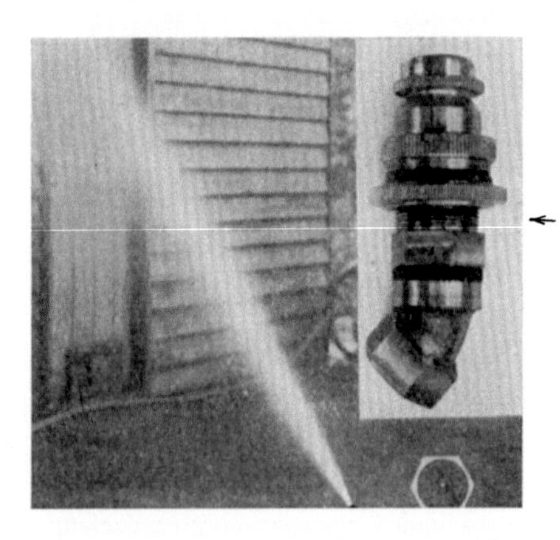

Fig. 251. Adjustment as in fig. 250 but with a fine disc opening giving a narrow angle or cone of spray, powerful but relatively coarse.

Fig. 252. Collar of nozzle screwed fully down. Disc opening large. This gives a very wide cone or angle of spray relatively fine and misty, but with little carrying power.

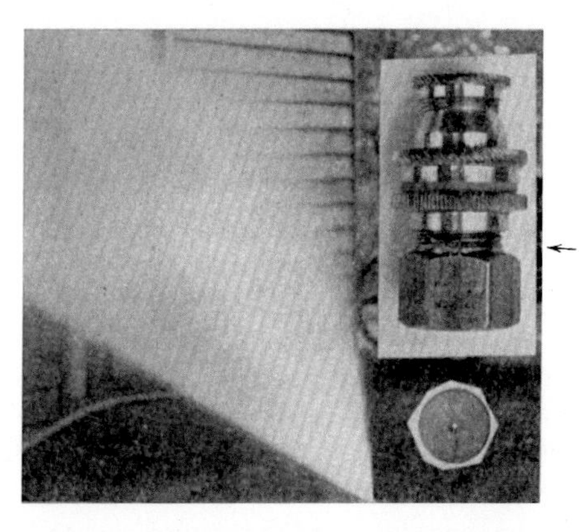

Fig. 253. Adjustment as in fig. 252 but disc opening very small. The cone or angle of spray is narrower but a very fine misty spray is produced.

independent of the fatigue of the workmen. A great deal of failure in spraying is undoubtedly traced to lack of adequate pressure on the pump. The figures (255 and 256) show the coarseness of the spray produced with different adjustments of the nozzle when the power is inadequate.

### C. A suitable type of machine.

Various forms of machines will be described in the following chapter. It is only necessary here to point out that it is better for a grower to ave a machine of a larger type than he could manage with, than the

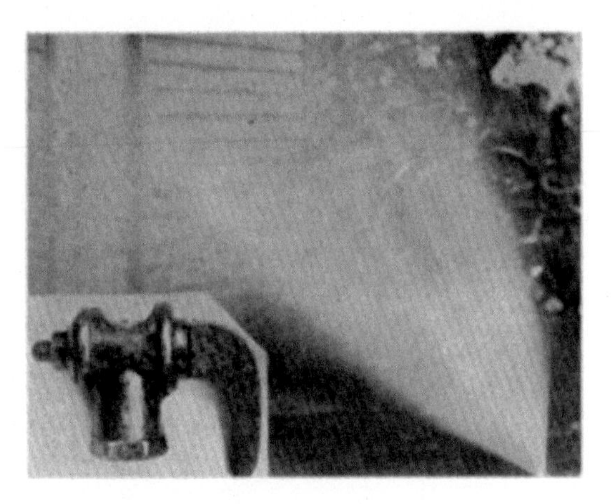

Fig. 254. "Seneca" nozzle for lime spraying.

reverse to be the case. Wherever a plantation exceeds say 20 acres, a power plant of some description will pay to instal, and will give far more efficient service than any other kind of hand or "traction" sprayer. An exception to this is in the case of the first few years of a freshly planted orchard, when the work can be done quite successfully with hand sprayers.

Other considerations in regard to spraying appliances are:

1. The pump and tank should be made of SUITABLE MATERIALS.
2. There should be an EFFICIENT AGITATOR to keep the spray liquid properly mixed.

Fig. 255. Nozzle with same adjustment as fig. 250 but with a low pump pressure. Showing coarseness of spray.

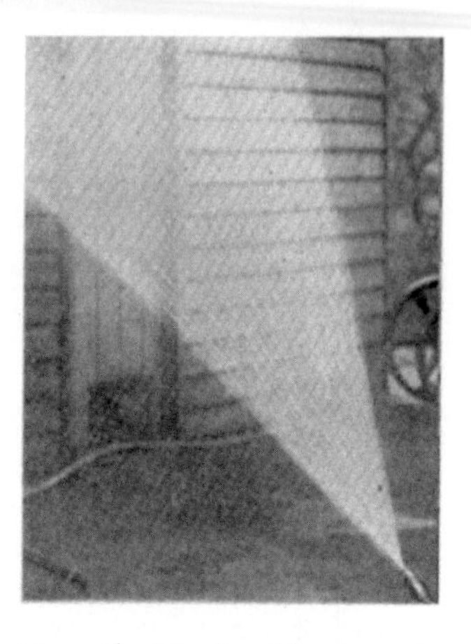

Fig. 256. Nozzle with same adjustment as fig. 252 but with a low pump pressure. Showing coarseness of spray.

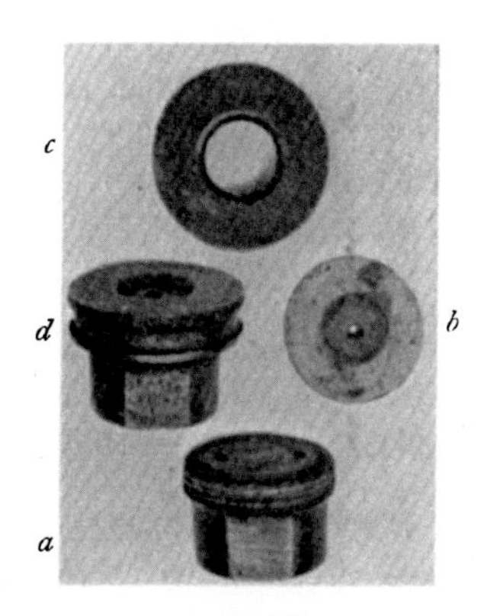


Fig. 257. "Mistry" nozzle for producing a fine misty spray, not adjustable. *a*, Body and cap with oblique holes. *b*, Disc, with fine jet opening. *c*, Cap for screwing jet in position. *d*, Complete nozzle.

Fig. 258. The "mistry" nozzle in action, showing the fine character of the spray produced.

3. A suitable AIR CHAMBER should be provided to keep the pressure steady.
4. The VALVES of the pump should be of an efficient non-choking type.
5. The pump should be capable of being DISMANTLED rapidly and easily in case of a breakdown.
6. The parts should be readily REPLACEABLE and of a standard pattern.
7. It should be possible to readily renew the PACKING of the piston, glands, etc.

1. Probably brass, bronze, or copper are the most suitable all-round materials for spray pumps and tanks. Nozzles should be of a heavy brass, as iron corrodes very readily. For lime-sulphur, which attacks brass and copper though not very seriously, wooden tanks are the best, and these are also suitable for Bordeaux. (Lime-sulphur has no effect on iron.) Lead-coated iron plate is also suitable for most materials. In any case, the sprayer must be well rinsed out after use and properly looked after when not in service.

2. An agitator for the tank is necessary for spraying with lead arsenate, Bordeaux, and similar washes. It should work automatically off some part of the gear and not require operating separately.

3. The size of the air chamber is important. If large, a steady pressure can be much more easily obtained. This applies to hand pumps; power pumps are usually provided with suitable air chambers.

4. Of the different kinds of valves, poppet, swing, check, ball, etc., the one which gives the most satisfaction is the ball valve. The action of the liquid tends to rotate the ball and so produce even wear. Trouble is however given with sprays which contain *insufficient soap*, but this applies more or less to all types of valves.

5. There will be occasions when it is necessary to examine the valves or other working parts of the pump and these should be readily accessible, otherwise much time will be wasted probably at the very period when it is most valuable.

6. It is well only to purchase a machine of which the parts are of standard pattern and can be replaced at once. For this reason, foreign pumps should on no account be bought unless it is certain that spare parts are stocked in this country and readily obtainable.

7. Stuffing boxes will require frequent re-packing and should be readily accessible.

# CHAPTER 44

## Types of Spraying Machinery

### 1. Syringe Sprayers

These are in the form of a garden syringe, with rose cap and also spraying nozzles. They are usually made with a ball valve for filling the syringe by suction. They represent the smallest type of sprayer, and are only suitable for small private gardens or greenhouses. The spray or "atomisation" obtained is fairly fine provided that plenty of pressure is given. It becomes very tedious in use after a short time (fig. 259).

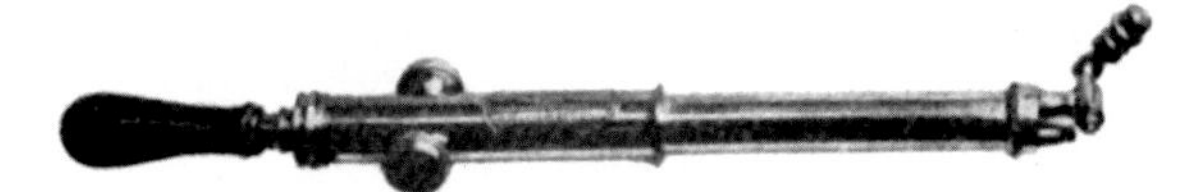

Fig. 259.

Fig. 260[1].

### 2. Pneumatic Hand Sprayer

This is a useful type for greenhouse work. The air pressure is obtained by means of an ordinary bicycle pump, and the sprayer can then be operated by pressing the thumb lever (fig. 260).

[1] Supplied by Messrs Weeks, Maidstone.

### 3. Bucket Pumps

These are also suitable for work in a small garden. In some patterns there is a rest provided for the foot and the pump is placed inside the bucket, the plunger being then operated by one hand while with the other the spray is directed. The constant moving of the bucket to fresh positions makes it unsuitable for any but small jobs.

Fig. 261 shows a very compact self-contained outfit marketed by an American company[1]. In this pattern an efficient agitator is worked with the pump handle, and the pump is secured to the bucket by means of wing nuts. Price about £6 complete.

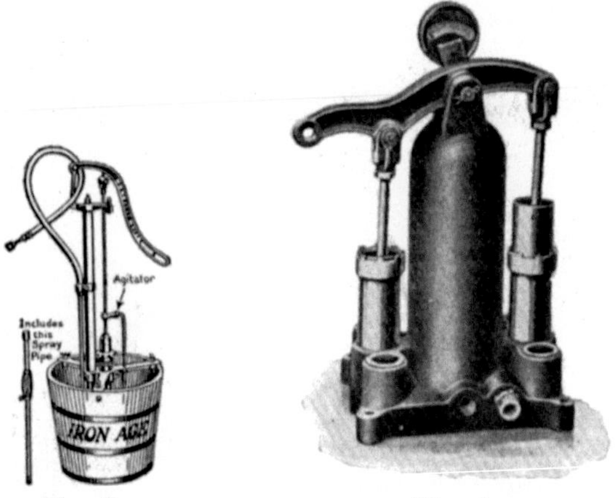


Fig. 261. Fig. 262.

### 4. Knapsack Sprayers

These are very handy for young fruit trees and for small plantations and experimental purposes. They are bean-shaped and designed to carry on the back with shoulder straps, one hand being used to work the pump and the other directing the spray. Their capacity is about three gallons. There are two types, the internal type fitted with rubber valves and diaphragms (like the original "Vermorel" pattern, fig. 263), and an external pump pattern with

[1] The Bateman Manufacturing Company, Grenloch, N.J., U.S.A.

metal valves, fig. 264. The latter type is probably the more suitable for all round work. Price about £4.

Fig. 263.

Fig. 264[1].

## 5. Automatic Sprayers

These are small compressed air sprayers, consisting of steel cylinders provided with an air pump and pressure gauge. The liquid is run in and the cylinders sealed up and charged with air to the desired pressure by means of the pump. The sprayer is then carried round to wherever required and the liquid sprayed by opening the tap on the lance. They are largely replacing the knapsack sprayers for similar uses, and have a working capacity of 1 to 3 gallons (fig. 265).

Fig. 265[1].

[1] Supplied by Messrs Weeks, Maidstone.

**6. Barrel Sprayer**

This is a popular type in America. It is mounted on a timber base and the pump handle has a backwards-and-forwards movement, which uses the back and hip muscles and can be worked for longer without fatigue than the up-and-down type. It is usually operated and carried on a farm cart or wagon. Fig. 266 shows a compact outfit manufactured by the Deming Co., and costing about $72 (at normal rate of exchange=£15).

Fig. 266.

**7. Barrow Sprayers** (figs. 267, 268)

This is a very popular hand-type of sprayer and is suitable for small plantations. The pump is worked by a vertically operated lever handle, and the tank, which may be of iron or wood, has a capacity of from 12 gallons in the smallest size to 50 gallons in the largest kind made. Tanks may also be had of iron, lead coated, and this is the most suitable material for general purposes. They are supplied with good lengths of hose and long lances, so as to reach high standard trees. The labour required is three men, one to work the pump and two to apply the spray, with two branch hoses and lances. Price about £24.

A similar type of machine is also made to be horse-drawn with a tank capacity of 100—110 gallons.

In America, a very popular form is barrel-shaped and mounted upon wheels.

Fig. 267.

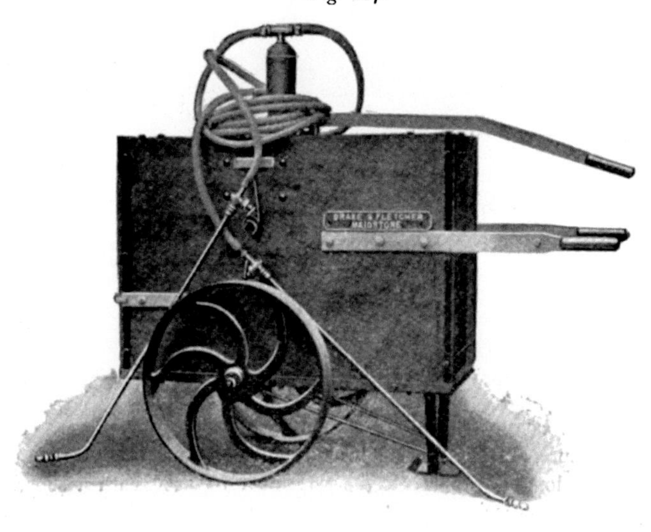

Fig. 268.

**8. Two-manual Hand Sprayers**

These are usually of the vertically operated type, and have handles on each side of the sprayer for two labourers. They may be obtained with or without tanks. In one popular type[1], the tank is in the form of a 10 gallon barrel, mounted on the carriage, alongside the pump (fig. 269). Price about £33.

Fig. 269.

**9. Horizontally Operated Pump**

This is the type shown in the illustration (fig. 270) and may be worked by one or two men. The advantage claimed is—as also in the case of No. 5—that it is much easier to maintain a good pressure, and the muscles used in operating are those which can work longer without fatigue. It must be used with a separate tank, as this is not supplied. Rotary, hand power sprayers are

[1] Manufactured by Messrs Weeks, Maidstone.

also made and the same advantage is claimed from them, viz. a minimum of fatigue.

Fig. 270[1].

## 10. Traction Pumps

These sprayers are operated by chains from the wheels on which the tank and pump are mounted. They are much used for spraying hops, but are generally unsuitable for trees, unless a hand action on the pump is also supplied. They are made for draught with two or three horses, according to the nature of the ground. A large number of nozzles, adjustable to any style, are supplied which project out on each side of the sprayer (see fig. 271)[2]. Price about £110. Similar designs operated by power are now manufactured and are much to be preferred since they are more regular in action (see fig. 272)[3]. Price about £200.

A similar pattern of sprayer is also used for spraying ground crops, such as potatoes, etc. In this case the nozzles are arranged on a large horizontal pipe at the back of the sprayer. For potato spraying, pairs of nozzles are arranged to spray each side of the potato hills; with other crops, the spray points downwards on to the rows. All nozzles are adjustable for different distances of rows (fig. 273)[2]. Price about £51.

[1] Sold by Messrs Craven, Evesham.
[2] Manufactured by Messrs Weeks, Maidstone.
[3] Manufactured by Messrs Drake and Fletcher, Maidstone.

## 11. Power Sprayers

All growers with plantations exceeding 20 acres will be well advised to invest in a power sprayer for their work. There is

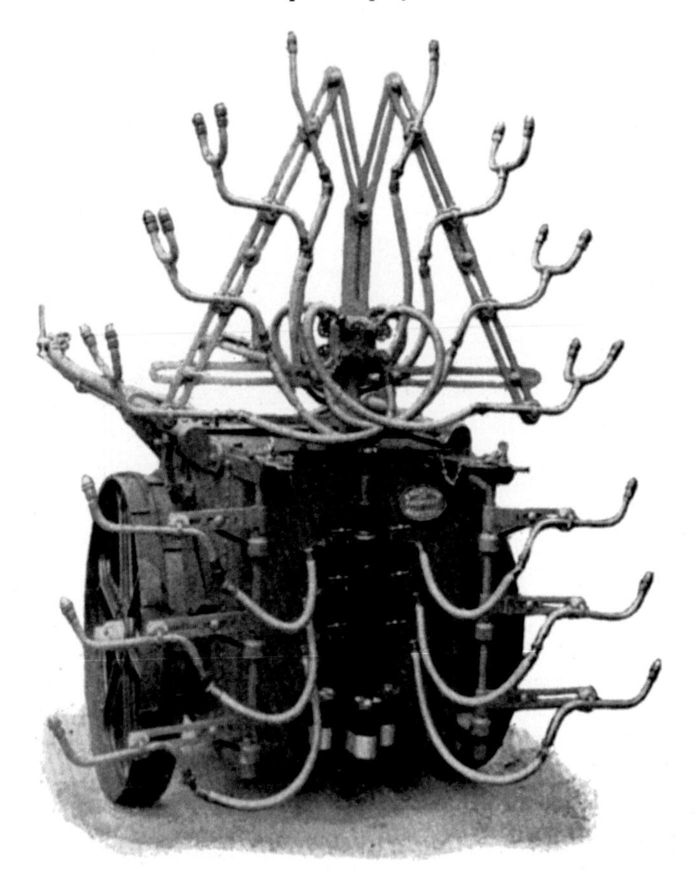

Fig. 271.

a great economy of time and labour and an enormous increase in efficiency—every practical foreman knows how the pressure drops on hand-operated sprayers when his back is turned. The time factor however is of the *greatest* importance—when bad attacks of pests occur every minute is golden, and this is when the power plant scores. Another great advantage of power plant is the very high pressure that can be maintained—up to 200 lbs. per square inch

Fig. 272.

and over. This ensures the spray reaching the highest trees and produces, with correct nozzle adjustment, the mistiest of sprays. Excellent power sprayers are procurable in this country, and American machines can also be obtained.

The grower is, however, advised to make quite sure that spare parts can be readily obtained for these latter before purchasing. Nothing is more annoying, and in fact often disastrous, than to find an essential part of the sprayer requires replacement at the last minute, when such parts are not obtainable without considerable delay.

The figures show a few standard types of sprayers suitable for all sizes of plantations. The capacity ranges from 4 to 12 gallons of spray fluid per minute.

Fig. 273[1].

A. Small Power.

These are designed tor the small grower who requires an inexpensive type as an alternative to hand labour. In the present high labour costs, they should prove a very attractive alternativ to the latter.

[1] Manufactured by Messrs Weeks, Maidstone.

I. Fig. 274 shows a low-priced, but thoroughly efficient, form of this type. It is practically an engine-operated hand sprayer of the barrow type, but the pump is superior to those supplied on such sprayers. The engine is $1\frac{1}{2}$ to 2 horse-power and will easily supply two nozzles. It is very portable, can be moved about by hand, and

Fig. 274[1].

has all the advantages of the hand type. It would undoubtedly be of great advantage to very many small growers to use a power sprayer of this type. It is also useful for small isolated plantations or orchards.

The present price of the sprayer complete is about £71. 10*s*.

[1] Manufactured by Messrs Weeks, Maidstone.

2. The figure (275) shows a small portable type, supplied by an American firm[1]. It may readily be conveyed through the plantations on a hand-cart. The engine is a first-rate make of 1½ H.P., water cooled. It is coupled to the pump by means of toothed gear wheels, and the pump is of the duplex type operated by a "walking beam" as shown in fig. 262. The tank is a 50 gallon barrel with large filling hopper, and is provided with a mechanical paddle. The frame is of channel steel, and the mechanical parts of first-rate workmanship throughout. Approximate price $295 (at normal rate of exchange=about £61).

Fig. 275.

B. MEDIUM POWER.

(*a*) The large figure (276[2]) shows a typical medium powered plant, priced complete at about £125. It is of stout build and workmanship, has 2-throw high pressure ram pumps and is mounted on a steel frame with an iron tank. It is capable of supplying 4 to 8 nozzles at high pressure, being supplied with a 3 H.P. (or 1½ H.P.) petrol or oil, water-cooled engine. This is probably the most suitable type for the average grower. It can also be used as a stationary engine to pump the spray fluid through the main pipes and branches.

(*b*) The American type[3] (fig. 277) is also a most satisfactory sprayer

[1] Manufactured by the Hayes Pump and Planter Co., Galva, Illinois, U.S.A.

[2] Manufactured by Messrs Drake and Fletcher, Maidstone.

[3] Manufactured by the Deming Co., Salem, U.S.A.

Fig. 276.

for the average sized plantation. A platform for spraying large trees is provided over the tank and engine and a very efficient mechanical agitator is fitted. The pump is a duplex type similar to that shown on fig. 278, and is coupled by belt with the engine (2 horse-power). It will maintain a pressure of 250 lbs. per square

Fig. 277.

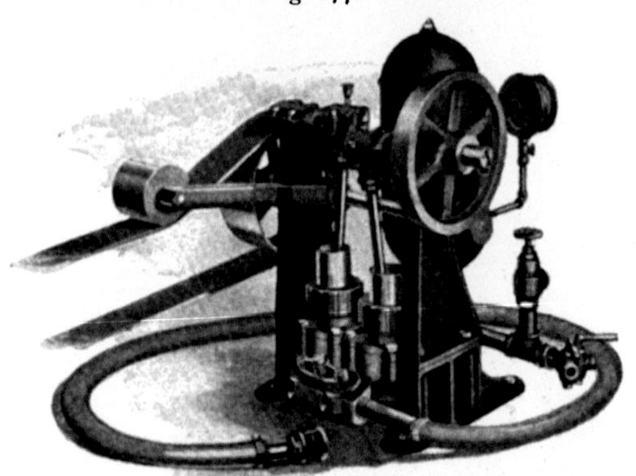

Fig. 278.

inch, and will supply two nozzles with large hole discs and four with small discs. Price approximately $300 (at normal rate of exchange = about £65).

(*c*) Another American medium powered sprayer is shown in fig. 279[1]. It is fitted with a 2-throw pump and Fairbanks-Morse 1½ H.P. water-cooled kerosene engine.

[1] Manufactured by the Hayes Pump and Planter Co., Galva, Illinois.

This is coupled to the pump by means of toothed gear wheels and the whole forms a very well-designed and compact outfit. The tank is of 100 or 150 gallons capacity, and the top acts as a platform for spraying. It is mechanically agitated and is guaranteed

Fig. 279.

for a pressure of 300 lbs. per square inch. It delivers $3\frac{1}{2}$ gallons per minute. Price approximately $450 (at normal rate of exchange = about £94).

C. High Power.

The high powered plant, suitable for large acreages of fruit are generally worked in connection with a permanent system of iron

Fig. 280. Spraying standard trees with a medium power petrol-driven outfit.

Fig. 281.

piping, laid so as to supply large areas of orchards from a common centre. This is undoubtedly the most efficient and economical method. The engine and pump is then placed in the most convenient situation for the water supply and for mixing the wash, and this is pumped through the permanent mains to wherever it is required to be sprayed. The alternative method is to move the machine along the main roads of the estate and employ a system of portable iron and indiarubber piping to reach down the rows of fruit. This is certainly a less efficient method, and has the great drawback that the plant has to be kept fed with water, but it has

Fig. 282[1].

to be adopted where a permanent pipe system is not installed. The following are three types of high power plant:

1. This plant (fig. 281[1]) has a maximum capacity of 1200—1900 gallons of liquid per hour. It is fitted with 5 H.P. or larger engine, and a set of 3-throw plunger pumps of gun metal, with a separate tank mounted on a carriage for the spraying mixture (if desired). It is capable of supplying 12 to 16 spraying nozzles at one time. The engine in this case is a Blackstone. Price of the 5 H.P. is approximately £308, and of the 3 H.P. £239.
2. The figure (282[2]) shows a similar type of plant of which the entire

[1] Messrs Weeks, Maidstone.
[2] Messrs Drake and Fletcher, Maidstone.

machine is manufactured by the same firm. Any other make of oil engine may be supplied if desired. The power ranges from 3—12 H. P. and the pumps are of the 3-throw plunger type[1], working at high pressure. The plant is adequate for the largest fruit acreages. The price of the 5 H. P. is approximately £275.

3. The figure (283) shows a powerful American type[2]. It is mounted on a carriage with a 200 gallon tank, and is specially suitable when very tall trees have to be dealt with. The pumps are capable of delivering at a pressure of 300 lbs. per square inch. The engine

Fig. 283.

uses gasoline as fuel, and is of $4\frac{1}{2}$ H. P. (a larger type has a 10 H. P. engine). The cover over the engine acts as a platform for spraying, and the working parts can be protected by waterproof blinds. The approximate price of the outfit is (without carriage) $516 and $725 (at normal rate of exchange about £108 and £151).

## 12. Dusting or blowing appliances

Like the liquid sprayers, these vary in size from the hand-operated type to the large capacity traction machines. For nurseries and

[1] See illustration of this pump, Appendix XI, p. 709.
[2] Supplied by the Field Force Pump Co., Elmira, N.Y.

small growers the hand type or, better, the drum or knapsack type are suitable (fig. 284).

For sulphuring hops, more complicated sulphurating machines are necessary. These are horse drawn, and the sulphur is mechanically distributed by means of machinery worked by gear from the wheels. (See fig. 285. Price about £32[1].)

Fig. 284.

Another type of powder distributor[2] adapted for standard fruit when dry sprays are used is that shown in fig. 286. (See also Appendix VIII, page 706.)

[1] Messrs Weeks, Maidstone.

[2] Supplied by Messrs Craven, Evesham.

Fig. 285.

Fig. 286.

# CHAPTER 45

## Spraying Methods

For small plantations, the portable hand sprayers are provided with a suitable length of delivery hose and are then pushed along the rows, so as to serve the trees on each side corresponding to the length of hose.

Although most of the power sprayers are mounted on a carriage of some sort, and therefore portable, it is, as previously mentioned (page 656), of great advantage to have steel mains running along each orchard from end to end.

At suitable points cocks (taps) are placed to which the lengths of hose may be attached for spraying trees in that section. A safety valve must be provided so that no damage occurs if all the cocks happen to be shut off.

When this system is used, the engine stands on the main cart track, and can be served with material very conveniently. It is usual to arrange so that the machine pumps its own water from a ditch or river if possible.

### Lances

These are rigid lengths of pipe made in bamboo or thin brass tube, and may be any length (or screwed in sections) so as to reach the highest trees.

### Hose

This should be only the *best 4-ply* so that no risk of bursts at inconvenient times may be incurred.

### Connections

A variety of connections made of brass are available, which may be employed to suit various purposes.

### Strainers

Fine brass wire strainers should always be employed on the tanks. There is nothing so annoying as stopped-up nozzles and this is bound to happen with many materials, especially those containing lime, if not carefully strained.

## Applying the Spray

A good deal of experience is necessary to apply the spray in an efficient and yet economical manner. Plenty of spray may be given with CONTACT WASHES and it is not of great importance, except from an economical point of view, if the spray is in excess and drips off the leaf. When spraying with Bordeaux, arsenate of lead and lime-sulphur, in summer, however, it is essential that no excess is sprayed and that the trees receive an even coat of the material, but no dripping off the leaves occurs. A very fine adjustment of the nozzle must be obtained and an actual *fog* produced as near as is possible. Also the operator must not stand too near the tree. It is well to thoroughly educate a few men on the right manner of spraying, and form a standing gang of these, so that they grow in experience each year. A strict check on the amount of material used should be kept and orchards inspected thoroughly after each spraying.

## Points to Remember

That in the case of a soap spray, the *soap must be in excess* (seen by the "head of lather" produced) otherwise the liquid will be sticky and will cause trouble with the valves.

That arsenates, Bordeaux and lime sprays must be kept constantly stirred.

That it pays to use only thoroughly efficient materials for spraying, even if the cost is considerable.

That delays are dangerous.

That the sooner treatment is applied, the less damage is done by the pest.

That the annual overhaul of the spraying plant previous to spraying in the spring should not be left till the last minute, so that, if spare parts are required, there is a margin of time for delivery before spraying becomes urgent.

That plant should be thoroughly cleansed and oiled before being put under cover for the slack periods.

The too much force in spraying hops, especially during the last period when the "pin" is forming, will result in damage. Powerful insecticides should also be reduced in percentage in the wash, and too large an excess of soap should also be avoided during this period.

# CHAPTER 46

## Combined Sprays

In cases where there is a great shortage of labour it is often preferable to use a **combined wash** for two or three kinds of pests rather than to only wash once for one of them. This is, however, with a few exceptions by no means an advisable practice for the following reasons:

(*a*) Each type of remedy generally requires application to the plant in a *special manner* and in particular with a varying adjustment of the spraying nozzle.

(*b*) Each pest has its own *particular period* when it is most open to treatment.

(*c*) Many of the substances used for spraying *react with one another chemically*, whereby the efficiency of one or both may be reduced, or risk of injury to the leaves may be incurred.

(*a*) Thus, arsenate of lead, and other caterpillar sprays, require a VERY FINE MISTY SPRAY. Nicotine, on the contrary, requires, especially for caterpillar, a *relatively coarse*, penetrating spray, provided by a coarse adjustment of the nozzle. Further, it is nothing like so efficient in its action when used apart from soap, while the latter should not be used with arsenate of lead, as, not only is there danger of scorching[1], but it will be found that the arsenate will tend to run off the leaves instead of adhering to them. The same remarks apply to Bordeaux used with nicotine.
In the case of the use of Bordeaux with arsenate of lead, this objection does not apply.

(*b*) An instance of this is in the case of apple aphis, and psylla (sucker). These should be sprayed before they enter the buds, while they are still exposed and just after hatching (see fig. 193). This is, of course, *much too early* to spray for caterpillar, so that a combined arsenate-nicotine wash will waste either the nicotine or the arsenate if applied for both caterpillar and aphis.

[1] See remarks on page 665.

In the case of nicotine used with lime-sulphur, the wash is usually applied too early for the nicotine to have any effect (winter strength) or too late to be of any use (summer strength).

The use of a combined wash of arsenate of lead and lime-sulphur is also inadvisable, as besides any chemical action which may be set up, the lime-sulphur is much less efficient at the weak strength

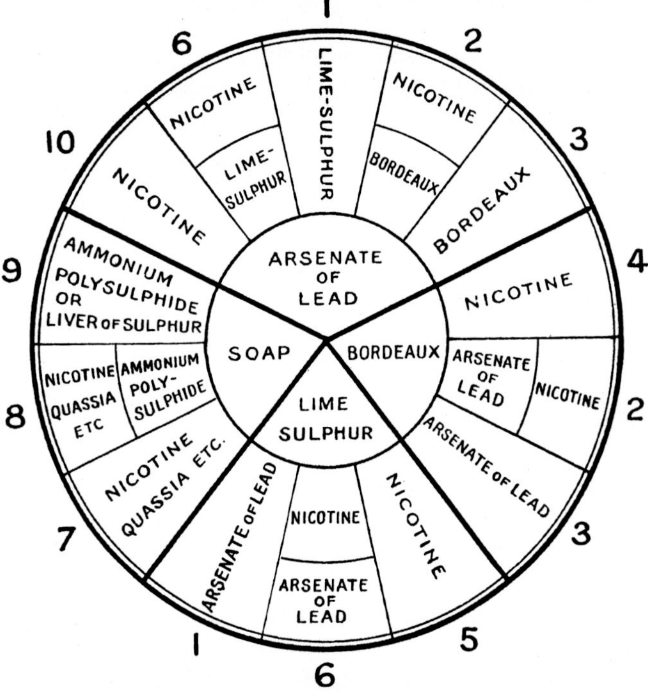


Fig. 287. Diagram showing materials which can be used together as a combined spray without undergoing chemical change.

at which it must be used when the trees are best open to deal with a caterpillar attack.

(*c*) Many growers have tried using soap with various other sprays, but in most cases the soap is either curded out altogether or *chemical action is set up* which results in severe leaf scorching. Thus lime-sulphur and Bordeaux destroy the soap, the former completely, the latter partially, while, although some samples of arsenate of lead do not curd out the soap—these are free from excess of lead—the soluble alkali salt of the arsenate is liable to be formed which is certain to cause injury to the plants. (See below.) The attached diagram (fig. 287) shows the combined sprays that it is possible to use without one or other of the constituents being altered or destroyed. The following brief comments may be useful:

1. **Arsenate of lead—lime-sulphur**

   See remarks in § *b* above. In America better results have been obtained by using calcium arsenate in place of the lead salt in this mixture.

2. **Arsenate of lead—Bordeaux—nicotine**

   The nicotine is much less powerful when used apart from soap (see § *a* above). This should, nevertheless, be a good combination to spray if used at exactly the right time.

3. **Arsenate of lead—Bordeaux**

   As far as is known this is an entirely unobjectionable combination and may be very successfully employed when apple scab is present as well as caterpillar.

4. **Bordeaux—nicotine**

   The same remarks apply as in the case of 2 above.

5. **Lime-sulphur—nicotine**

   Remarks as for 2 and 4 above.

6. **Lime-sulphur—nicotine—arsenate of lead**

   Remarks as for 1 and 2 above.

7. **Soap—nicotine; soap—quassia, etc.**

   These are normal mixtures and not really "combined" sprays.

8. **Soap—ammonium polysulphide—nicotine, quassia, etc.**

   This should form a very excellent combination for fungus (especially mildew) spraying and sucking insects. It must be remembered, however, that "A.P.S." is still in its experimental stages, and should be used cautiously as it has marked scorching tendencies on young foliage.

9. **Soap—"A.P.S."; soap—liver of sulphur**

   Normal mixtures.

**10. Arsenate of lead—nicotine**

Much employed by growers as an early spring wash. Has a good deal to be said for it as a labour-saving combination, but, as before stated, the nicotine is much less powerful than if used by itself with soft soap. Further, the nicotine should be applied earlier in the season in the case of attack by psylla (apple sucker), see § *b* above.

As regards the advisability of using arsenate of lead with soft soap, from the standpoint of possible damage to the leaves of the tree, this is a chemical question. Statements which are apparently very conflicting have been made by various authorities from time to time. Some have advised its use and found no scorching while others have strongly condemned the combination both on the grounds stated above and on account of the resulting damage.

The author has made careful investigations on the subject and the following is a summary of the results obtained:

1. The scorching action is largely dependent on the chemical composition of the soap used. If this is free from any trace of *caustic alkali* there is little or no chemical action on the arsenate of lead and thus no scorching results. If, on the other hand, there is free caustic alkali present—which often happens in commercial soaps—there is soluble arsenic formed which produces severe leaf-damage.
2. All makes of arsenate of lead are not the same chemically. Those which contain most of the *basic* arsenate are least likely to cause scorching even with soaps which contain some free alkali.
3. The degree of chemical action depends upon the *length of time* during which the soap is allowed to act upon the arsenate of lead before the mixture is sprayed. A greater degree of damage therefore occurs the larger the spray mixture is kept before using.
4. In the case of some brands of arsenate of lead which contained soluble lead acetate, the soap was curdled and rendered useless for spraying. Such varieties of arsenate are unsuitable for the purpose.

Summarising these results, if a grower decides to use soap with his arsenate, it is necessary in order to avoid less damage to obtain a *neutral* soap (guaranteed to contain no free alkali) and to use this with arsenate of lead which the manufacturers will guarantee is suitable for the purpose. Further, it is advisable to spray the mixed wash into the trees as quickly as possible.

# SECTION X

## SPRAYING CALENDAR

# SECTION X

## SPRAYING CALENDAR

The following calendar has been carefully compiled to serve as an index to the rest of the book, and it gives the approximate spraying operations for each month of the year. Practically all the pests are dealt with, and short directions are given, *but reference in all cases should be made to the pages indicated.*

This is absolutely necessary because it is not possible to give details with the limits fixed by a table, and further, there are often special precautions to be observed—e.g. not to spray with arsenical poisons when the fruit is ripening, etc.

As the seasons are liable to considerable variation from year to year, only a general indication can be given, and a good look-out should be kept for the **first appearance of the pest.**

It cannot be too strongly emphasised, as frequently stated in other parts of the book, that it is of the utmost importance to avoid delay, and to deal with each pest before it either has time to find shelter inside the buds, or to do permanent and irreparable damage to the tree.

## JANUARY

| Pest | Stage | Treatment: Preventive | Treatment: Remedies | Time of application | Page reference: Pest | Page reference: Remedy |
|---|---|---|---|---|---|---|
| Bud Moth ... | Caterpillar ... | ... ... ... ... | Lime-sulphur ... | Any time during month | 65 | 430 |
| Buff-Tip Moth ... | Pupa ... | Poultry in orchards ... | ... ... ... | All month ... ... | 69 | |
| Clouded Drab Moth | ,, ... | ,, ,, ... | ... ... ... | ,, ,, ... ... | 84 | |
| Codling Moth ... | Caterpillar ... | ... ... ... ... | Caustic Emulsion ... | End of month ... ... | 88 | 413 |
| ,, ,, ... | ,, ... | ... ... ... ... | *or* Lime-sulphur ... | ,, ,, ... ... | 88 | 430 |
| Currant Clearwing Moth | ,, ... | Prune off attacked shoots | ... ... ... | Any time during month | 81 | |
| Dot Moth ... | Pupa ... | Poultry in orchards ... | ... ... ... | All month ... ... | 96 | |
| Ermine (small) Moth | Caterpillar ... | ... ... ... ... | Caustic Soda ... | End of month ... ... | 100 | 413 |
| ,, ,, ,, | ,, ... | ... ... ... ... | *or* Lime-sulphur ... | ,, ,, ... ... | 100 | 430 |
| Magpie Moth ... | ,, ... | ... ... ... ... | Sticky bands ... | Middle to end of month | 134 | 404 |
| ,, ,, ... | ,, ... | ... ... ... ... | Lime ... ... | ,, ,, ,, | 134 | 420 |
| ,, ,, ... | ,, ... | ... ... ... ... | *or* Soot ... ... | ,, ,, ,, | 134 | 457 |
| Peppered Moth ... | Pupa ... | Turn over ground ... | ... ... ... | End of month ... ... | 147 | |
| Nut Weevil ... | Grub ... | Cultivate soil ... | ... ... ... | Any time during month | 198 | |
| Raspberry Weevil | ,, ... | Remove rubbish ... | ... ... ... | ,, ,, ,, | 202 | |
| Red-Legged Weevil | ,, ... | ,, ,, ... | ... ... ... | ,, ,, ,, | 206 | |
| Apple Sawfly ... | Maggot ... | Remove surface soil ... | Dig in soil insecticides | Middle to end of month | 237 | 432 |
| Apple-Oat Aphis... | Egg ... ... | Destroy prunings ... | ... ... ... | Any time during month | 281 | |
| Green Apple Aphis | ,, ... ... | ,, ,, ... | ... ... ... | ,, ,, ,, | 283 | |
| Strawberry Aphis | ,, ... ... | Cut off old leaves and burn | ... ... ... | ,, ,, ,, | 325 | |
| Woolly Aphis ... | All ... ... | ... ... ... ... | Caustic Emulsion ... | ,, ,, ,, | 327 | 413 |
| ,, ,, ... | ,, ... ... | ... ... ... ... | Inject Carbon Bisulphide | ,, ,, ,, | 327 | 410 |

## JANUARY (*continued*)

| Pest | Stage | Treatment: Preventive | Treatment: Remedies | Time of application | Page reference: Pest | Page reference: Remedy |
|---|---|---|---|---|---|---|
| Brown Scale ... | Louse ... | ... ... ... ... | Lime-sulphur ... | Any time during month | 361 | 430 |
| Mussel Scale ... | Egg ... ... | ... ... ... ... | Caustic Soda ... | ,, ,, ,, | 355 | 413 |
| ,, ,, ... | ,, ... ... | ... ... ... ... | *or* Lime-sulphur ... | ,, ,, ,, | 355 | 430 |
| ,, ,, ... | ,, ... ... | ... ... ... ... | *or* Paraffin ... | ,, ,, ,, | 355 | 441 |
| Apple Blossom Wilt | Pustules ... | Cut off cankered spurs | ... ... ... | ,, ,, ,, | 494 | |
| ,, Scab ... | Winter (ascigerous) | Cut off diseased shoots | ... ... ... | Toward end of month | 510 | |
| Brown Rot ... | On fruit ... | Remove mummified fruit | ... ... ... | First two weeks ... | 496 | |
| Cherry Leaf Curl | Mycelium ... | Prune off affected branches | ... ... ... | Before last week ... | 521 | |
| ,, ,, Scorch | Winter (ascigerous) | Remove dead leaves ... | ... ... ... | Any time during month | 523 | |
| Cherry Witches'-Brooms | Mycelium ... | Cut out brooms ... | ... ... ... | ,, ,, ,, | 528 | |

## FEBRUARY

| Pest | Stage | Treatment | | Time of application | Page reference | |
|---|---|---|---|---|---|---|
| | | Preventive | Remedies | | Pest | Remedy |
| Bud Moth ... ... | Caterpillar | ... ... ... ... | Lime-sulphur ... | Any time during month | 65 | 430 |
| Buff-Tip Moth ... | Pupa ... | Poultry in orchards ... | ... ... ... | All month ... ... | 69 | |
| Case Bearer Moth ... | Caterpillar | ... ... ... ... | Lime-sulphur ... | Any part of month ... | 74 | 430 |
| Clearwing (currant) Moth | ,, | Prune off affected shoots | ... ... ... | ,, ,, ... | 81 | |
| Codling Moth ... ... | ,, | ... ... ... ... | Lime-sulphur ... | ,, ,, ... | 88 | 430 |
| Ermine (small) Moth ... | ,, | ... ... ... ... | Caustic Soda ... | ,, ,, ... | 100 | 413 |
| ,, ,, ,, ... | ,, | ... ... ... ... | *or* Lime-sulphur ... | ,, ,, ... | 100 | 430 |
| Figure of Eight Moth | Eggs ... | Remove shoots with eggs | ... ... ... | ,, ,, ... | 105 | |
| Magpie Moth ... ... | Caterpillar | ... ... ... ... | Sticky bands ... | ,, ,, ... | 134 | 404 |
| ,, ,, ... ... | ,, | ... ... ... ... | Lime ... ... | ,, ,, ... | 134 | 420 |
| ,, ,, ... ... | ,, | ... ... ... ... | *or* Soot ... ... | ,, ,, ... | 134 | 457 |
| Peppered Moth ... | Pupa ... | Turn over ground ... | ... ... ... | ,, ,, ... | 147 | |
| Apple Sawfly ... ... | Maggot | ... ... ... ... | Dig in soil insecticides | Up to middle of month | 237 | 432 |
| Pear Midge ... ... | Pupa ... | Poultry in orchards ... | ... ... ... | Any time during month | 263 | |
| Currant-Lettuce Aphis | Eggs ... | Prune and destroy prunings | ... ... ... | ,, ,, ,, | 302 | |
| Raspberry Aphis ... | ,, | Destroy winter prunings | ... ... ... | ,, ,, ,, | 321 | |
| Strawberry Aphis ... | ,, | Cut off old leaves and burn | ... ... ... | ,, ,, ,, | 325 | |
| Brown Scale ... ... | Louse ... | ... ... ... ... | Lime-sulphur ... | ,, ,, ,, | 361 | 430 |
| Mussel Scale ... ... | ,, | ... ... ... ... | Caustic Soda ... | ,, ,, ,, | 355 | 413 |
| ,, ,, ... ... | ,, | ... ... ... ... | *or* Lime sulphur ... | ,, ,, ,, | 355 | 430 |
| ,, ,, ... ... | ,, | ... ... ... ... | *or* Paraffin ... | ,, ,, ,, | 355 | 441 |
| Gooseberry Red Spider | Various ... | ... ... ... ... | Paraffin ... ... | ,, ,, ,, | 377 | 441 |
| ,, ,, ,, | ,, | ... ... ... ... | and Soft Soap ... | ,, ,, ,, | 377 | 452 |
| ,, ,, ,, | ,, | ... ... ... ... | *or* Liver of Sulphur ... | ,, ,, ,, | 377 | 431 |
| ,, ,, ,, | ,, | ... ... ... ... | *or* Nicotine ... | ,, ,, ,, | 377 | 436 |
| Apple Canker ... ... | Mycelium | ... ... ... ... | Lime-sulphur ... | ,, ,, ,, | 501 | 430 |
| Peach Leaf Curl ... | ... ... | ... ... ... ... | Burgundy Spray ... | Between middle and end of month | 556 | 610 |
| ,, ,, ... | ... ... | ... ... ... ... | *or* Bordeaux Spray ... | Between middle and end of month | 556 | 605 |

## MARCH

| Pest | Stage | Treatment: Preventive | Treatment: Remedies | Time of application | Page reference: Pest | Page reference: Remedy |
|---|---|---|---|---|---|---|
| Bud Moth ... ... | Caterpillar... | ... ... ... | Arsenate of Lead ... | End of month ... | 65 | 397 |
| Buff-Tip Moth ... | Pupa ... | Poultry in orchards | ... ... ... | All month ... ... | 69 | |
| Gold-Tail Moth ... | Caterpillar... | ... ... ... | Lime-sulphur ... | 2nd or 3rd week ... | 112 | 430 |
| Lappet Moth ... ... | ,, ... | Hand pick caterpillars | Arsenate of Lead ... | Any time during month | 127 | 397 |
| March Moth ... ... | Egg ... | Destroy egg bands... | ... ... ... | ,, ,, ,, | 139 | |
| Raspberry Moth ... | Caterpillar... | ... ... ... | Arsenate of Lead ... | Twice during month | 157 | 397 |
| Winter Moth ... ... | Egg ... | ... ... ... | Lime salt wash ... | Middle of month ... | 176 | 430 |
| Apple-Blossom Weevil | Beetle ... | ... ... ... | Lime-sulphur ... | End of month ... | 190 | 430 |
| ,, ,, ,, | ,, ... | ... ... ... | *or* Lime ... ... | ,, ,, ... | 190 | 428 |
| ,, ,, ,, | ,, ... | ... ... ... | *or* Nicotine Soap ... | ,, ,, ... | 190 | 436 |
| Pear Midge ... ... | Fly ... | ... ... ... | Nicotine Soap ... | ,, ,, ... | 263 | 436 |
| Blue Apple Aphis ... | Egg ... | ... ... ... | Lime on shoots ... | Beginning of month | 276 | 428 |
| ,, ,, ... | Lice ... | ... ... ... | *or* Nicotine Soap ... | End of month ... | 276 | 436 |
| Plum Leaf-Curling Aphis | Stem-Mother | ... ... ... | Nicotine Soap ... | Middle of month ... | 313 | 436 |
| ,, ,, ,, | ,, ,, | ... ... ... | *or* Lime ... ... | ,, ,, ... | 313 | 428 |
| Big Bud Mite ... ... | Mites ... | ... ... ... | Carbolic and Soap ... | Beginning of month | 367 | 410 |
| Leaf Blister Mite ... | ,, | ... ... ... | Lime-sulphur ... | End of month ... | 372 | 430 |
| Gooseberry Red Spider | Spider ... | ... ... ... | Nicotine ... ... | Beginning of month | 377 | 436 |
| ,, ,, ,, | ,, ... | ... ... ... | and Liver of Sulphur ... | ,, ,, | 377 | 431 |
| ,, ,, ,, | ,, ... | ... ... ... | and Paraffin ... | ,, ,, | 377 | 441 |
| Apple Blossom Wilt ... | Pustules ... | ... ... ... | Ammonium Polysulphide | Middle of month ... | 494 | 601 |
| ,, Brown Rot ... | On fruit ... | ... ... ... | Lime-sulphur ... | Beginning of month | 497 | 612 |
| ,, Mildew... ... | Mycelium ... | ... ... ... | ,, ,, ... | Middle of month ... | 507 | 612 |
| ,, ,, ... ... | ,, ... | ... ... ... | *or* Bordeaux ... | ,, ,, ... | 507 | 605 |
| ,, Scab ... ... | Various ... | ... ... ... | Full strength Lime-sulphur | Early in month ... | 510 | 612 |
| ,, ,, ... ... | ,, ... | ... ... ... | *or* Bordeaux ... | ,, ,, ... | 510 | 605 |
| Cherry Witches'-Broom | Spores ... | ... ... ... | Bordeaux ... ... | ,, ,, ... | 528 | 605 |
| Pear Leaf Spot ... | Conidial ... | ... ... ... | Lime-sulphur ... | End of month ... | 561 | 612 |
| ,, ,, | ,, ... | ... ... ... | *or* Bordeaux ... | ,, ,, ... | 561 | 605 |
| Pear Scab ... ... | Various ... | ... ... ... | Full strength Lime-sulphur | Early in month ... | 564 | 612 |
| Plum Scab ... ... | ,, ... | ... ... ... | ,, ,, ,, | ,, ,, ... | 569 | 612 |

## APRIL

| Pest | Stage | Treatment | | Time of application | Page reference | |
|---|---|---|---|---|---|---|
| | | Preventive | Remedies | | Pest | Remedy |
| Blister Moth ... | Egg ... ... | ... ... ... ... | Arsenate of Lead ... | Middle to end of month | 58 | 397 |
| Brown-Tail Moth | Caterpillar | Destroy tents ... ... | ... ... ... | Any time during month | 61 | |
| Clouded Drab Moth | ,, ... | ... ... ... ... | Arsenate of Lead ... | ,, ,, ,, | 84 | 397 |
| Codling Moth ... | ,, ... | ... ... ... ... | ,, ,, ,, ... | Directly blossoms fall | 88 | 397 |
| Figure of Eight Moth | ,, ... | ... ... ... ... | ,, ,, ,, ... | Any time during month | 105 | 397 |
| Lackey Moth ... | ,, ... | Destroy tents ... ... | ... ... ... | ,, ,, ,, | 121 | |
| Lappet Moth ... | ,, ... | Hand pick larvæ ... | Arsenate of Lead ... | Till middle of month ... | 127 | 397 |
| Leaf Miner Moth ... | ,, ... | Hand pick infested leaves | ... ... ... | Any time during month | 132 | |
| March Moth ... | ,, ... | ... ... ... ... | Arsenate of Lead ... | End of month ... | 139 | 397 |
| Mottled Umber Moth | ,, ... | ... ... ... ... | ,, ,, ,, ... | Any time during month | 143 | 397 |
| Tortrix Moth ... | ,, ... | ... ... ... ... | ,, ,, ,, ... | Early in month ... | 166 | 397 |
| Vapourer Moth ... | ,, ... | ... ... ... ... | ,, ,, ,, ... | Any time during month | 172 | 397 |
| Winter Moth ... | ,, ... | ... ... ... ... | ,, ,, ,, ... | Early in month ... | 176 | 397 |
| Twig Cutting Weevil | Beetle ... | Jarring on to tarred boards | ,, ,, ,, ... | Middle of month ... | 207 | 397 |
| Raspberry Beetle ... | ,, ... | ... ... ... ... | ,, ,, ,, ... | End of month ... ... | 214 | 397 |
| Gooseberry Sawfly ... | Grub ... | ... ... ... ... | ,, ,, ,, ... | ,, ,, ... ... | 242 | 397 |
| ,, ,, ,, ... | ,, ... ... | ... ... ... ... | *or* Pyrethrum ... | ,, ,, ... ... | 242 | 444 |
| ,, ,, ,, ... | ,, ... ... | ... ... ... ... | *or* Hellebore ... | ,, ,, ... ... | 242 | 421 |
| Plum Fruit Sawfly ... | Fly ... ... | ... ... ... ... | Arsenate of Lead ... | ,, ,, ... ... | 253 | 397 |
| Pear Midge ... ... | Maggot ... | ... ... ... ... | If bad destroy blossoms with Paris Green | Beginning of month ... | 263 | 442 |
| Blue Apple Aphis ... | Lice ... ... | ... ... ... ... | Nicotine and Soap | ,, ,, ... | 276 | 436 |
| Green Apple Aphis | ,, ... ... | ... ... ... ... | ,, ,, ,, ... | Regularly during month | 283 | 436 |
| Currant-Lettuce Aphis | Stem-Mother and Lice | ... ... ... ... | Soft Soap ... | End of month ... ... | 302 | 452 |
| ,, ,, ,, | ,, ,, ... | ... ... ... ... | and Nicotine ... | ,, ,, ... ... | 302 | 436 |
| ,, ,, ,, | ,, ,, ... | ... ... ... ... | and Paraffin ... | ,, ,, ... ... | 302 | 441 |

## APRIL (*continued*)

| Pest | Stage | Treatment | | Time of application | Page reference | |
|---|---|---|---|---|---|---|
| | | Preventive | Remedies | | Pest | Remedy |
| Strawberry Aphis ... | Lice ... | ... ... ... ... | Soap and Nicotine | Beginning of month ... | 325 | 436 |
| Apple Sucker ... | Nits ... | ... ... ... ... | Lime coat buds ... | Middle of month ... | 337 | 428 |
| Capsid Bug ... ... | Louse ... | ... ... ... ... | Thick Lime wash ... | End of month ... ... | 342 | 428 |
| Leaf Hoppers ... | Adults ... | Knocking off on to tarred boards | Soft Soap and Nicotine | Middle to end of month | 350 | 436 |
| ,, ,, ... ... | ,, ... | ... ... ... ... | and Paraffin Emulsion | ,, ,, ,, | 350 | 441 |
| Big Bud Mite ... | Mite ... | Pick off swollen buds | ... ... ... | End of month ... ... | 367 | |
| Leaf Blister Mite ... | ,, ... | Destroy affected leaves | ... ... ... | Early in month ... | 372 | |
| Apple Bitter Rot ... | Conidial | ... ... ... ... | Bordeaux spray | Every two weeks from middle to end of month | 491 | 605 |
| Apple Blossom Wilt | On flowers ... | Remove wilting trusses | ... ... ... | End of month ... ... | 494 | |
| Apple Mildew ... | Conidial ... | Cut away diseased tips | ... ... ... | Middle of month ... | 507 | |
| Cherry Leaf Scorch | ,, ... | ... ... ... ... | Bordeaux spray ... | Early in month ... | 523 | 605 |
| Currant Leaf Spot ... | Acigerous ... | ... ... ... ... | ,, ,, ... | ,, ,, ... ... | 533 | 605 |
| ,, ,, ,, ... | ,, ... | ... ... ... ... | *or* Dilute Lime-sulphur | ,, ,, ... ... | 533 | 612 |
| ,, ,, ,, ... | ,, ... | ... ... ... ... | *or* Ammonium Polysul- | ,, ,, ... ... | 533 | 601 |
| ,, ,, ,, ... | ,, ... | ... ... ... ... | phide and Soap ... | ,, ,, ... ... | 533 | 452 |
| Die-Back of Gooseberry | ... ... | ... ... ... ... | Copper Sulphate ... | ,, ,, ... ... | 547 | 611 |
| Plum Leaf Rust ... | ... ... | ... ... ... ... | Lime-sulphur *or* Bordeaux ... | ,, ,, ... | 567 | 612 / 605 |

## MAY

| Pest | Stage | Treatment | | Time of application | Page reference | |
|---|---|---|---|---|---|---|
| | | Preventive | Remedies | | Pest | Remedy |
| Clearwing (Apple) Moth | Pupa ... | Apply sticky composition to bark | ... ... ... | Beginning of month ... | 78 | |
| Ermine (small) Moth ... | Caterpillar | Cut out and burn nests | Arsenate of Lead ... | Beginning to middle of month | 100 | 397 |
| Goat Moth ... ... | ,, ... | ... ... ... ... | Place lump of Cyanide of Potash in hole and seal | Any time during month | 108 | 423 |
| Gold-Tail Moth ... | ,, ... | ... ... ... ... | Arsenate of Lead ... | ,, ,, ,, | 112 | 397 |
| Lackey Moth ... ... | ,, ... | Destroy tents ... ... | ... ... ... | ,, ,, ,, | 121 | |
| Magpie Moth ... ... | ,, ... | ... ... ... ... | Arsenate of Lead ... | ,, ,, ,, | 134 | 397 |
| Pith Moth ... ... | ,, ... | Hand pick dead shoots | ... ... ... | End of month ... ... | 151 | |
| Tortoiseshell Butterfly | Egg ... | Destroy egg bands ... | ... ... ... | Beginning of month ... | 163 | |
| Apple-Blossom Weevil | Grub *or* Pupa | Shake off and burn withered buds | ... .. ... | ,, ,, ,, ... | 190 | |
| Leaf Weevils ... ... | Beetle ... | Jar beetles off at night on to sticky boards | ... ... ... | Middle of month ... | 196 | 404 |
| Nut Weevil ... ... | ,, ... | ... ... ... ... | Arsenate of Lead ... | Early in month ... | 198 | 397 |
| Raspberry Weevil ... | ,, ... | Jar beetles off at night on to sticky boards | ... ... ... | Middle of month ... | 202 | 404 |
| Red-Legged Weevil ... | ,, ... | ,, ,, ,, ,, | ... ... ... | ,, ,, ... | 206 | 404 |
| Twig Cutting Weevil ... | ,, .. | Jar off beetles on dull days on to tarred boards | ... ... ... | Beginning of month ... | 207 | 417 |
| Ground Beetles ... | ,, ... | Trap in jam jars, sunk in ground and covered with straw | ... ... ... | End of month ... ... | 214 | |
| Raspberry Beetles ... | ,, ... | Jar off on dull days ... | ... ... ... | Middle of month ... | 215 | |
| Cockchafer ... ... | ,, ... | ,, ,, ,, ... | Arsenate of Lead ... | Beginning of month ... | 222 | 397 |
| Rose Chafer ... ... | ,, ... | ,, ,, ,, ... | ,, ,, ,, ... | ,, ,, ,, ... | 226 | 397 |
| Apple Sawfly ... ... | Maggot ... | Pick and destroy attacked fruitlets | ... ... ... | Any time during month | 237 | |
| Nut Sawfly ... ... | ,, ... | ... ... ... ... | Arsenate of Lead ... | On first appearance on leaf | 250 | 397 |
| Plum Fruit Sawfly ... | ,, ... | ... ... ... ... | ,, ,, ,, ... | Beginning of month ... | 253 | 397 |

## MAY (*continued*)

| Pest | Stage | Treatment: Preventive | Treatment: Remedies | Time of application | Page reference: Pest | Page reference: Remedy |
|---|---|---|---|---|---|---|
| Plum Leaf Sawfly ... | Maggot ... | ... ... ... ... | Arsenate of Lead ... | Middle of month ... | 256 | 397 |
| Social Pear Sawfly ... | ,, ... | ... ... ... ... | ,, ,, ,, ... | ,, ,, ... | 258 | 397 |
| Pear Midge ... ... | ,, ... | Pick off infected fruit | ... ... ... | Beginning of month ... | 263 | |
| Black Cherry Aphis ... | Lice ... | ... ... ... ... | Soft Soap and Insecticide | Any time during month | 290 | 452 |
| Woolly Aphis ... ... | ,, ... | ... ... ... ... | Nicotine Wash ... | Early in month ... | 327 | 436 |
| Apple Sucker ... ... | Louse ... | ... ... ... ... | ,, ,, ... | ,, ,, ... | 337 | 436 |
| Capsid Bug ... ... | Bug ... | ... ... ... ... | ,, ,, ... | End of month ... ... | 342 | 436 |
| Apple Bitter Rot ... | Conidial ... | ... ... ... ... | Bordeaux Spray ... | Every two weeks ... | 491 | 605 |
| ,, Brown Rot ... | On fruit ... | Remove infected fruit | ... ... ... | Middle of month ... | 497 | |
| ,, Leaf Spot ... | Conidial ... | ... ... ... ... | Summer Strength Bordeaux | Early in month ... | 505 | 605 |
| ,, Mildew... ... | ,, ... | ... ... ... ... | Soft Soap and Ammonium Polysulphide | Middle of month ... | 507 | 601 |
| ,, Scab ... ... | ,, ... | ... ... ... ... | Summer Strength Bordeaux | When leaves show fungus | 510 | 605 |
| ,, ,, ... ... | ,, ... | ... ... ... ... | Summer Strength Lime-sulphur | ,, ,, ,, ,, | 510 | 612 |
| Cherry Leaf Curl ... | Spores ... | ... ... ... ... | Bordeaux Mixture ... | Middle of month ... | 521 | 605 |
| ,, Scab ... ... | ... ... | ... ... ... ... | As for Apple Scab ... | ... ... ... ... | 526 | 605 |
| ,, Witches'-Broom | Spores ... | ... ... ... ... | Bordeaux Mixture ... | Early in month ... | 528 | 605 |
| Gooseberry Black Knot | Fruit ... | Remove wilting branches | ... .. ... | Any time during month | 545 | |
| Pear Leaf Spot ... | Conidial ... | ... ... ... ... | {Dilute Bordeaux<br>{,, Lime-sulphur | Middle of month}<br>,, ,, } ... | 561 | {605<br>{612 |
| Pear Scab ... ... | ... ... | ... ... ... ... | See Apple Scab ... | ... ... ... ... | 564 | |
| Plum Leaf Rust ... | ... ... | ... ... ... ... | {Lime-sulphur}<br>{Bordeaux} ... | End of month ... ... | 567 | {612<br>{605 |
| Raspberry Rust ... | ... ... | ... ... ... ... | Ammonium Polysulphide | Any time when Rust appears | 583 | 601 |
| Strawberry Leaf Spot | ... ... | ... ... ... ... | Summer Strength Lime-sulphur | Any time during month | 587 | 612 |

# JUNE

| Pest | Stage | Treatment | | Time of application | Page reference | |
|---|---|---|---|---|---|---|
| | | Preventive | Remedies | | Pest | Remedy |
| Blister Moth ... ... | Caterpillar | Hand pick infested leaves | ... ... ... ... | Anytime during month | 58 | |
| Tortoiseshell Butterfly | ,, | Destroy young colonies | Arsenate of Lead ... | Early in month ... | 164 | 397 |
| Wood Leopard Moth ... | ,, | ... ... ... | Put in piece of Cyanide of Potash and seal | Middle of month ... | 181 | 423 |
| Apple-Blossom Weevil | Beetle ... | Jar on to tarred boards | ... ... ... ... | Beginning of month | 190 | 417 |
| Leaf Weevils ... ... | ,, ... | ... ... ... | Arsenate of Lead ... | ,, ,, | 196 | 397 |
| Nut Weevils ... ... | ,, ... | Jar off beetles ... | ... ... ... ... | ,, ,, | 198 | |
| Raspberry Weevils ... | ,, ... | ... ... ... | Arsenate of Lead ... | ,, ,, | 202 | 397 |
| Red-Legged Weevils ... | ,, ... | ... ... ... | ,, ,, ... | ,, ,, | 206 | 397 |
| Gooseberry Sawfly ... | Maggot ... | ... ... ... | Spray with Arsenate of Lead | Anytime during month | 242 | 397 |
| | | | *or* Spray with Hellebore | ... ... ... | | 421 |
| | | | *or* Spray with Pyrethrum | ... ... ... | | 444 |
| Slugworm ... ... | ,, ... | ... ... ... | Spray with Arsenate of Lead *or* Hellebore | Early in month ... | 247 | 397<br>421 |
| Plum Leaf Sawfly ... | ,, ... | Grease band and jar trees | Arsenate of Lead ... | ,, ,, ... | 256 | 404<br>397 |
| Social Pear Sawfly ... | ,, ... | Cut and destroy tents | ... ... ... ... | ,, ,, ... | 258 | |
| Pear Midge ... ... | ,, ... | Run fowls over ground | Dress ground with Kainit | ,, ,, ... | 263 | 426 |
| Leaf Blister Currant Aphis | Lice ... | ... ... ... | Nicotine Soap wash ... | ,, ,, ... | 297 | 436 |
| Hop Aphis ... ... | ,, ... | ... ... ... | Soft Soap ... ... | ,, ,, ... | 305 | 452 |

## JUNE (*continued*)

| Pest | Stage | Treatment | | Time of application | Page reference | |
|---|---|---|---|---|---|---|
| | | Preventive | Remedies | | Pest | Remedy |
| Mealy Plum Aphis ... | Lice ... | ... ... ... | Soft Soap and Nicotine ... | Early in month ... | 317 | 436 |
| Raspberry Aphis ... | ,, ... | ... ... ... | ,, ,, ,, ... | ,, ,, ... | 321 | 436 |
| Hop Red Spider ... | Spider ... | ... ... ... | Sulphuring ... ... | Middle of month ... | 375 | 452 |
| | | | Soft Soap—Nicotine ... | ... ... ... | | |
| | | | Liver of Sulphur ... | ... ... ... | | |
| Gooseberry Red Spider | ... ... | ... ... ... | A compound wash ... | Anytime during month | 377 | |
| Eelworm ... ... | ... ... | ... ... ... | Soil insecticides ... | ,, ,, ,, | 383 | 433 |
| Snails and Slugs ... | ... ... | Ducks among plants | ,, ,, ... | ,, ,, ,, | 384 | 433 |
| ,, ,, ... | ... ... | ... ... ... | Soot and Lime ... ... | ,, ,, ,, | 384 | 457 & 428 |
| Apple Leaf Spot ... | Conidial | ... ... ... | Summer strength Bordeaux | Middle of month ... | 505 | 605 |
| Cherry Leaf Scorch ... | ,, | ... ... ... | ,, ,, ,, | Early in month ... | 523 | 605 |
| Die-back of Gooseberry | ,, | ... ... ... | ,, ,, ,, | ,, ,, ... | 547 | 605 |
| American Gooseberry Mildew | ,, | ... ... ... | Lime-sulphur ... ... | Anytime during month | 539 | 612 |
| | | | *or* Ammonium Polysulphide | | | 601 |
| European Gooseberry Mildew | ,, | .. ... ... | Various, see p. 544 ... | ,, ,, ,, | 543 | 544 |
| Hop Mildew ... ... | ... ... | ... ... ... | Flowers of Sulphur ... | ,, ,, ,, | 551 | |
| Pear Leaf Spot ... | Conidial | ... ... ... | Dilute Bordeaux ... | Middle of month ... | 561 | 605 |
| | | | *or* Dilute Lime-sulphur ... | | | 612 |
| Plum Wither-Tip ... | Mycelium | Remove diseased tips | ... ... ... ... | Anytime during month | 578 | |

## JULY

| Pest | Stage | Treatment | | Time of application | Page reference | |
|---|---|---|---|---|---|---|
| | | Preventive | Remedies | | Pest | Remedy |
| Buff-Tip Moth ... | Caterpillar | ... ... ... | Arsenate of Lead ... | End of month ... | 69 | 397 |
| Codling Moth ... | ,, | Apply sack bands to trunks | *or* Grease-bands ... | Early in month ... | 88 | 404 |
| Hawk (Eyed) Moth | ,, | Hand pick caterpillars | Arsenate of Lead ... | End of month ... | 117 | 397 |
| Plum Fruit Sawfly ... | Maggot ... | Destroy infected fruit | ... ... ... | Early in month ... | 253 | |
| Mussel Scale ... | Lice ... | ... ... ... | Nicotine Soap ...<br>*or* Paraffin Emulsion | ,, ,, ... | 355 | 436<br>417 |
| Apple Blossom Wilt | On fruit ... | Remove rotten apples | ... ... ... | ,, ,, ... | 494 | |
| ,, Brown Rot ... | On spurs | Cut off infected spurs | ... ... ... | ,, ,, ... | 497 | |
| Plum Wither-Tip ... | Mycelium | Remove diseased tips | ... ... ... | Middle of month ... | 578 | |

## AUGUST

| Pest | Stage | Treatment | | Time of application | Page reference | |
|---|---|---|---|---|---|---|
| | | Preventive | Remedies | | Pest | Remedy |
| Bud Moth ... ... | Caterpillar ... | ... ... ... ... | Arsenate of Lead ... | Any time during month | 65 | 397 |
| Buff-Tip Moth ... | ,, ... | Jar off caterpillars ... | Grease band trees ... | End of month ... ... | 69 | 404 |
| Clearwing (Currant) Moth | ,, ... | Prune off attacked shoots | ... ... ... | Any time during month | 81 | |
| Dot Moth ... ... | ,, ... | Hand pick caterpillars | Arsenate of Lead ... | Early in month ... | 96 | 397 |
| Mottled Umber Moth | Pupa ... | Turn over ground under trees | ... ... ... | Any time during month | 143 | |
| Peppered Moth ... | Caterpillar ... | Hand pick caterpillars | Arsenate of Lead ... | ,, ,, ,, | 147 | 397 |
| Plum Fruit Moth ... | ,, ... | Shake trees and destroy plums which fall | ... ... ... | ,, ,, ,, | 154 | |
| Nut Weevil ... ... | Grub ... | Remove and burn fallen nuts | ... ... ... | ,, ,, ,, | 198 | |
| Rose Chafer ... | ,, ... | Hand pick grubs ... | ... ... ... | Early in month ... | 226 | |
| Leaf Hoppers ... | Louse ... | ... ... ... ... | Nicotine and Soap | Any time during month | 350 | 436 |
| Pear Leaf Spot ... | Acigerous ... | Destroy fallen leaves ... | ... ... ... | ,, ,, ,, | 561 | |

## SEPTEMBER

| Pest | Stage | Treatment | | Time of application | Page reference | |
|---|---|---|---|---|---|---|
| | | Preventive | Remedies | | Pest | Remedy |
| Case-Bearer Moth ... | Caterpillar ... | ... ... ... ... | Arsenate of Lead ... | Middle of month ... | 74 | 397 |
| Gold-Tail Moth ... | ,, ... | Place sacking around trees | ... ... ... | Early in month ... | 112 | |
| Magpie Moth ... | ,, ... | ... ... ... ... | Arsenate of Lead ... | Any time during month | 134 | 397 |
| March Moth ... | Pupa ... | Turn over ground ... | ... ... ... | ,, ,, ,, | 139 | |
| Plum Fruit Moth ... | Caterpillar ... | Place sacking around trees | *or* apply sticky bands | End of month ... ... | 154 | 404 |
| Swift Moth ... ... | ,, ... | Hoe ground ... ... | Apply soil insecticides | Middle of month ... | 160 | 433 |
| Winter Moth ... | ... ... | ... ... ... ... | Grease band trees ... | End of month ... ... | 176 | 404 |
| Social Pear Sawfly ... | Maggot ... | ... ... ... ... | Apply soil insecticides | ,, ,, ... ... | 258 | 433 |
| Black Cherry Aphis | Female and Male | ... ... ... ... | Soft Soap and Insecticide *or* Paraffin Emulsion | Middle of month ... | 290 | 452, 417 |
| Plum Leaf Curling Moth | Winged female | ... ... ... ... | Soft Soap and Insecticide | ,, ,, ... | 313 | 452 |
| Apple Bitter Rot ... | Acigerous ... | Remove mummified apples, cut out canker on branches | ... ... ... | Any time during month | 491 | |
| Apple Canker ... | ,, ... | Remove cankered branches | Lime-sulphur ... | Early in month ... | 501 | 612 |

## OCTOBER

| Pest | Stage | Treatment: Preventive | Treatment: Remedies | Time of application | Page reference: Pest | Page reference: Remedy |
|---|---|---|---|---|---|---|
| Brown-Tail Moth ... | Caterpillar ... | ... ... ... | Winter strength Lime-sulphur | End of Month ... ... | 61 | 430 |
| | | | *or* Strong Caustic Soda | ... ... ... ... | | 413 |
| Clouded Drab Moth | Pupa ... ... | Fowls in orchard ... | ... ... ... | All month and succeeding ones | 84 | |
| Codling Moth ... | Caterpillar ... | ... ... ... | Lime-sulphur ... | End of month ... ... | 89 | 430 |
| Dot Moth ... ... | Pupa ... ... | Fowls in orchard ... | ... ... ... | All month and succeeding ones ... ... ... | 96 | |
| Ermine (small) Moth | Caterpillar ... | ... ... ... | Lime-sulphur ... | End of month ... ... | 100 | 430 |
| | | | *or* Caustic Emulsion ... | | | 413 |
| Figure of Eight Moth | Moth ... ... | Use light traps ... | ... ... ... | Early in month ... ... | 105 | |
| Mottled Umber Moth | Moth ... ... | ... ... ... | Apply grease bands ... | Middle of month ... | 143 | 404 |
| Raspberry Moth ... | Caterpillar ... | Hoe ground ... | Soil insecticides ... | Early in month ... ... | 157 | 433 |
| Vapourer Moth ... | Egg ... ... | Remove egg masses | ... ... ... | ,, ,, ... ... | 172 | |
| Raspberry Weevil ... | Grub ... ... | ... ... ... | Hoe in soil insecticides | ,, ,, ... ... | 202 | 433 |
| Cockchafer ... ... | Grub ... ... | ... ... ... | ,, ,, ,, | ,, ,, ... ... | 222 | 433 |
| Gooseberry Sawfly ... | Maggot ... | ... ... ... | ,, ,, ,, | ,, ,, ... ... | 242 | 433 |
| Slugworm ... ... | ,, ... | ... ... ... | ,, ,, ,, | ,, ,, ... ... | 247 | 433 |
| Social Pear Sawfly ... | ,, ... | ... ... ... | ,, ,, ,, | ,, ,, ... ... | 258 | 433 |
| Blue Apple Aphis ... | Male & female | ... ... ... | Nicotine Soap ... | Middle of month ... | 276 | 436 |
| | | | *or* Paraffin Emulsion | | | 417 |
| Green Apple Aphis ... | Winged ... | ... ... ... | Nicotine Wash ... | Any time during month ... | 283 | 436 |
| Apple Sucker ... | ,, ... | ... ... ... | Nicotine Wash ... | ,, ,, ,, ... | 337 | 436 |
| | | | *or* Paraffin Emulsion | ... ... ... ... | | 417 |

## NOVEMBER

| Pest | Stage | Treatment | | Time of application | Page reference | |
|---|---|---|---|---|---|---|
| | | Preventive | Remedies | | Pest | Remedy |
| Blister Moth ... | Pupa ... | ... ... ... ... | Lime-sulphur ... | End of month ... ... | 58 | 430 |
| Lackey Moth ... | Eggs ... | Remove branches with eggs | ... ... ... | Any time during month | 121 | |
| Plum Fruit Moth | Caterpillar | ... ... ... ... | Lime-sulphur ... | ,, ,, ,, | 154 | 430 |
| Raspberry Beetle | Pupa ... | Prune canes and destroy cuttings | ... ... ... | ,, ,, ,, | 215 | |

## DECEMBER

| Pest | Stage | Treatment | | Time of application | Page reference | |
|---|---|---|---|---|---|---|
| | | Preventive | Remedies | | Pest | Remedy |
| Pith Moth ... | Caterpillar | Prune off attacked shoots | ... ... ... | Any time during month | 151 | |
| Raspberry Beetle | Pupa ... | Hoe in Soot or Lime ... | ... ... ... | ,, ,, ,, | 215 | 427, 428 |
| Shot-Borer Beetle | Beetle ... | Cut down and burn attacked trees | ... ... ... | ,, ,, ,, | 219 | |

# SECTION XI

## TABLES AND APPENDICES

# TABLE I

## WEIGHTS AND MEASURES IN COMMON USE

### WEIGHTS (English)

1 ounce = 437½ grains = 28·35 grammes.

1 pound (lb.) = 16 ounces (oz.) = 7000 grains = 453·6 grammes, or 0·454 kilogramme.

1 hundred weight (cwt.) = 4 quarters (qr.) = 8 stones = 112 lbs. = 50·80 kilogrammes.

1 ton = 20 cwt. = 2240 lbs. = 1016·65 kilogrammes.

### WEIGHTS (American)

As for English, but tons are usually "short" tons of 2000 lbs. = 0·892 of an English ton.

### WEIGHTS (Metric)

1 gramme (g.) = 15·432 grains.

1 kilogramme (kg.) = 1000 g. = 2·2046 lbs.

1 metrical ton = 1000 kg. = 0·9842 English ton = 1·023 American tons (of 2000 lbs.).

1 cubic foot of water = 62·5 lbs. (nearly).

1 cwt. of water occupies 1·79 cu. feet.

1 ton of water occupies 35·7 cu. feet.

### MEASURES (English)

1 fluid ounce = $\frac{1}{20}$ pint = 1·733 cu. inches = 28·35 cubic centimetres.

1 pint = 20 ozs. = 34·66 cu. inches = 567 cubic centimetres.

1 quart = ¼ gallon = 2 pints = 40 ozs. = 2½ lbs. water.

1 gallon = 4 quarts = 10 lbs. of water = 70,000 grains = 4·536 litres (kilogrammes water) = 277·28 cu. inches.

1 peck = 2 gallons = 20 lbs. water.

1 bushel = 4 pecks = 8 gallons.

1 quarter = 8 bushels = 32 pecks = 64 gallons.

### MEASURES (American)

As for English but U.S. gallon = 8·345 lbs. water = 3·7854 litres

### MEASURES (Metric)

1 litre = 1000 cubic centimetres = 0·2201 English gallon.

# TABLE II

Amounts of solid or liquid substances to be taken to obtain 1,000,000 gallons of wash of different percentage strengths.

(Divide by specific gravity to obtain liquids in measured ounces)

| Percentage Strength | In 1 gallon | | | | In 10 gallons | | | | In 100 gallons | | | |
|---|---|---|---|---|---|---|---|---|---|---|---|---|
| | ozs. | | | | ozs. | | | | ozs. | | | |
| 0·0156 | — | | | | ¼ | | | | 2½ | | | |
| 0·0313 | — | | | | ½ | | | | 5 | | | |
| 0·0625 | — | | | | 1 | | | | 10 | | | |
| 0·094 | — | | | | 1½ | | | | 15 | | | |
| 0·100 | — | | | | 1 3/5 | | | | 16 | | | |
| | | | | | | | | | lbs. | ozs. | pts. | ozs. |
| 0·125 | — | | | | 2 | | | | 1 | 4 | 1 | 0 |
| 0·157 | ¼ | | | | 2½ | | | | 1 | 9 | 1 | 5 |
| 0·188 | 3/10 | | | | 3 | | | | 1 | 14 | 1 | 10 |
| 0·20 | ⅓ | | | | 3 1/5 | | | | 2 | 0 | 1 | 12 |
| 0·25 | 2/5 | | | | 4 | | | | 2 | 8 | 2 | 0 |
| 0·31 | ½ | | | | 5 | | | | 3 | 2 | 2 | 10 |
| 0·38 | 3/5 | | | | 6 | | | | 3 | 12 | 3 | 0 |
| 0·50 | 4/5 | | | | 8 | | | | 5 | 0 | 4 | 0 |
| 0·63 | 1 | | | | 10 | | | | 6 | 4 | 5 | 0 |
| 0·75 | 1 1/5 | | | | 12 | | | | 7 | 8 | 6 | 0 |
| 0·88 | 1 2/5 | | | | 14 | | | | 8 | 12 | 7 | 0 |
| | | | | | | | | | | | galls. | pts. |
| 1·00 | 1 3/10 | | | | 16 | | | | 10 | 0 | 1 | 0 |
| | | | | | lbs. | ozs. | pts. | ozs. | | | | |
| 1·25 | 2 | | | | 1 | 4 | 1 | 0 | 12 | 8 | 1 | 2 |
| 1·50 | 2 2/5 | | | | 1 | 8 | 1 | 4 | 15 | 0 | 1 | 4 |
| 1·88 | 3 | | | | 1 | 14 | 1 | 10 | 18 | 12 | 1 | 7 |
| 2·00 | 3 1/5 | | | | 2 | 0 | 1 | 12 | 20 | 0 | 2 | 0 |
| 2·50 | 4 | | | | 2 | 8 | 2 | 0 | 25 | 0 | 2 | 4 |
| 3·00 | 4 4/5 | | | | 3 | 0 | 2 | 8 | 30 | 0 | 3 | 0 |
| 3·75 | 6 | | | | 3 | 12 | 3 | 0 | 37 | 8 | 3 | 12 |
| 5 | 8 | | | | 5 | 0 | 4 | 0 | 50 | 0 | 5 | 0 |
| 6 | 9½ | | | | 6 | 0 | 4 | 16 | 60 | 0 | 6 | 0 |
| 10 | 16 | | | | 10 | 0 | 8 | 0 | 100 | 0 | 10 | 0 |
| | lbs. | ozs. | pts. | ozs. | | | | | | | | |
| 15 | 1 | 8 | 1 | 4 | 15 | 0 | 12 | 0 | 150 | 0 | 15 | 0 |
| 20 | 2 | 0 | 1 | 12 | 20 | 0 | 16 | 0 | 200 | 0 | 20 | 0 |

# TABLE III

## TO FIND THE CAPACITY OF A TANK, ETC.

To find the capacity of a *square or oblong tank*, multiply the length by the width in feet, and the result by the height in feet, and multiply this figure by $6\frac{1}{4}$ (6·25). This gives the capacity in gallons.

If the weight is required, this is obtained in lbs. by multiplying the number of gallons by 10 (if the liquid is water or the same specific gravity) or by 10 times the specific gravity if not the same as water.

Or the capacity in gallons may be divided by 11·2 giving the result in cwts.

If the tank is *cylindrical*, the following table may be used. This shows the capacities for a tank 1 foot in length and various diameters. To obtain the capacity for other heights, multiply the figure opposite the correct diameter by the height of the vessel.

| Diameter ft. | Capacity in gallons | Weight of water | | |
|---|---|---|---|---|
| 1 | 4·9 | | | 49 lbs. |
| 2 | 19·7 | | 1 cwt. | 85 lbs. |
| 3 | 44·0 | | 3 cwts. | 104 lbs. |
| 4 | 78·1 | | 6 cwts. | 109 lbs. |
| 5 | 123 | | 10 cwts. | 110 lbs. |
| 6 | 176 | | 15 cwts. | 80 lbs. |
| 7 | 240 | 1 ton | $1\frac{1}{2}$ cwts. | |
| 8 | 328 | 1 ton | $9\frac{1}{4}$ cwts. | |
| 9 | 396 | 1 ton | $15\frac{1}{2}$ cwts. | |
| 10 | 491 | 2 tons | $2\frac{3}{4}$ cwts. | |
| 11 | 597 | 2 tons | $13\frac{1}{4}$ cwts. | |
| 12 | 709 | 3 tons | $3\frac{1}{4}$ cwts. | |

# TABLE IV

## WEIGHTS OF COMMON METALS

The following are the weights of one square foot of Metal Sheet of various thicknesses:

| Thick-ness | Wrought Iron | Cast Iron | Steel | Copper | Brass | Lead | Zinc |
|---|---|---|---|---|---|---|---|
| inch | lb. | lb. | lb. | lb. | lb. | lb. | lb. |
| $\frac{1}{16}$ | 2·53 | 2·34 | 2·55 | 2·89 | 2·73 | 3·71 | 2·34 |
| $\frac{1}{8}$ | 5·05 | 4·69 | 5·10 | 5·78 | 5·47 | 7·42 | 4·69 |
| $\frac{3}{16}$ | 7·58 | 7·03 | 7·66 | 8·67 | 8·20 | 11·13 | 7·03 |
| $\frac{1}{4}$ | 10·10 | 9·38 | 10·21 | 11·76 | 10·94 | 14·83 | 9·38 |
| $\frac{5}{16}$ | 12·63 | 11·72 | 12·76 | 14·45 | 13·67 | 18·54 | 11·72 |
| $\frac{3}{8}$ | 15·16 | 14·06 | 15·31 | 17·34 | 16·41 | 22·25 | 14·06 |
| $\frac{7}{16}$ | 17·68 | 16·41 | 17·87 | 20·23 | 19·14 | 25·96 | 16·41 |
| $\frac{1}{2}$ | 20·21 | 18·75 | 20·42 | 23·13 | 21·88 | 29·67 | 18·75 |
| $\frac{9}{16}$ | 22·73 | 21·09 | 22·97 | 26·02 | 24·61 | 33·38 | 21·09 |
| $\frac{5}{8}$ | 25·27 | 23·44 | 25·52 | 28 91 | 27·34 | 37·08 | 23·44 |
| $\frac{11}{16}$ | 27·79 | 25·78 | 28·07 | 31·80 | 30·08 | 40·79 | 25·78 |
| $\frac{3}{4}$ | 30·31 | 28·13 | 30·63 | 34·69 | 32·81 | 44·50 | 28·13 |
| $\frac{13}{16}$ | 32·84 | 30·47 | 33·18 | 37·58 | 35·55 | 48·21 | 30·47 |
| $\frac{7}{8}$ | 35·87 | 32·81 | 35·73 | 40·47 | 38·28 | 51·92 | 32·81 |
| $\frac{15}{16}$ | 37·90 | 35·16 | 38·28 | 43·36 | 41·02 | 55·63 | 35·16 |
| 1 | 40·42 | 37·50 | 40·83 | 46·25 | 43·75 | 59·33 | 37·50 |

# APPENDIX I

## CALENDAR OF INSECT PESTS

These diagrams are summaries of the calendars given at the end of each description on insect pests, and show the approximate duration of each stage of the insect. They will repay careful study.

Incidentally they show how very varied is the life-history of the pests concerned. It will be seen, for instance, that the winter is passed by many insects in the larval stage (caterpillar, grub, etc.), but many also hibernate in the pupal stage, a few in the egg stage, and one or two winter as adults, sheltered in more or less secure retreats.

| | |
|---|---|
| BUD | Bud Moth |
| BL | Blister |
| BT | Brown-Tail |
| BF | Buff-Tip |
| CB | Case-bearers |
| ACW | Apple Clearwing |
| CCW | Currant Clearwing |
| CD | Clouded Drab |
| C | Codling |
| DE | December |
| DO | Dot |
| E | Ermine (small) |
| F | Figure of Eight |
| G | Goat |
| GT | Gold-Tail |
| H | Hawk |
| L | Lackey |
| LP | Lappet |
| LM | Leaf Miner |
| M | Magpie |
| MR | March |
| MU | Mottled Umber |
| PP | Peppered |
| PT | Pith |
| PF | Plum Fruit |
| R | Raspberry |
| S | Swift |
| T | Tortoiseshell |
| TX | Tortrix |
| V | Vapourer |
| W | Winter |
| WL | Wood Leopard |
| ABW | Apple Blossom Weevil |
| NW | Nut Weevil |
| RW | Raspberry Weevil |
| RB | Raspberry Beetle |
| SB | Shot-borer Beetle |
| CK | Cockchafer |
| RC | Rose Chafer |
| AS | Apple Sawfly |
| GCS | Gooseberry & Currant Sawfly |
| SW | Slug Worm |
| NS | Nut Sawfly |
| PFS | Plum Fruit Sawfly |
| SPS | Social Pear Sawfly |
| PM | Pear Midge |

## Duration of Egg Stage of Insects

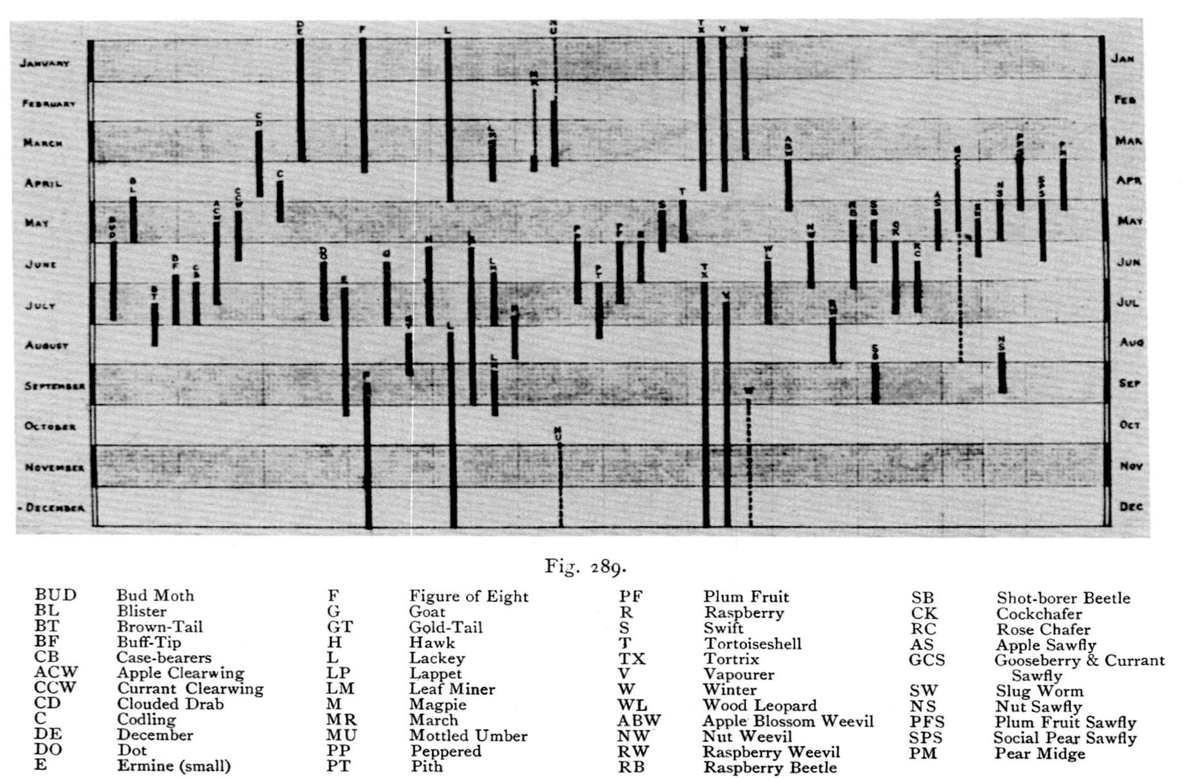


Fig. 289.

| | | | | | | | |
|---|---|---|---|---|---|---|---|
| BUD | Bud Moth | F | Figure of Eight | PF | Plum Fruit | SB | Shot-borer Beetle |
| BL | Blister | G | Goat | R | Raspberry | CK | Cockchafer |
| BT | Brown-Tail | GT | Gold-Tail | S | Swift | RC | Rose Chafer |
| BF | Buff-Tip | H | Hawk | T | Tortoiseshell | AS | Apple Sawfly |
| CB | Case-bearers | L | Lackey | TX | Tortrix | GCS | Gooseberry & Currant Sawfly |
| ACW | Apple Clearwing | LP | Lappet | V | Vapourer | | |
| CCW | Currant Clearwing | LM | Leaf Miner | W | Winter | SW | Slug Worm |
| CD | Clouded Drab | M | Magpie | WL | Wood Leopard | NS | Nut Sawfly |
| C | Codling | MR | March | ABW | Apple Blossom Weevil | PFS | Plum Fruit Sawfly |
| DE | December | MU | Mottled Umber | NW | Nut Weevil | SPS | Social Pear Sawfly |
| DO | Dot | PP | Peppered | RW | Raspberry Weevil | PM | Pear Midge |
| E | Ermine (small) | PT | Pith | RB | Raspberry Beetle | | |

## Duration of the Larva Stage of Insects

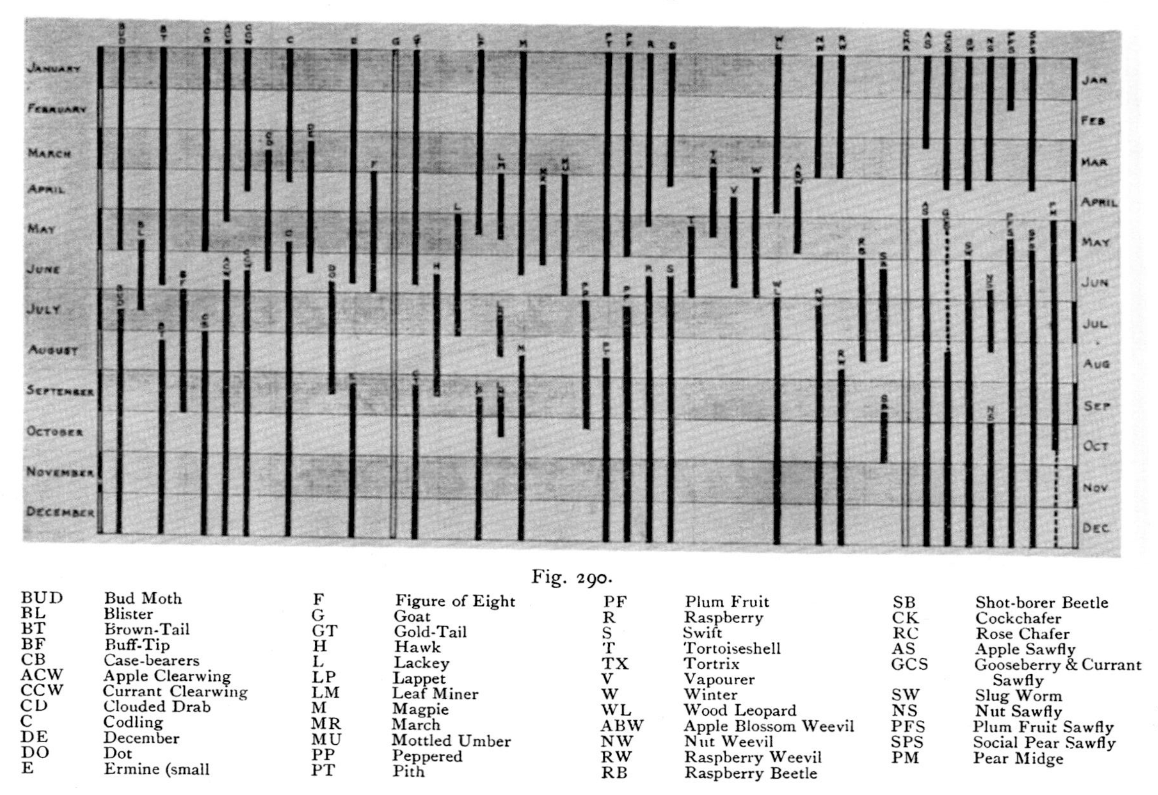


Fig. 290.

| | | | | | | | |
|---|---|---|---|---|---|---|---|
| BUD | Bud Moth | F | Figure of Eight | PF | Plum Fruit | SB | Shot-borer Beetle |
| BL | Blister | G | Goat | R | Raspberry | CK | Cockchafer |
| BT | Brown-Tail | GT | Gold-Tail | S | Swift | RC | Rose Chafer |
| BF | Buff-Tip | H | Hawk | T | Tortoiseshell | AS | Apple Sawfly |
| CB | Case-bearers | L | Lackey | TX | Tortrix | GCS | Gooseberry & Currant Sawfly |
| ACW | Apple Clearwing | LP | Lappet | V | Vapourer | | |
| CCW | Currant Clearwing | LM | Leaf Miner | W | Winter | SW | Slug Worm |
| CD | Clouded Drab | M | Magpie | WL | Wood Leopard | NS | Nut Sawfly |
| C | Codling | MR | March | ABW | Apple Blossom Weevil | PFS | Plum Fruit Sawfly |
| DE | December | MU | Mottled Umber | NW | Nut Weevil | SPS | Social Pear Sawfly |
| DO | Dot | PP | Peppered | RW | Raspberry Weevil | PM | Pear Midge |
| E | Ermine (small | PT | Pith | RB | Raspberry Beetle | | |

## Duration of the Pupa Stage of Insects

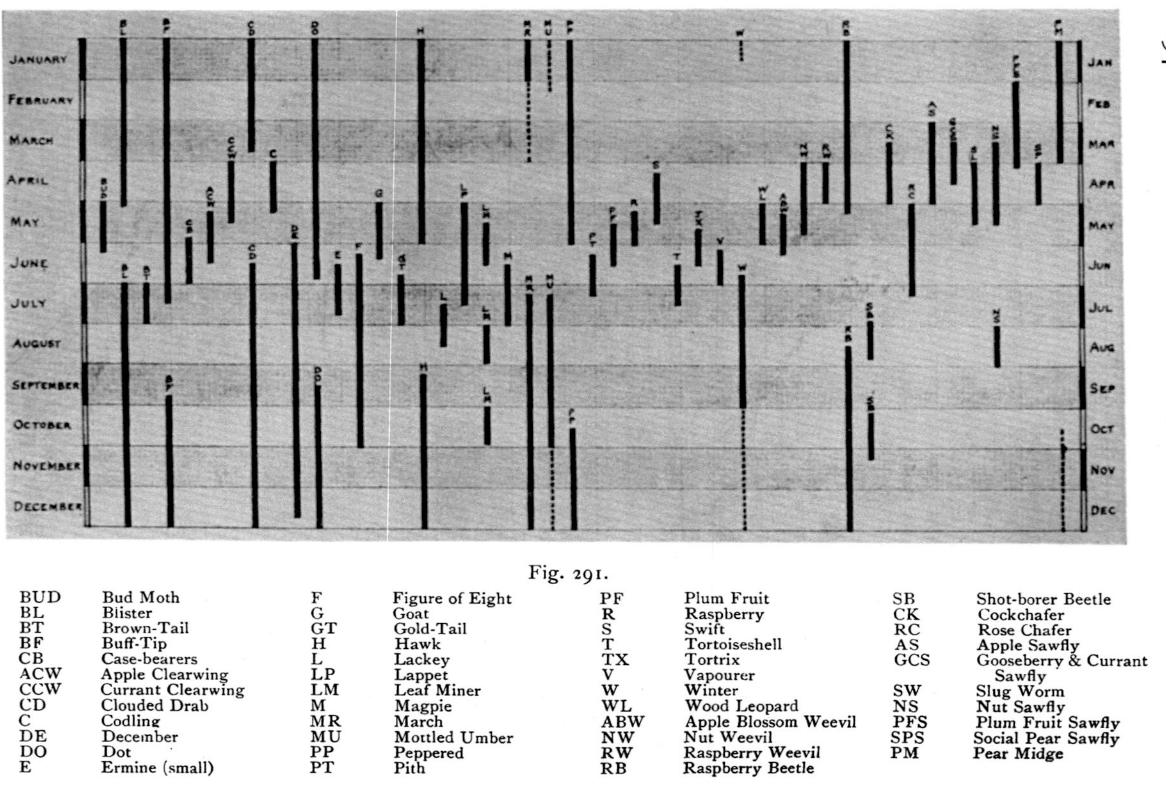


Fig. 291.

| | |
|---|---|
| BUD | Bud Moth |
| BL | Blister |
| BT | Brown-Tail |
| BF | Buff-Tip |
| CB | Case-bearers |
| ACW | Apple Clearwing |
| CCW | Currant Clearwing |
| CD | Clouded Drab |
| C | Codling |
| DE | December |
| DO | Dot |
| E | Ermine (small) |
| F | Figure of Eight |
| G | Goat |
| GT | Gold-Tail |
| H | Hawk |
| L | Lackey |
| LP | Lappet |
| LM | Leaf Miner |
| M | Magpie |
| MR | March |
| MU | Mottled Umber |
| PP | Peppered |
| PT | Pith |
| PF | Plum Fruit |
| R | Raspberry |
| S | Swift |
| T | Tortoiseshell |
| TX | Tortrix |
| V | Vapourer |
| W | Winter |
| WL | Wood Leopard |
| ABW | Apple Blossom Weevil |
| NW | Nut Weevil |
| RW | Raspberry Weevil |
| RB | Raspberry Beetle |
| SB | Shot-borer Beetle |
| CK | Cockchafer |
| RC | Rose Chafer |
| AS | Apple Sawfly |
| GCS | Gooseberry & Currant Sawfly |
| SW | Slug Worm |
| NS | Nut Sawfly |
| PFS | Plum Fruit Sawfly |
| SPS | Social Pear Sawfly |
| PM | Pear Midge |

## Duration of the Adult Stage of Insects

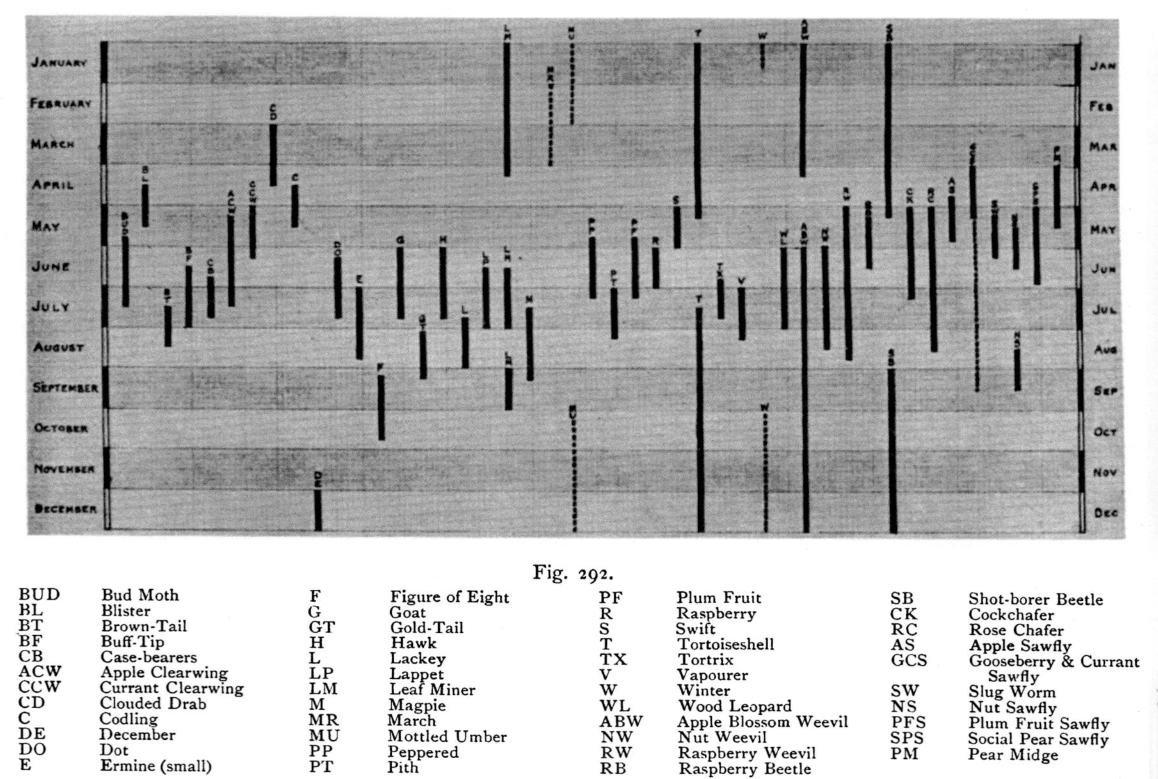


Fig. 292.

| | | | | | | | |
|---|---|---|---|---|---|---|---|
| BUD | Bud Moth | F | Figure of Eight | PF | Plum Fruit | SB | Shot-borer Beetle |
| BL | Blister | G | Goat | R | Raspberry | CK | Cockchafer |
| BT | Brown-Tail | GT | Gold-Tail | S | Swift | RC | Rose Chafer |
| BF | Buff-Tip | H | Hawk | T | Tortoiseshell | AS | Apple Sawfly |
| CB | Case-bearers | L | Lackey | TX | Tortrix | GCS | Gooseberry & Currant Sawfly |
| ACW | Apple Clearwing | LP | Lappet | V | Vapourer | | |
| CCW | Currant Clearwing | LM | Leaf Miner | W | Winter | SW | Slug Worm |
| CD | Clouded Drab | M | Magpie | WL | Wood Leopard | NS | Nut Sawfly |
| C | Codling | MR | March | ABW | Apple Blossom Weevil | PFS | Plum Fruit Sawfly |
| DE | December | MU | Mottled Umber | NW | Nut Weevil | SPS | Social Pear Sawfly |
| DO | Dot | PP | Peppered | RW | Raspberry Weevil | PM | Pear Midge |
| E | Ermine (small) | PT | Pith | RB | Raspberry Beetle | | |

# APPENDIX II

## THE GIPSY MOTH

### *Ocneria (Lymantia) dispar*

This is not a pest in England, but has caused an enormous amount of damage in the States.

It was introduced into Massachusetts about the year 1868, and has cost many thousands of dollars to keep it in check. In 1891, a commission was established for its extermination, and the reports issued make interesting reading. It is perhaps the most dangerous pest ever introduced into the States, and unless the present campaign against it is maintained, it will threaten the entire crops of the country. The caterpillar feeds upon an enormous number of different plants, and is especially difficult to control.

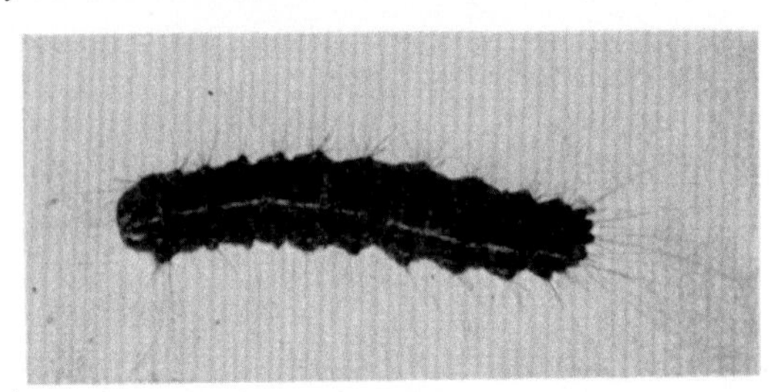

Fig. 293. Caterpillar of the Gipsy Moth. Natural size.

The adult insect is shown in fig. 295. The females are much larger and of stouter build than the males. They are creamy white in colour with irregular transverse lines of grey.

The eggs are laid in masses of 400—500 in a great diversity of situations, and are covered with yellow hair and scales from the abdomen. They are found in July onwards until late in September, and remain throughout the winter, the caterpillars hatching out early the following year (April—June).

The caterpillar (fig. 293) is about 1½ inches long when mature and is of cream colour, so thickly spotted with black as to appear brown in tint. There is a broken cream line along the middle of the body, and long tufts of hairs project from each segment. It pupates during July and September, forming a chocolate brown pupa (see fig. 294) with a very incomplete cocoon, which serves to hold it in place. The moths emerge in a few days.

The remedies adopted are:—

1. Spraying the plants in spring and summer with Arsenate of Lead to kill the caterpillars.
2. Hunting out and destroying the egg masses during the autumn and winter.

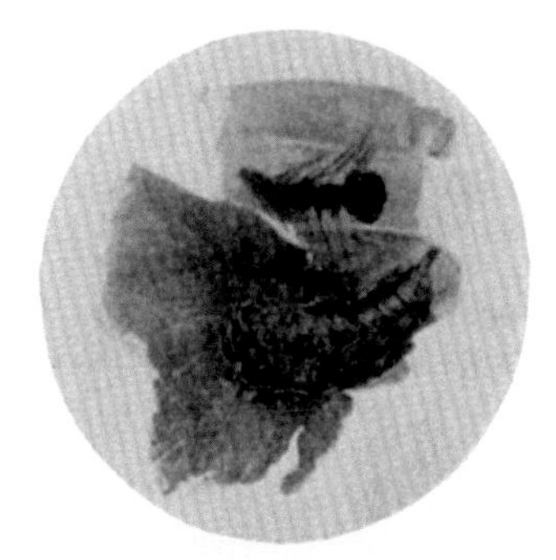

Fig. 294. Pupa of the Gipsy Moth with shrivelled larva skin. Natural size.

Fig. 295. Male (above) and female (below) of Gipsy Moth. Natural size

# APPENDIX III

## THE STRUCTURE OF INSECTS

The subjoined figure is from a photograph of a wasp (*Vespa crabro*) belonging to the family Hymenoptera, carefully dissected to show clearly the various parts of the body. The Cockchafer (*Melolontha vulgaris*), family Coleoptera, is shown similarly dissected in fig. 297. In each case the insect is shown complete, viewed from above and below. In the centre is a complete dissection of the body parts, while on the right the leg parts are shown separated from one another.

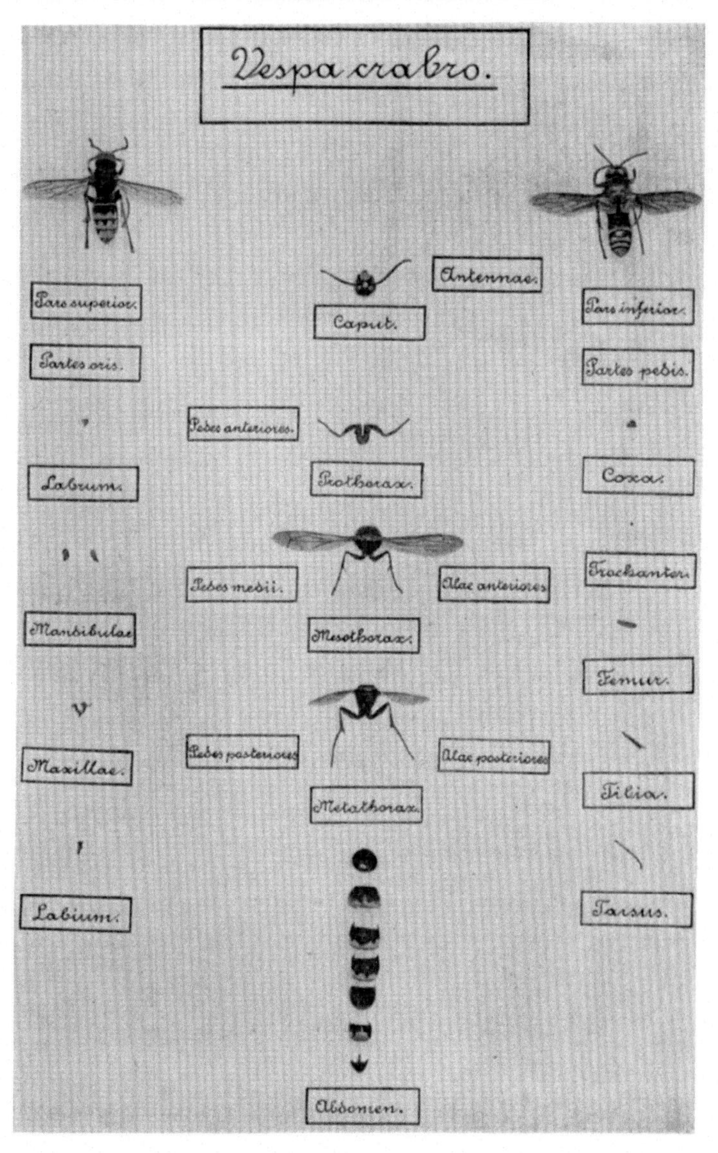


Fig. 296. The Common Wasp carefully dissected to show all the body parts.

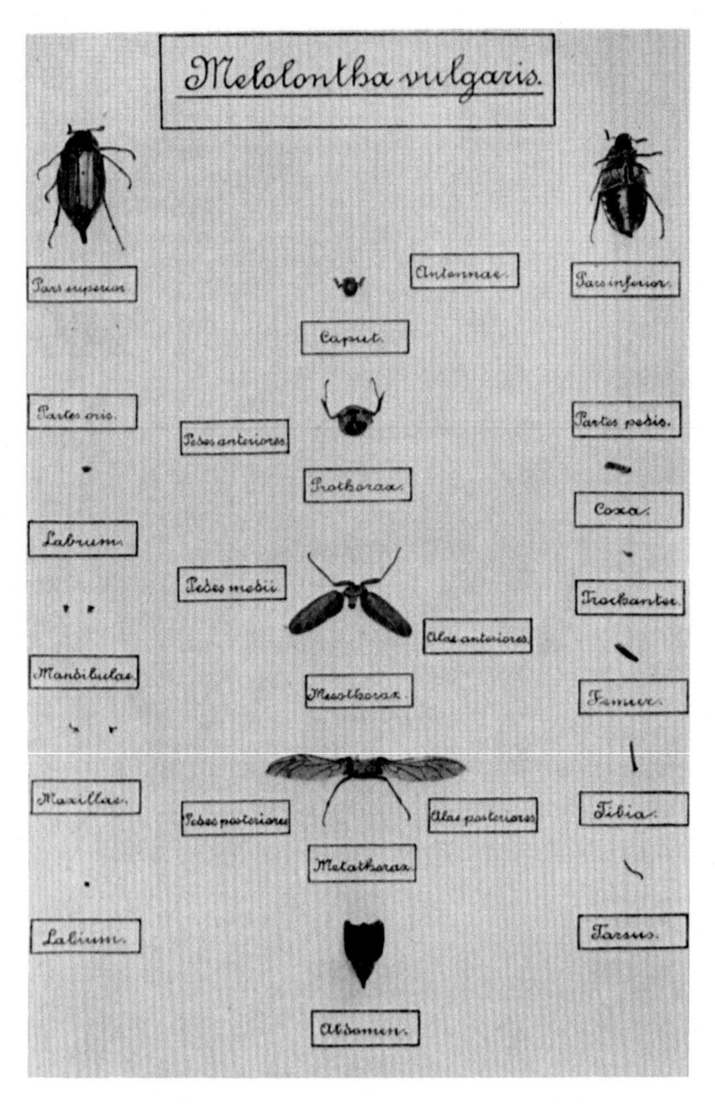


Fig. 297. The Cockchafer Beetle dissected to show the body parts.

# APPENDIX IV

## "THE HOP DOG"

### *Dasychira (Orygia) pudibunda*

This very hairy caterpillar has been found on hops from time to time, but is seldom a serious pest. It was reported in Herefordshire as attacking apples to a serious extent in 1913.

Fig. 298. Caterpillar of "Hop Dog," *Dasychira pudibunda*. Natural size.

The moth generally appears in May and June, but frequently occurs earlier. Eggs are laid in groups of about 50. The caterpillars are found from July to October. They are about 1½ to 2 inches in length when mature.

# APPENDIX V

## Refractive Indices of Aqueous Solutions of Nicotine (Fryer)

| Per Cent. Nicotine | Refractive Index, 15° C | Difference | Per Cent. Nicotine | Refractive Index, 15° C. | Difference | Per Cent. Nicotine | Refractive Index, 15° C. | Difference | Per Cent. Nicotine | Refractive Index, 15° C. | Difference |
|---|---|---|---|---|---|---|---|---|---|---|---|
| | | | | | 18 | | | 22 | | | 21 |
| 100 | 1·5300 | 7 | 74 | 1·4932 | 19 | 48 | 1·4369 | 22 | 22 | 1·3803 | 22 |
| 99 | 1·5293 | 8 | 73 | 1·4913 | 20 | 47 | 1·4347 | 22 | 21 | 1·3781 | 21 |
| 98 | 1·5285 | 9 | 72 | 1·4893 | 20 | 46 | 1·4325 | 22 | 20 | 1·3760 | 22 |
| 97 | 1·5276 | 10 | 71 | 1·4873 | 21 | 45 | 1·4303 | 22 | 19 | 1·3738 | 21 |
| 96 | 1·5266 | 11 | 70 | 1·4852 | 21 | 44 | 1·4281 | 22 | 18 | 1·3717 | 22 |
| 95 | 1·5255 | 11 | 69 | 1·4831 | 22 | 43 | 1·4259 | 22 | 17 | 1·3695 | 21 |
| 94 | 1·5244 | 12 | 68 | 1·4809 | 22 | 42 | 1·4237 | 22 | 16 | 1·3674 | 22 |
| 93 | 1·5232 | 13 | 67 | 1·4787 | 22 | 41 | 1·4215 | 22 | 15 | 1·3652 | 21 |
| 92 | 1·5219 | 13 | 66 | 1·4765 | 22 | 40 | 1·4193 | 22 | 14 | 1·3631 | 22 |
| 91 | 1·5206 | 13 | 65 | 1·4743 | 22 | 39 | 1·4171 | 22 | 13 | 1·3609 | 21 |
| 90 | 1·5193 | 14 | 64 | 1·4721 | 22 | 38 | 1·4149 | 22 | 12 | 1·3588 | 21 |
| 89 | 1·5179 | 14 | 63 | 1·4699 | 22 | 37 | 1·4127 | 22 | 11 | 1·3567 | 21 |
| 88 | 1·5165 | 15 | 62 | 1·4677 | 22 | 36 | 1·4105 | 22 | 10 | 1·3546 | 21 |
| 87 | 1·5150 | 16 | 61 | 1·4655 | 22 | 35 | 1·4083 | 22 | 9 | 1·3525 | 21 |
| 86 | 1·5134 | 16 | 60 | 1·4633 | 22 | 34 | 1·4061 | 22 | 8 | 1·3504 | 21 |
| 85 | 1·5118 | 16 | 59 | 1·4511 | 22 | 33 | 1·4039 | 21 | 7 | 1·3483 | 20 |
| 84 | 1·5102 | 16 | 58 | 1·4589 | 22 | 32 | 1·4018 | 22 | 6 | 1·3463 | 20 |
| 83 | 1·5086 | 17 | 57 | 1·4567 | 22 | 31 | 1·3997 | 21 | 5 | 1·3443 | 21 |
| 82 | 1·5069 | 17 | 56 | 1·4545 | 22 | 30 | 1·3976 | 22 | 4 | 1·3422 | 20 |
| 81 | 1·5052 | 17 | 55 | 1·4523 | 22 | 29 | 1·3954 | 21 | 3 | 1·3402 | 20 |
| 80 | 1·5035 | 17 | 54 | 1·4501 | 22 | 28 | 1·3933 | 22 | 2 | 1·3382 | 20 |
| 79 | 1·5018 | 17 | 53 | 1·4479 | 22 | 27 | 1·3911 | 22 | 1 | 1·3362 | 21 |
| 78 | 1·5001 | 17 | 52 | 1·4457 | 22 | 26 | 1·3889 | 21 | 0 | 1·3341 | |
| 77 | 1·4984 | 17 | 51 | 1·4435 | 22 | 25 | 1·3868 | 22 | | | |
| 76 | 1·4967 | 17 | 50 | 1·4413 | 22 | 24 | 1·3846 | 22 | | | |
| 75 | 1·4950 | 18 | 49 | 1·4391 | 22 | 23 | 1·3824 | 21 | | | |

REFRACTIVE INDICES OF AQUEOUS SOLUTIONS OF PYRIDINE (FRYER)

| Per Cent. Pyridine | Refractive Index, 15° C. | Difference | Per Cent. Pyridine | Refractive Index, 15° C. | Difference | Per Cent. Pyridine | Refractive Index, 15° C. | Difference | Per Cent. Pyridine | Refractive Index, 15° C. | Difference |
|---|---|---|---|---|---|---|---|---|---|---|---|
| | | | | | 17 | | | 20 | | | 20 |
| 100 | 1·5136 | 8 | 74 | 1·4782 | 18 | 48 | 1·4277 | 19 | 22 | 1·3770 | 20 |
| 99 | 1·5128 | 9 | 73 | 1·4764 | 18 | 47 | 1·4258 | 19 | 21 | 1·3750 | 19 |
| 98 | 1·5119 | 10 | 72 | 1·4746 | 18 | 46 | 1·4239 | 20 | 20 | 1·3731 | 19 |
| 97 | 1·5109 | 10 | 71 | 1·4728 | 18 | 45 | 1·4219 | 19 | 19 | 1·3712 | 20 |
| 96 | 1·5099 | 11 | 70 | 1·4710 | 18 | 44 | 1·4200 | 19 | 18 | 1·3692 | 20 |
| 95 | 1·5088 | 11 | 69 | 1·4792 | 19 | 43 | 1·4181 | 19 | 17 | 1·3672 | 20 |
| 94 | 1·5077 | 12 | 68 | 1·4673 | 19 | 42 | 1·4162 | 20 | 16 | 1·3652 | 20 |
| 93 | 1·5065 | 12 | 67 | 1·4654 | 19 | 41 | 1·4142 | 19 | 15 | 1·3632 | 20 |
| 92 | 1·5053 | 12 | 66 | 1·4635 | 19 | 40 | 1·4123 | 20 | 14 | 1·3612 | 20 |
| 91 | 1·5041 | 13 | 65 | 1·4616 | 19 | 39 | 1·4103 | 19 | 13 | 1·3592 | 20 |
| 90 | 1·5028 | 13 | 64 | 1·4597 | 20 | 38 | 1·4084 | 20 | 12 | 1·3572 | 20 |
| 89 | 1·5015 | 13 | 63 | 1·4577 | 20 | 37 | 1·4064 | 19 | 11 | 1·3552 | 20 |
| 88 | 1·5002 | 14 | 62 | 1·4557 | 20 | 36 | 1·4045 | 20 | 10 | 1·3532 | 20 |
| 87 | 1·4988 | 14 | 61 | 1·4537 | 21 | 35 | 1·4025 | 20 | 9 | 1·3512 | 20 |
| 86 | 1·4974 | 14 | 60 | 1·4516 | 20 | 34 | 1·4005 | 19 | 8 | 1·3492 | 19 |
| 85 | 1·4960 | 15 | 59 | 1·4496 | 21 | 33 | 1·3986 | 20 | 7 | 1·3473 | 19 |
| 84 | 1·4945 | 15 | 58 | 1·4475 | 21 | 32 | 1·3966 | 19 | 6 | 1·3454 | 19 |
| 83 | 1·4930 | 15 | 57 | 1·4454 | 20 | 31 | 1·3947 | 20 | 5 | 1·3435 | 19 |
| 82 | 1·4915 | 15 | 56 | 1·4434 | 20 | 30 | 1·3927 | 20 | 4 | 1·3416 | 19 |
| 81 | 1·4900 | 16 | 55 | 1·4414 | 20 | 29 | 1·3907 | 19 | 3 | 1·3397 | 19 |
| 80 | 1·4884 | 17 | 54 | 1·4394 | 19 | 28 | 1·3888 | 20 | 2 | 1·3378 | 19 |
| 79 | 1·4867 | 17 | 53 | 1·4375 | 19 | 27 | 1·3868 | 19 | 1 | 1·3359 | 18 |
| 78 | 1·4850 | 17 | 52 | 1·4356 | 19 | 26 | 1·3849 | 20 | 0 | 1·3341 | |
| 77 | 1·4833 | 17 | 51 | 1·4337 | 20 | 25 | 1·3829 | 20 | | | |
| 76 | 1·4816 | 17 | 50 | 1·4317 | 20 | 24 | 1·3809 | 19 | | | |
| 75 | 1·4799 | 17 | 49 | 1·4297 | 20 | 23 | 1·3790 | 20 | | | |

# APPENDIX VI

## "REVERSION," "GOING WILD" OR "NETTLE HEAD OF CURRANTS"

For a good number of years now it has been noticed that in certain plantations the foliage of currants, especially black currants, has undergone a curious modification. This consists of the growth of narrow pointed leaves, more "saw-toothed" than normal (see fig. 299). Together

Fig. 299.

with this alteration in the character of the leaf, there usually occurs a great increase in the numbers of lateral shoots. Of late years this trouble has been on the increase, and is now apparently responsible for as much damage as the "big bud" disease. In many cases "reversion" is associated with the latter disease, but in many others there is no evidence of its presence at all. Thus it is open to grave doubt whether big bud has any action in causing reversion—indeed, the opposite might be the case, and the mite attack be consequent upon the reversion.

In many cases abnormal growth of laterals was shown to be the result of the injury to the terminal bud. It is now proved, however, that even cutting off the terminal bud does not produce reversion. Experiments are now being undertaken to secure mite-free reverted stock and to investigate the whole subject. This work is being undertaken by Professor Lees at Bristol.

# APPENDIX VII

## ARSENIC POISONING

If arsenic compounds have been taken into the mouth, the following mixture is very valuable. It is well for all users of arsenic compounds to prepare a quantity of this, enough for a few doses in case of emergency (a still better antidote may be bought at any druggist's stores under the name of "liquor ferri dialysate"—dose 10—30 minims).

Take 4 parts of ferric sulphate solution (liquor ferri persulphatis B.P.) and mix with 12½ parts of water, keep in a stoppered bottle. Rub one part of magnesia with cold water to a thin smooth cream, dilute to 75 parts and transfer to another stoppered bottle. When required for use, shake the magnesia up well, and add gradually to the ferrous sulphate, shaking at intervals. Dose is 4 fluid ounces.

After either of these has been administered, the stomach must be emptied and stimulants and warmth applied. In any case, lose no time in calling in medical aid.

# APPENDIX VIII

## POWDER ("DRY") SPRAYING

The figure shows a new method of powder spraying, using a portable air compressor and special powder distributor. Powder spray has been employed to a considerable extent in the United States, and while in some cases it has proved equally as effective as wet spraying, it is, on the average, much less efficient, while the costs are about the same. It is therefore unlikely to displace wet spraying, except where suitable water is not available, or in very hilly districts, where water carriage is difficult.

The most popular powder mixture for dry spraying is composed of about 85 parts of sulphur (pure sulphur, dried lime-sulphur or similar substance) and the remainder Arsenate of Lead. Mixtures of Bordeaux Powder and Arsenates (Lead or Calcium) are also used with or without a diluent such as chalk powder.

Fig. 300. Powder spraying by means of portable air-compressor.

# APPENDIX IX

## THE USE OF BENEFICIAL INSECTS IN COMBATING INSECT PESTS

The artificial use of beneficial insects by collection or propagation for pest-destruction has not, as far as the author is aware, been tried to any extent in this country. In America, however, this method has been used in several instances with conspicuous success. It is mainly applicable in the case of insect pests imported from other countries. Frequently such pests increase in their new surroundings at an alarming rate, which is due to the fact that the natural enemies of the insect, which are sufficient to keep it in check in its native country, are absent. The remedy is to discover the identity of these "beneficial" insects, and to introduce them in the areas affected.

One well-known instance of this was the case of the "Fluted scale." The orange growers of California were attacked very severely with this pest, which threatened to ruin the industry. The then entomologist to the U. S. Department of Agriculture (C. V. Riley) traced the origin of this pest to Australia. In 1888 an entomologist was sent to discover the natural enemies of the pest in Australia. As a result the lady-bird beetle, known as *Vedalia cardinalis*, was collected and brought over to California and distributed to growers. In less than 18 months the beetles had practically rid the country of the pest.

In other districts the localities where lady-birds hibernate are noted (usually in the mountains), and they are collected in large quantities in the winter, kept in cold storage, and sent in boxes of several thousands to growers when attacks of aphis grow serious.

Although the concentration of certain fruit in definite areas is of great assistance to such a method in America, while in England the scattered orchards offer less favourable conditions, it might be well worth while to make experiments in this direction.

Besides enemies of their own class, many insect pests are subject to bacterial and fungoid diseases which are very contagious and rapidly fatal. It might be possible to propagate suitable bacteria and fungi and produce infection of the pest by spraying cultures of these, or in other ways. Even in America such methods are only in their infancy, and there is a large field open for future research.

# APPENDIX X

## CHERRY AND PLUM LEAF BLIGHT

(*Cylindrosporium padi*)

This disease appears to be slightly on the increase, especially in the United States. It occurs mainly on young stock. The disease appears in late spring, as pale spots on the leaves, sometimes with a reddish tint. The diseased areas spread and the attacked portion dies and falls out, leaving holes in the leaves. Severe attacks cause a leaf-fall early in the season. The best control treatment is to spray with Bordeaux in spring, followed by a second treatment after a month's interval.

# APPENDIX XI

## SPRAYING PUMP

(*Referred to on page* 657)

Fig. 301. Three-throw high pressure ram pump for large spraying outfits.

# APPENDIX XII

## PLUM LEAF-CURLING APHIS

Fig. 302. Commencement of attack on young branch of plum. (See page 313.)

## APPENDIX XII—*continued*

Fig. 303. Plum leaf-curling aphides magnified (× 15). Inset, natural size. (See page 313.)

# APPENDIX XIII

## APPLE SUCKER (Psylla)

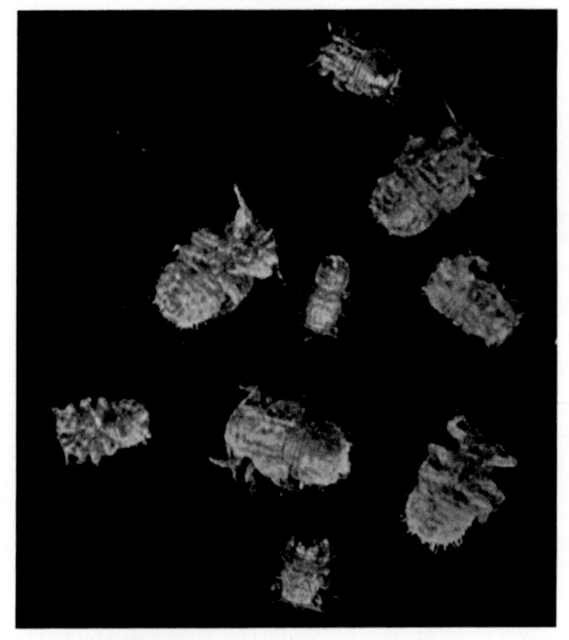

Fig. 304. Larval and nymph stages of Apple Sucker (Psylla) highly magnified (× 15). (See page 337.)

# APPENDIX XIV

## LACE-WING FLY

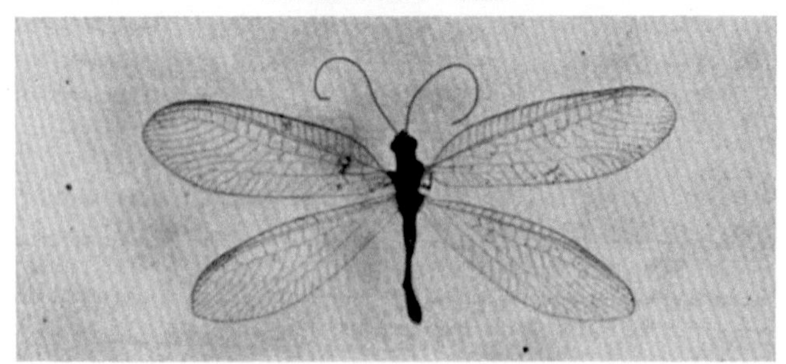

Fig. 305. Adult of the Lace-wing Fly, the larva of which is termed the "Aphis lion." (See page 469.)

# GENERAL INDEX

Printed in the United States
By Bookmasters